Applied Weed and Her

Kassio Ferreira Mendes
Antonio Alberto da Silva
Editors

Applied Weed and Herbicide Science

Editors
Kassio Ferreira Mendes
Department of Agronomy
Federal University of Viçosa
Viçosa, Minas Gerais, Brazil

Antonio Alberto da Silva
Department of Agronomy
Federal University of Viçosa
Viçosa, Minas Gerais, Brazil

ISBN 978-3-031-01940-1 ISBN 978-3-031-01938-8 (eBook)
https://doi.org/10.1007/978-3-031-01938-8

© The Editor(s) (if applicable) and The Author(s), under exclusive license to Springer Nature Switzerland AG 2022
This work is subject to copyright. All rights are solely and exclusively licensed by the Publisher, whether the whole or part of the material is concerned, specifically the rights of translation, reprinting, reuse of illustrations, recitation, broadcasting, reproduction on microfilms or in any other physical way, and transmission or information storage and retrieval, electronic adaptation, computer software, or by similar or dissimilar methodology now known or hereafter developed.
The use of general descriptive names, registered names, trademarks, service marks, etc. in this publication does not imply, even in the absence of a specific statement, that such names are exempt from the relevant protective laws and regulations and therefore free for general use.
The publisher, the authors and the editors are safe to assume that the advice and information in this book are believed to be true and accurate at the date of publication. Neither the publisher nor the authors or the editors give a warranty, expressed or implied, with respect to the material contained herein or for any errors or omissions that may have been made. The publisher remains neutral with regard to jurisdictional claims in published maps and institutional affiliations.

This Springer imprint is published by the registered company Springer Nature Switzerland AG
The registered company address is: Gewerbestrasse 11, 6330 Cham, Switzerland

The authors dedicate this book to the research, teaching and extension group in Integrated Weed Management (from the Portuguese, Manejo Integrado de Plantas Daninhas—MIPD) at the Federal University of Viçosa, MG, Brazil.

Preface

Weeds are species that damage crops or human activities directly or indirectly. These plants have rapid growth and high competitive capacity, in addition to high seed production and the ability to germinate in different environments. However, some species, in addition to sexual reproduction, are able to reproduce by vegetative propagation.

The survival strategies of weeds allow perpetuation of the species in the environment, even in adverse situations. These strategies consist of high seed production in order to maximize the chances of survival. In some species, the seeds have structures that aid in dispersal, promoting the occupation of new areas. Another form of dispersion vegetative propagation structures, which also serve as a source of reserves, contributes to the persistence of these species even under unfavourable conditions.

Because of the survival strategies of weeds, identifying the species and knowing their biology is important to define management methods that allow control without causing damage to the environment and must be economically viable. In this book, the survival mechanisms of weeds are discussed.

However, the chemical control of weeds is the most used technique worldwide and very frequently in modern agriculture. In addition, knowing the retention, absorption and translocation of the herbicide is fundamental for it to reach the target and have an efficient control, although weeds can have target-site (TSR) and non-target-site resistance (NTSR) mechanisms to herbicides. All the parameters of the evolution of weed resistance and genetically modified crops resistant to herbicides are covered in this book.

Viçosa, Minas Gerais, Brazil — Kassio Ferreira Mendes

Acknowledgement

The authors thank the Graduate Program in Crop Science from the Department of Agronomy, the Federal University of Viçosa, MG, Brazil.

Contents

About the Editors

Kassio Ferreira Mendes is a Professor of Biology and Integrated Management of Weeds, Department of Agronomy, Federal University of Viçosa (UFV), Brazil. He graduated from the State University of Mato Grosso, Brazil, and received a master's in Agronomy from UFV. He did his PhD and post-doc in Nuclear Energy in Agriculture at the Center of Nuclear Energy in Agriculture, University of São Paulo (USP), Brazil, with a research fellowship at the University of Minnesota (U of M) and United States Department of Agriculture-Agricultural Research Service (USDA-ARS). Dr. Mendes is a member of the Brazilian Society of Weed Science.

Antonio Alberto da Silva holds a PhD in Soils and Plant Nutrition from the University of São Paulo (USP). He is currently full professor at the Federal University of Viçosa (UFV) and 1B fellow in research productivity by CNPq. He published 380 articles in scientific journals, published/organized 11 books and 57 book chapters. He supervised seven post-doctorates and supervised 56 PhD, 69 master's and 36 scientific initiation students. He is the leader of a CNPq research group and participates in several others. He is a consultant in the Graduate Programs of Crop Science and Agrochemistry of the UFV and a member of CTNBio. He acts mainly on the following topics: environmental herbicide interaction, herbicide resistance and integrated weed management.

Aspects of Biology and Ecophysiology, Survival Mechanisms, and Weed Classifications

Adalin Cezar Moraes de Aguiar, Kassio Ferreira Mendes, Lucas Heringer Barcellos Júnior, Elisa Maria Gomes da Silva, Laryssa Barbosa Xavier da Silva, and Antonio Alberto da Silva

Abstract After the emergence of agriculture, the environment became populated not only by cultivated species, but also by species with undesirable characteristics, called weeds. These spontaneous plants have characteristics that differentiate them from other plants and allow them to successfully invade, establish, and persist in agricultural environments. The ability to reproduce sexually through seeds and/or asexually through vegetative structures has allowed weeds to reproduce easily, increasing their ability to exploit agricultural environments. Moreover, the high production of propagules, the presence of dispersal mechanisms, and the dormancy of these reproductive structures have allowed weeds to persist and overcome adversities caused by man and the environment. Knowing the morphophysiological characteristics of weed species and the possibility of grouping species according to their similarities, such as taxonomic classifications, as to habitat, life cycle, and growth habit, helps in the positioning of management techniques. Even if their weed communities have been altered by human action and the environment, weeds will always be present in agricultural environments, so there is a need to live with their presence, avoiding their negative effects, but trying to make the most of their contributions.

Keywords Evolutionary strategies · Weed reproduction · Propagule dispersal · Dormancy · Seed bank · Taxonomy · Habitat · Life cycle · Growth habit

---

A. C. M. de Aguiar (✉) · L. H. Barcellos Júnior · E. M. G. da Silva · L. B. X. da Silva
Department of Agronomy, Federal University of Viçosa, Viçosa, Minas Gerais, Brazil

K. F. Mendes · A. Alberto da Silva
Department of Agronomy, Federal University of Viçosa, Viçosa, Minas Gerais, Brazil
e-mail: kfmendes@ufv.br; aasilva@ufv.br

© The Author(s), under exclusive license to Springer Nature Switzerland AG 2022
K. F. Mendes, A. Alberto da Silva (eds.), *Applied Weed and Herbicide Science*, https://doi.org/10.1007/978-3-031-01938-8_1

1 Introduction

With the emergence of agriculture approximately 12,000 years ago, humans began to create new fertile habitats not only for cultivated species that were intentionally sown, but also for undesirable species that adapted in order to exploit this new environment. In some cases, these weed species are closely related to agricultural crops, in which one can include wild species as well as wild descendants of crops, which have evolved through de-domestication (Fig. 1) (Guo et al. 2018).

Over time, weeds have developed mechanisms that provide them with the ability to survive in environments subject to various types and intensities of limitations. Among the main survival mechanisms of weeds are the ability to compete for resources, capacity to produce propagules, disuniformity of the germination process, viability of propagules under unfavorable conditions, alternative mechanisms of reproduction, ease of dissemination of propagules and rapid growth and initial development (Brighenti and Oliveira 2011).

In an agroecosystem, weeds generally cause negative impacts on human activities, whether agricultural, forestry, livestock, ornamental, among others. Despite this, weeds also provide beneficial effects to humans, animals, and the environment, such as the production of medicinal compounds, soil and water conservation, and

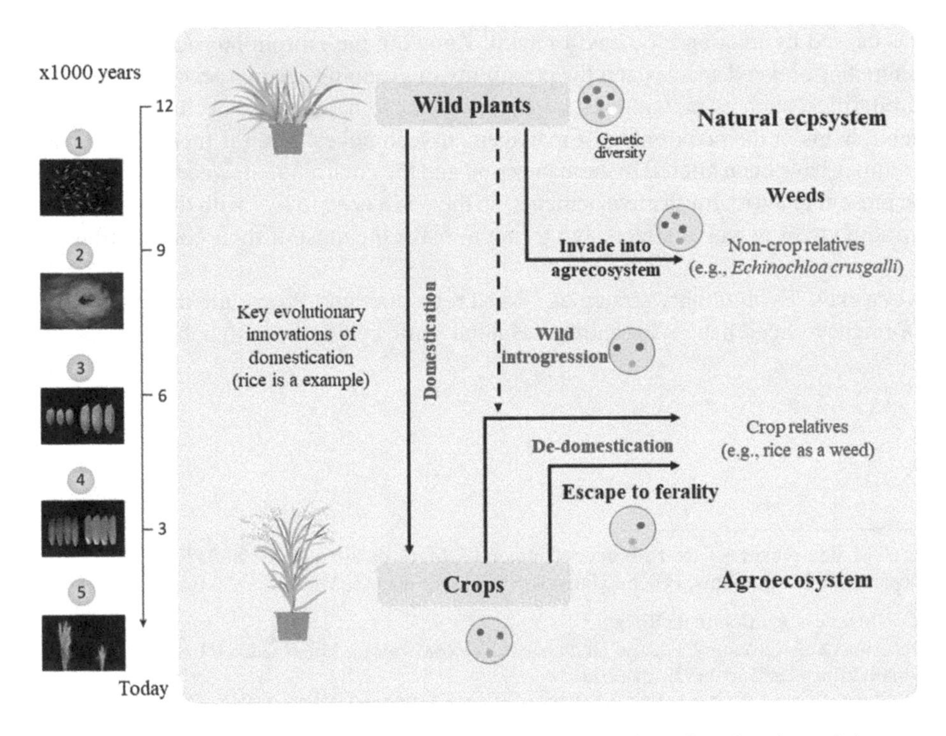

Fig. 1 Evolutionary relationships between wild plants, crops, and weeds, using rice and ricegrass species as an example. (Source: Guo et al. (2018))

maintenance of natural enemies. In this sense, the sustainable management of weeds becomes important in order to avoid damage and maintain the balance of the agroecosystem.

To manage weeds efficiently, it is important to know their behavior inside and outside cultivated areas. Factors such as reproduction, dispersal, dormancy, longevity, and viability of the seed/propagule bank (disseminules or diaspores) in the soil are important to understand the dynamics of weeds within the agroecosystem (Barrat-Segretain 1996, Vivian et al. 2008, Chauhan and Abugho 2013, Hossain and Begum 2015, Saeed et al. 2020).

Besides knowing the biology and ecophysiology, it is important to understand some characteristics that allow grouping weeds. Classification is essential for their science and control, using classification keys that take into account characteristics of the adult plant, habitat, cycle, and growth habit (Brighenti and Oliveira 2011). In certain cases, the management techniques and selectivity of some herbicides are based on specific characteristics of weeds, such as morphological and physiological differences between species.

This chapter covers the conceptual aspects of weeds, origin, and evolution of species, as well as aspects of reproduction, dispersal, dormancy, and seed bank of weeds. In addition, the chapter deals with the classification of weeds, with the aim of helping farmers and researchers with information that contributes to the science and management of weeds.

2 Weed Concept and Characteristics

The term weed is used to refer to certain plants that become undesirable at certain times and places, related to human activity. It is often common to find other terms to refer to an undesirable plant, such as: invasive plant, weed, adventitious plant, weed, as well as colloquial terms like weed, marsh, and inço. Pitelli (2015) proposes the use of the term weed plant to designate any higher plant that interferes with human interests and the environment, and the term weed community to designate the set of weed plants that inhabit a given environment, including the site of infestation, when it is necessary to be more specific.

A priori, no plant species can be considered a weed. Some weeds can be useful when present, such as soil erosion containment, nutrient recycling, medicinal properties, environmental phytodecontaminants, nectar supply for bees, among others.

Weeds can be grouped into **common** and **true**. **Common weeds** are those that do not have the capacity to grow and develop under unfavorable conditions. These plants occur most often in the crop succession system in the same area, such as the corn/soybean system, where corn plants also known as volunteer plants or tiger plants emerge within the soybean cycle, causing damage to the crop (López-Ovejero et al. 2016). Another example is the cotton/soybean system, where the regrowth of cotton plants inside the soybean causes negative effects to the crop. These weeds are becoming increasingly important, mainly due to herbicide resistance through genetically modified organisms (GMOs), which will be better addressed in Chapter 8.

Table 1 Characteristics that confer success to weeds in invading, establishing, and persisting in agricultural environments

1.	Germination requirements met in many environments
2.	Discontinuous (internally controlled) germination and great seed longevity
3.	Rapid growth through the vegetative phase to flowering
4.	Continuous seed production as long as growing conditions allow
5.	Self-compatibility, but not autogamy or complete apomixis
6.	Cross-pollination, when it occurs, by non-specialized visitors or wind
7.	High seed production under favorable environmental circumstances
8.	Production of some seed in a wide range of environmental conditions, and tolerance and plasticity
9.	Adaptations for short and long distance dispersal
10.	When perennial, vigorous vegetative reproduction or regeneration of fragments
11.	When perennial, fragile so that it is not easily removed from the soil
12.	Ability to compete interspecifically, by special means (rosette, smother growth, and release of allelochemicals)

Source: Baker (1974) and Monaco et al. (2002)

True weeds have characteristics that differentiate them from other plants and allow them to successfully invade, establish, and persist in agricultural environments under adverse conditions of temperature, humidity, and light. These characteristics give these weeds an enormous capacity to survive in environments disturbed by man (Table 1).

3 Origin and Evolution of Weeds

Since the advent of agriculture, humans have encountered plants that have hindered their goal of managing the environment. Several terms referring to these plants are found in ancient books. In the Bible, the parable of the sower (Matthew 13:25–30), brings a reference to these plants.

> [...] [25] but while everyone was sleeping, his enemy came and sowed weeds among the wheat, and went away. [26] When the wheat sprouted and formed heads, then the weeds also appeared.[27] "The owner's servants came to him and said, 'Sir, didn't you sow good seed in your field? Where then did the weeds come from?"[28] "'An enemy did this,' he replied. "The servants asked him, 'Do you want us to go and pull them up?"[29] "'No,' he answered, 'because while you are pulling the weeds, you may uproot the wheat with them. [30] Let both grow together until the harvest. At that time, I will tell the harvesters: First collect the weeds and tie them in bundles to be burned; then gather the wheat and bring it into my barn'" (Matthew 13:25–30).

Today, these plants that interfere with man's agricultural activities are called weeds (Ellstrand et al. 2010). Weeds originated initially from natural disturbances such as glaciation, mountain collapses, and the action of rivers and seas (Musik 1970). Due to the very concept of weeds, they started to appear when man stopped being nomadic and started his agricultural activities in a fixed territory, separating the beneficial ones (cultivated plants) from the plants that were of no interest (weeds).

Some plant species become weeds because they are competitive and adaptable, able to exploit naturally disturbed or human-interfered habitats. As humans manipulate the environment to meet survival needs, they provide a suitable environment for certain plant species, which can thrive under these circumstances. Weeds are just opportunistic species that are well adapted to grow on sites that have suffered disturbance (Chandrasena 2014). Weeds became totally domesticated through an unconscious process, in which they occurred simultaneously with crops, benefiting from the growing conditions. And also because of their superior adaptability to the crop, they became the target of human selection. Weeds evolve within the man-made habitat in three main ways (De Wet and Harlan 1975):

1. Through wild colonizers, selection and adaptation to continuous habitat disturbance;
2. As derived from hybridization between wild and cultivated breeds of domestic species;
3. Domesticated species abandoned by man.

Weeds are often classified using ecological categories related to population behavior. They can be colonizers (plants that appear early in vegetation succession) or naturalized species (exotic species that form sustainable populations without direct assistance from humans) (Radosevich et al. 2007).

The process of invasion of an alien species can be divided into three phases: introduction, colonization, and naturalization (Cousens and Mortimer 1995). These three phases of invasion are defined as follows:

Introduction: As a result of dispersal, propagules reach a location beyond their previous geographic range and establish populations of adult plants.

Colonization: Plants from the founder population reproduce and increase in number to form a self-perpetuating colony.

Naturalization: The species establishes new, self-perpetuating populations, disperses widely, and is incorporated into the resident flora.

Weeds can be classified according to evolutionary strategies based on genetically determined patterns of carbon resource allocation (Radosevich et al. 2007).

There are two key external factors that limit the amount of plant material, which can accumulate in an area which are stress and disturbance (Grime 1979). When the extremes of these factors are considered, the following possible evolutionary development strategies appear according to Table 2.

Stress tolerant: These are plants that survive in unproductive environments by reducing their biomass allocation for vegetative growth and reproduction and increasing their allocation for maintenance and defense. They exhibit characteristics that ensure the resilience of relatively mature individuals in hostile and limited environments such as recurrent droughts or floods, and even biotic factors such as resource utilization by neighboring plants and pest attack.

Competitors: These are plants that evolved with characteristics that maximize the capture of environmental resources under productive conditions. These plants have extensive vegetative growth and are abundant during the early and intermediate stages of succession.

Table 2 Evolutionary strategies of weeds resulting from disturbance and stress

Disturbance intensity	Strength of stress	
	High	Low
High	Plant death	Ruderals
Low	Stress tolerant	Competitors

Ruderals: These are plants found in highly disturbed but potentially productive environments. These are generally plants, with a short life span, rapid growth, and high seed production. They occupy the first stages of succession.

Modifications in cropping systems over the years have made an important contribution to the evolution of weeds. An example of this is the adoption of the no-till farming system, due to its ability to promote changes in the population dynamics and composition of the weed community. These changes occur due to alterations in management conditions, such as not disturbing the soil, and the presence of plant residues predominating over the soil, exerting chemical, physical, and biological effects on weeds, besides the adoption of chemical control, used to replace the mechanical control, performed by plowing and harrowing in conventional cultivation (Gomes Júnior and Christoffoleti 2008).

The control method is a key determinant in the evolution of the weed community. The control methods in general (cultural, mechanical, physical, chemical, and biological) are all non-selective, especially when used on a weed species or a plant community with inter- and intra-specific variations in morphology, habit, and growth. Each species may or may not adapt to any of these forms of control. Because weeds show plasticity and differential responses when exposed to adverse environmental conditions, the adoption of a single, continuous control method over a long period of time causes species to migrate, flower, or die. The species that survive and manage to reproduce gradually become more suitable and adapted, changing the population number and diversity in the absence of others (Qasem 2013). The population of less adapted individuals, on the other hand, are restricted in number and also in growth until they are strongly suppressed and may even become extinct.

Another example of evolution is the emergence of herbicide-resistant weed biotypes. The selection pressure caused by intensive herbicide use has the ability to select biotypes within resistant weed populations that pre-exist in the environment. Resistance is the natural and heritable ability of some biotypes within a given plant population to survive and reproduce after exposure to a dose of an herbicide that would be lethal to a susceptible population. This is the result of the evolutionary process, which occurs through the wide genetic variability of weeds, which allows the adaptation and survival of these species in various environmental and agroecosystem conditions (Christoffoleti and López-Ovejero 2003).

Knowing the origin and evolution of weeds is important to understand how these species behave when faced with changes in the environment caused by man and also to try to predict what will happen to these plants in the future. It is also worth pointing out that not only do weeds have characteristics that cause damage to human activities, but depending on the location, season, and growing conditions, weeds can also have beneficial effects.

4 Positive and Negative Effects of Weeds

4.1 Positive Points

Despite the negative impacts caused by weeds in agricultural and non-agricultural areas, some species can provide beneficial effects to humans, animals, and the environment, obtained by knowledge and correct use of these plants. Among the positive aspects provided by weeds are as follows:

(a) Benefits to the soil in containing erosion, reducing raindrop impact, preventing surface overheating, retaining moisture, increasing organic matter content, nutrient cycling, and improving soil pH. The extract of gallant soldier (*Galinsoga parviflora*) plants was efficient in increasing the pH of the soils used in studies, and southern sandbur (*Cenchrus echinatus*) showed high nutrient cycling when present (Meda et al. 2002, Cury et al. 2012).
(b) Hosts of natural enemies of pests or pathogens, besides honey bees and other pollinating insects. In a study of bee hive analysis, a high pollen load was observed originating from flowers of the weed plant assa-peixe (*Vernonia polyanthes*) (Modro et al., 2011).
(c) Feeding of terrestrial and aquatic animals. Alexandergrass (*Urochloa plantaginea*), showed good capacity for supplementation of steers and Mexican teosinte (*Euchlaena mexicana*) was efficient in supplementation of grass carp associated with feed (Martins et al. 2000, Costa et al. 2008).
(d) Use as ornamental plants. Morning glories (*Ipomoea* spp.) and balloon vine (*Cardiospermum halicacabum*), are species with ornamental characteristics (Brighenti and Oliveira 2011).
(e) Assistance in phytoremediation processes, allowing the removal or degradation of residual compounds in the soil. The jack bean (*Canavalia ensiformis*) was efficient in the remediation of soil contaminated with the herbicide trifloxysulfuron-sodium (Santos et al. 2006).
(f) They have medicinal properties. The tropical whiteweed (*Ageratum conyzoides*) has antiseptic action and is used against skin diseases, and the arrowleaf sida (*Sida rhombifolia*), can be used in wound healing (Sahu 1984).

4.2 Negative Points

Despite the numerous uses of some weed species, the vast majority interfere negatively with human activities, generating various damages. The damage caused by weeds can be direct, that is, closely related to losses in productivity and product quality, and/or indirect, interfering in other agricultural practices.

4.2.1 Direct Damage

On average, about 20–30% of the cost of crop production is due to weed control (Silva et al. 2007). In addition to reducing crop productivity, weeds cause other direct losses such as:

(a) Reduction in the quality of the commercial product (weed tubers growing inside potato tubers; weed impurities in the cotton fiber) (Fig. 2a, b).
(b) They are responsible for the non-certification of seeds and crop seedlings, when these are harvested together with seeds of certain prohibited weed species. A major problem is the presence of weedy rice (*Oryza sativa*) in cultivated white

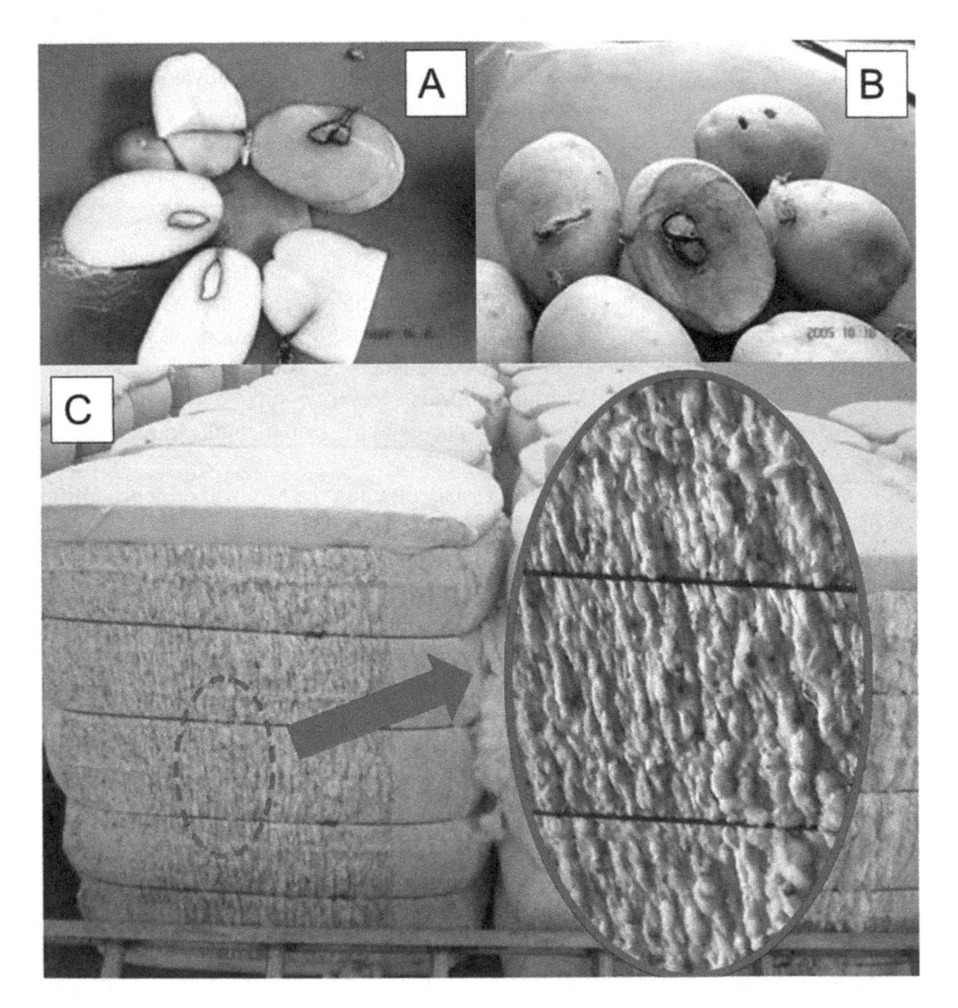

Fig. 2 Damage to English potatoes due to penetration and development of *Cyperus rotundus* tubers (**a, b**) and cotton bale with plant impurities from weeds present at harvest (**c**). (Source: Silva et al. (2007) and Ferreira et al. (2017))

rice (Marchezan 1994); and also the infestation of purple nutsedge (*Cyperus rotundus*) in citrus seedlings produced in nurseries.

(c) They can intoxicate domestic animals, when present in pastures, such as cafezinho (*Palicourea marcgravii*), among others, which when ingested, can cause the death of animals (D'Oliveira et al. 2018).

(d) Some species perform parasitism in major crops, fruit trees, ornamental plants, among others. Such as the mahogany mistletoe (*Phoradendron rubrum*) in citrus and the witchweed (*Striga lutea*) in corn, in addition to root system parasitic weeds, such as broomrape (*Orobanche* spp.), which can generate losses in tobacco crops (Cidasc 2018).

Other weed species can also reduce the value of the land, such as the *Cyperus rotundus* and *Artemisia verlotorum*. These species, when present in areas with crops that have little competitive capacity, such as vegetables in general, parks and gardens, have a very high cost of control, making it economically unviable (Silva et al. 2007).

4.2.2 Indirect Damage

Weeds can be alternative hosts for organisms that are harmful to cultivated plant species and can cause diseases, such as golden bean mosaic, sugarcane mosaic, and others. Weeds can also be hosts for nematodes. An example of this is black prickly pear, which has been reported to host the nematode *Meloidogyne mayaguensis* in the state of Paraná (Carneiro et al. 2006).

Some species, besides the direct damage they cause to crops, can also hinder or even prevent the execution of certain cultural practices and the harvest. Examples are some species of morning glories (*Ipomoea grandifolia, Ipomoea aristolochiaefolia, Ipomoea purpurea,* and others of this genus), which affect the cultivation of corn and sugarcane. The presence of these plants decreases the efficiency of the machines and increases the losses during the harvesting operation even when there is moderate infestation in the fields. The presence of morning glory (*I. purpurea*) reduced, in addition to productivity, pepper harvesting efficiency in Chile (Schutte 2017). Southern sandbur (*Cenchrus echinatus*), bristly starbur (*Acanthospermum hispidum*), cat's claw (*Acacia plumosa*), and other thorny plants can even impede the manual harvesting of crops.

Weeds are also inconvenient in non-cultivated areas: industrial areas, public roads, railroads, and oil refineries. Weeds can also generate problems in aquatic environments, such as water collection, fishing and increased irrigation costs, as well as the maintenance of dams and hydroelectric plants (Tanaka et al. 2002, Thomaz 2002).

5 Aggressiveness of Weeds

The characteristics of true weeds make them more aggressive in terms of development and rapid occupation of the soil, with this, they dominate the cultivated plants, if man does not interfere, using the available control methods. These aggressiveness characteristics are (Silva et al. 2007):

(a) High capacity of producing diaspores (seeds, bulbs, tubers, rhizomes, stolons, among others). For example, Redroot pigweed (*Amaranthus retroflexus*) produces up to 291,570 seeds per plant (Sellers et al. 2003).
(b) Maintenance of viability even under unfavorable conditions. For example, when passing through the digestive tract of sheep and goats, 18% of Wild poinsettia (*E. heterophylla*) seeds were not damaged (Lacey et al. 1992).
(c) Ability to germinate and emerge at great depths. For example, Wild poinsettia can germinate at depths greater than 15 cm (Scheren et al. 2013). This may be the cause of the failure of herbicides applied to the soil.
(d) Great unevenness in the germination process. This occurs due to the complex dormancy processes, one of the survival strategies of weeds. For example, Curly dock (*Rumex crispus*) seeds showed germination above 70% after being buried 20 cm deep for 17 years (Burnside et al. 1996).
(e) Alternative reproduction mechanisms. Many weed species present more than one reproduction mechanism. For example, johnsongrass (*Sorghum halepense*): reproduces by seeds and rhizomes; bermuda grass (*Cynodon dactylon*): by seeds and stolons; and purple nutsedge (*C. rotundus*): by seeds, tubers, basal bulb, and stolons (Lorenzi 2014).
(f) Ease of dispersion of propagules and seeds over great distances. This occurs by the action of water, wind, animals, man, machines, among others. For example, Horseweed, hairy fleabane or tall fleabane (*Conyza* spp.) are easily moved with distances of 72–145 km with light wind of about 18 km h^{-1} during a single flight and with higher speeds, of 72 km h^{-1}, horseweed seeds can reach 550 km of distance during a single flight (Shields et al. 2006).
(g) Rapid growth and early development. Many weeds grow and develop faster than many crops. For example, in onion cultivation, weeds germinate and grow much faster, easily dominating the crop, when it is conducted by direct sowing instead of transplanting (Pontes Jr et al. 2021).
(h) Great longevity of diaspores. For example, observations using 14C dating have shown that the seed of the sacred lotus (*Nelumbo nucifera*) can be viable for 1000 years, and that of the Common lambsquarters (*Chenopodium album*), for 1700 years (Shen-Miller et al. 1995). This great longevity is due to numerous and complex dormancy processes.

6 Reproduction, Dispersal, Dormancy and Weed Seed Bank

6.1 Reproduction

Weeds multiply and reproduce sexually and/or asexually (vegetatively). Sexual reproduction requires pollination and fertilization of an egg, this usually occurs through pollination of a flower, which subsequently produces seeds. The viable seed then has the potential to produce a new plant.

Seed production is considered the main form of weed reproduction, especially for annual plants. One of the characteristics of weed aggressiveness is the ability to produce propagules, by which the plants can perpetuate themselves. In general, most weeds produce a high amount of seeds, which ensures a high rate of dispersal and re-establishment of an infestation, but production can vary widely among species (Table 3).

The variability in weed seed production depends mainly on the characteristics of the species. Plants of common cocklebur (*Xanthium strumarium*), in general, produce a low quantity of seeds per plant (Table 3), different from that of the hairy fleabane (*Conyza bonariensis*), with production results of more than 800,000 seeds per plant. The variability, however, does not depend only on the species, but environmental factors such as climate, precipitation, and temperature; crop factors such as fertility, emergence time, cropping systems, plant density, and the presence of crops also influence the weed seed production potential.

Besides the quantity of seeds produced, the germination processes of the seeds are important for those working with weed management. The variability in weed emergence occurs due to inter and intrapopulation variability, adaptive capacity and environmental conditions that prevail during seed development, which can hinder the management (Eslami 2011). Moreover, the seed is one of the routes of entry of herbicides, in addition to parts of the seedlings, such as hypocotyl, radicle, and caulicle. Many herbicides act, that is, have their mechanisms of action linked to the germination process, preventing the plant from establishing itself.

If the seed is not in dormancy and there are favorable environmental conditions, such as adequate water supply, temperature, oxygen concentration and presence or absence of light, being positive or negative photoblastic, it will enter in the germination process (Mondo et al. 2010, Yamashita and Guimarães 2010, Saeed et al. 2020). Temperature is one of the most important factors in the germination process of weed seeds. Saeed et al. (2020) reported that seeds of common cocklebur (*X. strumarium*) had high influence of temperature on germination, with highest germination percentage near 30 °C (Fig. 3a).

Even under ideal germination conditions, the weeds have difficulty emerging, especially at greater depths in the soil profile. Plants of sheep tick were not able to emerge when buried at a depth of 15 cm below the soil surface (Fig. 3b).

Light is another important factor affecting the germination of weed seeds. Light-responsive plants are called photoblastic. When they germinate in the presence of light, they are called **positive photoblastics**; and when they germinate in the absence of light, i.e., in the dark, they are **negative photoblastics**. However, when they germinate both in the presence and absence of light, they are called **non-photoblastic** or **indifferent** plants (Carvalho 2013).

Table 3 Seed production capacity per plant in different weed species

Common name	Scientific name	Seeds number by plant	Local	Source
Common waterhemp	*Amaranthus rudis*	288.950	USA	Sellers et al. (2003)
Palmer amaranth	*Amaranthus palmeri*	250.700	USA	
Redroot pigweed	*Amaranthus retroflexus*	291.570	USA	
Smooth pigweed	*Amaranthus hybridus*	254.540	USA	
Spiny amaranth	*Amaranthus spinosus*	113.950	USA	
Tumble pigweed	*Amaranthus albus*	50.090	USA	
Common lambsquarters	*Chenopodium album*	8.749–371.199	United Kingdom	Grundy et al. (2004)
Bristly starbur	*Acanthospermurn hispidum*	166 a 815	India	Hosamani et al. (1971)
Common cocklebur	*Xanthium strumarium*	26 a 1.205	India	
		8.938	USA	Bararpour and Oliver (1998)
Hairy Beggarticks	*Bidens pilosa*	1.244–10.507	India	Shivakumar et al. (2014)
		687	Brazil	Santos et al. (2002)
	Bidens spp.	960–3.620	Brazil	Fleck et al. (2003)
Florida beggarweed	*Desmodium tortuosum*	714	Brazil	Santos et al. (2002)
Tropical whiteweed	*Ageratum conyzoides*	2.808	India	Shivakumar et al. (2014)
Hairy Fleabane Horseweed	*Conyza bonariensis*	76.408–121.482	Brazil	Piasecki et al. (2019)
		119.100	Australia	Wu et al. (2007)
		366.425–878.086	Brazil	Kaspary et al. (2017)
	Conyza canadensis	100.000–200.000	USA	Bhowmik and Bekech (1993)
Sowthistle	*Sonchus oleraceus*	8.289–46.050	Australia	Mobli et al. (2020)
Dandelion	*Taraxacum officinale*	2.170	United Kingdom	Bostock and Benton (1979)
Wild radish	*Raphanus raphanistrum*	503–1.250	Australia	Blackshaw et al. (2002)
		59–1.030	Australia	Reeves et al. (1981)
Arrowleaf sida	*Sida rhombifolia*	640–1.340	Brazil	Fleck et al. (2003)
Prickly Sida	*Sida spinosa*	1.231	USA	Walker and Oliver (2008)

Morning glory	*Ipomoea purpurea*	15.000–26.000	USA	Crowley and Buchanan (1982)
Cypress vine	*Ipomoea quamoclit*	9.000	USA	
Wild Poinsettia	*Euphorbia heterophylla*	478–488	Brazil	Santos et al. (2002)
		110–125	Brazil	Hassanpour-Bourkheili et al. (2020)
Sicklepod	*Senna obtusifolia*	11.420	USA	Bararpour and Oliver (1998)
		350	USA	Walker and Oliver (2008)
Curly dock	*Rumex crispus*	1.575–24.932	Czech Republic	Hejcman et al. (2012)
Windmill Grass	*Chloris polydactyla*	3.175–30.000	Brazil	Carvalh et al. (2005)
Benghal dayflower	*Commelina benghalensis*	1.600	Philippines	Pancho (1964)
Goosegrass	*Eleusine indica*	120.000	Brazil	Takano et al. (2016)
Junglerice Barnyardgrass	*Echinochloa colona*	2.200–3.100	Philippines	Chauhan and Johnson (2010)
		603–7.716	Philippines	Chauhan (2013)
	Echinochloa crus-galli	2.100–2.900	Philippines	Chauhan and Johnson (2010)
		682–2.977	Philippines	Chauhan (2013)
Itchgrass	*Rottboellia cochinchinensis*	460–560	Philippines	Chauhan (2013)
Johnsongrass	*Sorghum halepense*	28.000	Israel	Horowitz (1973)
Italian Ryegrass	*Lolium multiflorum*	633–1.382	Brazil	Vargas et al. (2005)
Sourgrass	*Digitaria insularis*	662–1.155	Brazil	Ferreira et al. (2018)

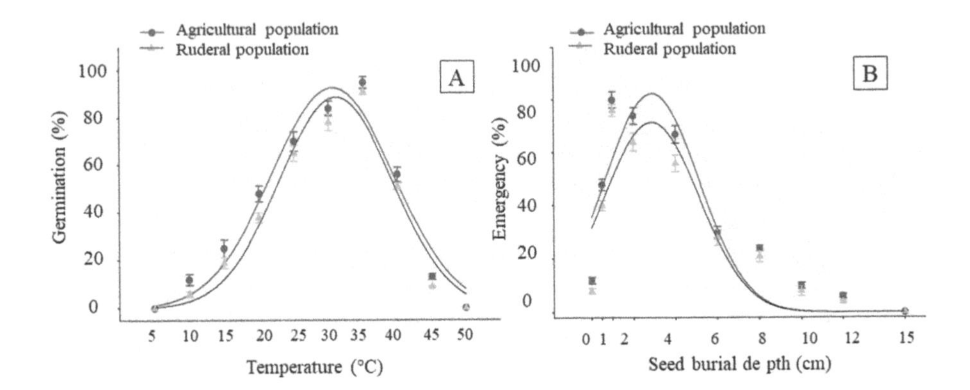

Fig. 3 Influence of temperature on seed germination percentage (**a**) and seeding depth (**b**) of populations acquired in agricultural and non-agricultural environments of *Xanthium strumarium*. (Source: Saeed et al. (2020))

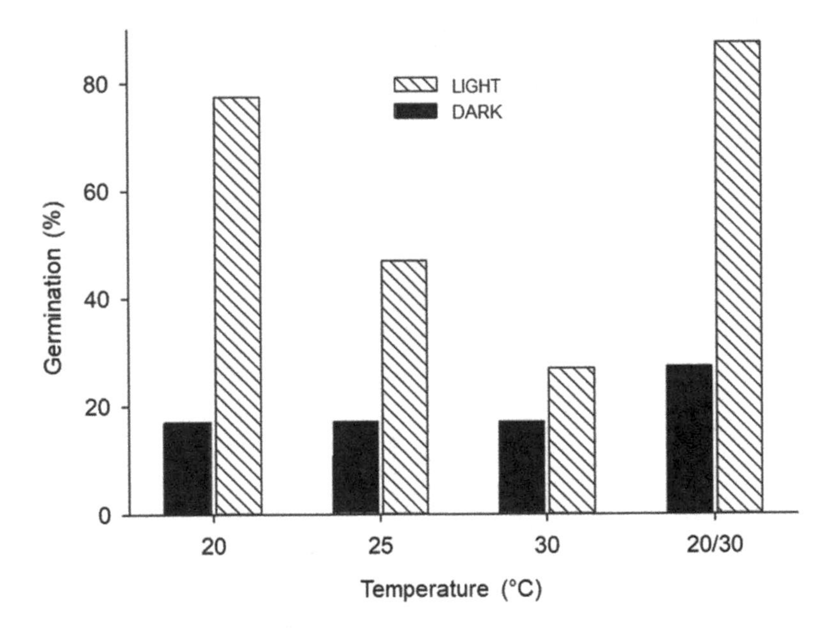

Fig. 4 Germination of *Conyza* spp. in four thermal regimes and two light levels, evaluated at 12 days after sowing. (Source: Adapted from Vidal et al. (2007))

Horseweed, hairy fleabane or tall fleabane (*Conyza* spp.) is a positive photoblastic weed that emerges in soils with low-disturbance production systems (no-till and fruit growing areas) (Fig. 4). In addition to the response to light, the weed has a high germination response in periods with milder temperatures during the year, i.e., during the desiccation season for planting winter crops and in citrus during the spring, which is evidenced by the higher germination of the weed at a temperature of 20 °C (Vidal et al. 2007).

Asexual reproduction (vegetative propagation) is a survival mechanism of great importance in perennial weeds. Propagules are classified as: stolons, rhizomes, tubers, bulbs, shoots, roots, stems or fragments of these organs, which have two essential characteristics: dormancy and food reserves. Thus, certain species such as johnsongrass (*Sorghum halepense*) and bermuda grass (*Cynodon dactylon*), which have, besides seeds, vegetative reproduction through rhizomes and stolons, respectively, are more competitive because they have a high reproductive capacity.

There are several types of vegetative propagation structures of weeds, which stand out (Fig. 5):

Bulbs: Bulbs are modified underground structures for reserve and vegetative reproduction, formed by part of the stem and modified leaves. For example, common wood sorrel (*Oxalis latifolia*).

Stolon: It is a type of stem that develops into adventitious roots and an aerial part in the region of the nodes, parallel to the ground. For example, bermuda grass (*Cynodon dactylon*) and rhodes grass (*Chloris gayana*).

Rhizome: It is a long underground stem that produces adventitious roots and aerial part, with reserve function and vegetative reproduction. For example, johnsongrass (*Sorghum halepense*) and sourgrass (*Digitaria insularis*).

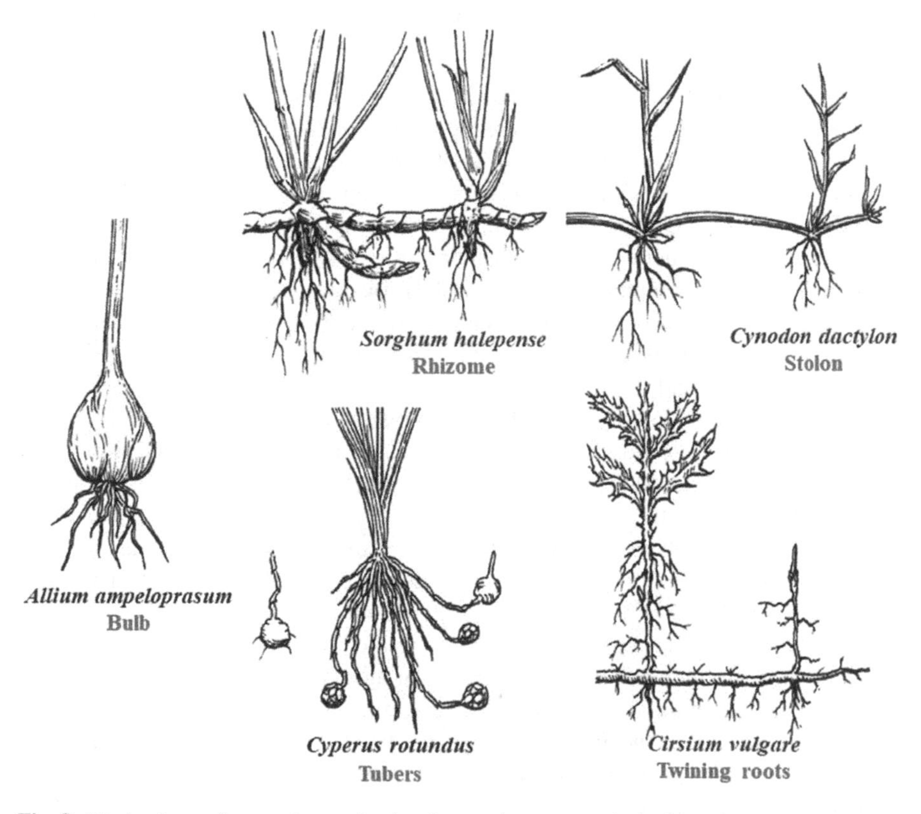

Fig. 5 Mechanisms of asexual reproduction (vegetative propagation) of weeds

Tuber: It is a type of underground stem, usually short and rounded, that has a large amount of reserves and buds. For example, Purple nutsedge (*Cyperus rotundus*) and yellow nutsedge (*Cyperus esculentus*).

Twin roots: These are underground roots, parallel to the soil surface, which have buds capable of regenerating the aerial part. For example, mama-cadela (*Brosimum gaudichaudii*) and musk thistle (*Carduus nutans*).

6.2 Dispersion

One of the main characteristics that gives weeds aggressiveness is the ability to spread their propagation structures, mainly seeds. Factors such as wind, water, animals, including man, are disseminators of weeds.

The dispersal of weed seeds and propagules can be carried out by their own means, i.e. from the mother plant, called **Autocory**, or by external dispersal agents, i.e. from the environment, called **Allocory**. In the case of autocoria, some species have the ability to release their seeds, such as wild poinsettia (*E. heterophylla*) and castor oil plant (*Ricinus communis*).

Weed dispersal by external agents is classified according to the dispersing agent (Fig. 6):

Anemocoria (Wind): This type of abiotic dispersal of weed seeds is the result of wind-driven aerial movement, where the seeds typically land on the ground or on obstacles consisting of vegetation or on soil microsites such as burlap. Dispersal ability in this case depends on wind speed, structure and size of the flight surface, and the size of the seed. One example is the horseweed (*Conyza canadensis*), native to North America that is expanding in many parts of the world due to its spatial dispersal potential (Benvenuti 2007). Besides all the plants of the genus Conyza, there are also species such as sourgrass (*Digitaria insularis*), dandelion (*Taraxacum officinale*), and sowthistle (*Sonchus oleraceus*) that also disperse with the aid of wind.

Hydrocoric (Water): In this mode of spatial movement of diaspores, the seeds are transported by water runoff or by flotation along drainage channels, rivers, and floods. In this case it is indispensable for the hydrocoric dispersion is the buoyancy of the seeds, the proximity of drainage ditches, rivers and also the water flow interferes with the speed and capacity of dispersion. Another factor is the time of seed production by plants and the period of greater rainfall. As an example, we have the curly dock (*Rumex* spp.) and tall windmill grass (*Chloris elata*).

Zoocoria (Animals): Dispersal by animals is divided into epizoocoria, which refers to dispersal through adherence to the skin of animals such as cattle, horses, and sheep, and wild animals such as rats and mice, as an example, the common cocklebur (*Xanthium strumarium*), the hairy beggarticks (*Bidens* spp.), and the bristly starbur (*Acanthospermum hispidum*), which have structures capable of adhering to animals. The endozoocoria is a form of dispersal that consists of the ingestion

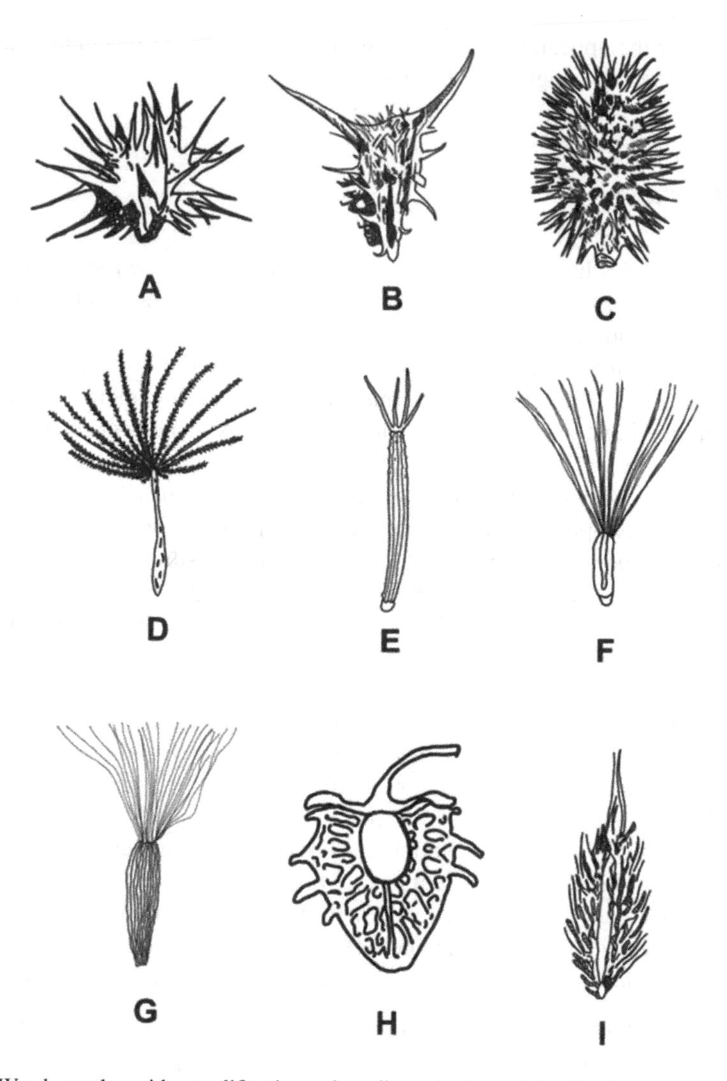

Fig. 6 Weed seeds with modifications for dissemination: (**a**) *Cenchrus echinatus*, (**b**) *Acanthospermum hispidum*, (**c**) *Xanthium strumarium*, (**d**) *Taraxacum officinale*, (**e**) *Bidens subalternans*, (**f**) *Conyza* spp., (**g**) *Sonchus oleraceus,* (**h**) *Rumex obtusifolius*, (**i**) *Digitaria insularis*. (Drawings: Courtesy of Antônio Vitor Braga Martins)

of the diaspore by the animals, and its movement and excretion of the diaspore by the surrounding fauna. Myrmecocoria is the dispersal performed by ants, which represents an example of mutualistic interaction between plant and animal species. Finally, anthropocoria is the dispersal of seeds by human activity.

Any agronomic activity, such as the seed trade, can intensify accidental seed dispersal or lead to the introduction of exotic weed species. An example of this is the occurrence of palmer amaranth (*Amaranthus palmeri*) in Brazil, where the main hypothesis for the appearance of this weed is by the entry of machinery contaminated with this weed, originating from other countries.

6.3 Dormancy

Some weed species have developed mechanisms that allow their survival. Among these is dormancy, which represents one of the main abilities of plant species to ensure survival and perpetuation.

There are two mechanisms of dormancy, the first linked to internal events of the seeds (embryo) and the second to external characteristics (tegument, endosperm or barriers imposed by the fruit) called endogenous and exogenous dormancy, respectively (Nikolaeva 1977, Vivian et al. 2008). From this, dormancy is considered as the seeds whose morphological and biochemical characteristics do not allow germination, even under optimal environmental conditions; and dormancy as the absence of germination due to one or more internal factors (understood as endogenous dormancy) that act in the germination process under appropriate water, thermal, and gas conditions. The overcoming is proven when the seed is able to soak up water and emit the primary root, that is, when germination occurs (Vivian et al. 2008).

It is important to emphasize the difference between dormancy and quiescence, because although they are difficult to separate conceptually, they are distinct events in practice, because the quiescence, is a state of rest in which, being viable the seed, is easily overcome with the provision of the necessary environmental conditions (Braccini 2011). The dormant seed, on the other hand, is not able to germinate, in a specified period of time, in the combinations of environmental conditions in which the seed would normally germinate if it were not as in the state of dormancy, that is, the dormant seed does not germinate in environmental conditions normally considered favorable or adequate (Baskin and Baskin 2004, Cardoso 2009).

Weed dormancy can be subdivided according to its mechanism of dormancy:

Physiological dormancy: It occurs when the embryo presents some specific physiological mechanism that prevents the production of the primary root, and can be subdivided into deep, intermediate or superficial physiological dormancy, according to the intensity. The deep dormancy occurs in cases where the seed needs, for example, a long period with low temperatures to overcome it. The intermediate and superficial dormancy is caused by the absence of embryo growth or by the generation of abnormal seedlings, even when the embryo is isolated from the seed. The intermediate and superficial dormancy are more common, being necessary the isolation of the embryo to generate normal seedlings (Vivian et al. 2008).

Morphological dormancy: It occurs in species that have a rudimentary or immature embryo, that is, seeds in which the embryo has not completed its growth or final development.

Physical Dormancy: Usually occurs in species that have large seeds, whose embryo stores a large part of the reserves. This mechanism is associated with impermeability to water, protected by layers of single or double lignified cells.

Chemical dormancy: This is imposed by the presence of inhibitors in the fruit pericarp, in which compounds are translocated to the seeds before dispersal or disconnection from the mother plant and prevent the growth of the embryo and germination. The main inhibitors are gibberellic acid, abscisic acid, indoleacetic acid, cytokinins, and ethylene.

Mechanical Dormancy: This is defined as the inhibition of germination by the presence of hard fruits or woody walls, usually attributed to the endocarp or extended to the mesocarp of species.

Seed dormancy is classified into two forms: *primary or natural dormancy* and *secondary or induced dormancy* (Benech-Arnold et al. 2000). *Primary dormancy* is an innate seed characteristic, developed while present in the mother plant and which remains after dispersal. *Secondary dormancy* refers to the state of induction of dormancy, under conditions not favorable to germination, in non-dormant seeds or in those whose primary dormancy has been overcome.

Some factors affect weed seed dormancy, such as temperature and water availability, light, nitrate, and other soil compounds (Vivian et al. 2008):

Temperature: Temperature is the main environmental factor that causes changes in seed dormancy, especially in temperate regions. The break of dormancy occurs when the temperature is adequate, in some cases, in addition to cooling or increasing the temperature, the fluctuation in daily temperature is even more important to overcome dormancy.

Water: The hydric state of the seeds is the second most important factor in inducing and overcoming dormancy, although water is interrelated with the other factors, such as temperature and light. The hydration and drying of seeds influences the increase or decrease in germination rates and percentage of weed emergence. For many species, the overcoming of seed dormancy is achieved after repeating these drying and hydration cycles.

Light: The response to light is distinct among species, being mainly related to phytochromes. The action of light in the red and far red region on the phytochrome promotes the alteration of its isomeric form allowing the balance between the active (far red phytochrome, Fvd—P730) and inactive form (red phytochrome, Fv—P660). When the Fvd/Fv ratio is high, a greater stimulus to germination occurs, however, when the ratio is lower, the predisposition to the dormant state of the seeds is greater (Figs. 7 and 8).

Nitrate: There are many compounds present in soils that have some effect on the seed bank. However, only nitrate and nitrite seem to significantly influence seed dormancy. The action of nitrate can be both favorable in inducing and overcoming dormancy for many species.

Some methods can be used to overcome weed seed dormancy. There are naturally occurring processes under environmental conditions that allow overcoming

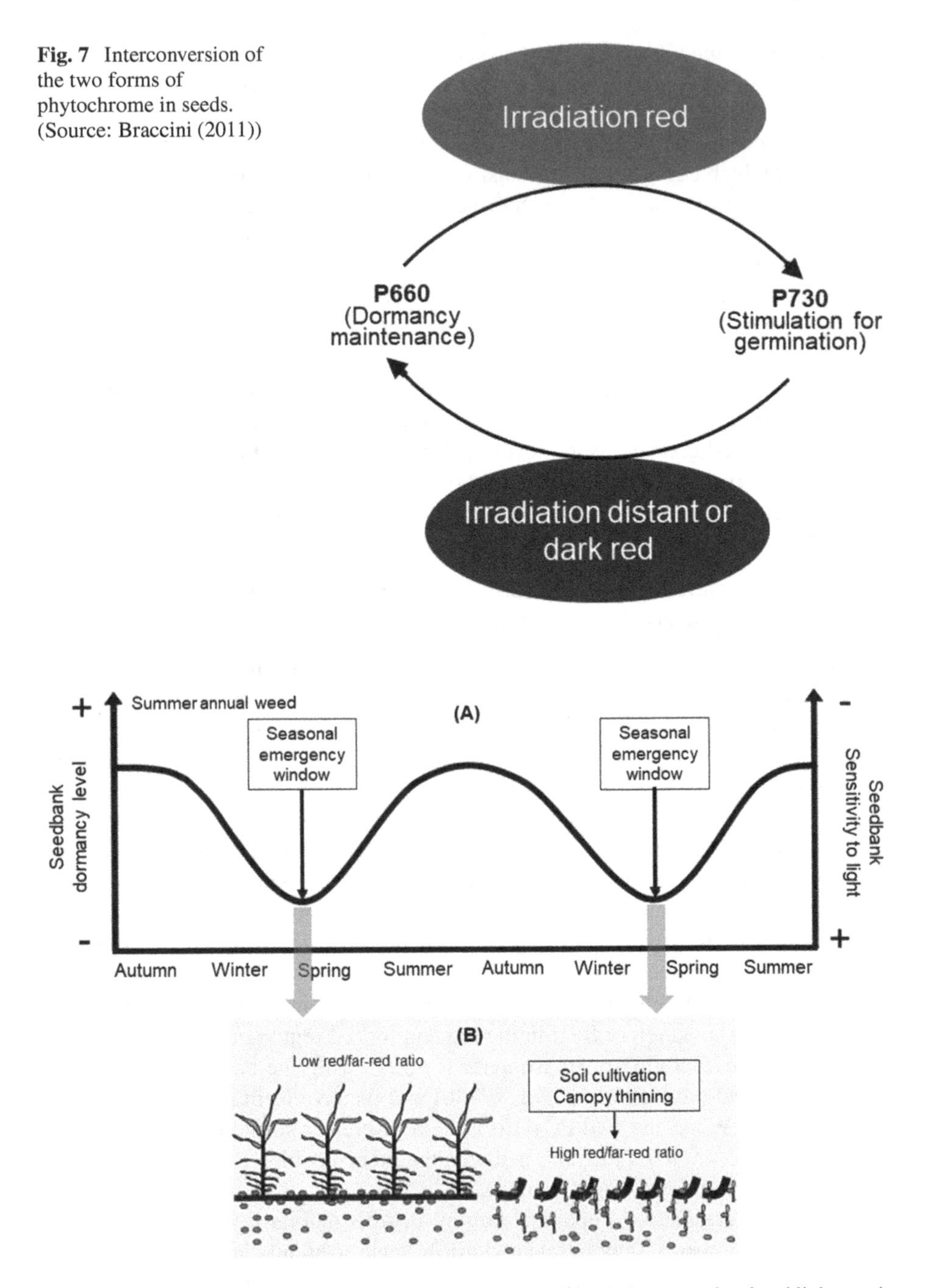

Fig. 7 Interconversion of the two forms of phytochrome in seeds. (Source: Braccini (2011))

Fig. 8 Schematic representation of seasonal changes in seed bank dormancy level and light sensitivity for a summer annual weed (**a**) and schematic representation of the termination of dormancy by light in sensitive seeds (seeds with low level of dormancy) (**b**). (Source: Adapted from Batlla and Benech-Arnold (2014))

Table 4 Methods to overcome the main mechanisms of seed dormancy

Dormancy type	Overcoming methods
Impermeability and mechanical constraints of the integument	Immersion in solvents (hot water, alcohol, acetone, etc.)
	Mechanical scarification
	Scarification with sulfuric acid
	Rapid cooling
	Exposure to high temperature
	Increased oxygen tension
	Shock or impact against hard surfaces
Dormant embryo	Stratification at low temperature Treatment with hormones (gibberellins or cytokinins)
Dormancy in Poaceae	Breaking up of the caryopsis
	Treatment with potassium nitrate
	Exposure to light
	Use of alternate temperatures
	Application of pre-chilling
	Increasing oxygen tension
	Treatment with hormones
	Germination at suboptimal temperature
Impermeable tegument combined with dormant embryo	Mechanical or sulfuric acid scarification followed by low temperature stratification
Double dormancy (dormant epicotyl and radicle)	Stratification at low temperatures, followed by favorable conditions for radicle and epicotyl growth, respectively

Source: Popinigis (1985)

Table 5 Percent germination of weed seeds submitted to treatment without mechanical (*Euphorbia heterophylla* and *Ipomoea nil*) and chemical (*Sida glaziovii* and *Urochloa plantaginea*) dormancy breaking

	Species			
Treatment	*E. heterophylla*	*I. nil*	*S. glaziovii*	*U. plantaginea*
No breakage	64,0 a	47,2 b	2,7 b	24,5 b
With breakage	66,2 a	71,5 a	19,5 a	45,2 a

Source: Salvador et al. (2007)

species dormancy, however, most studies for overcoming dormancy are conducted in the laboratory. Popinigis (1985) suggests some methods for overcoming seed dormancy according to the types of dormancy (Table 4).

The chemical and mechanical methods are the most used to overcome weed dormancy. The mechanical scarification with the help of sandpaper was efficient to break weed seed dormancy (*Ipomoea nil*), however, it was not efficient in overcoming dormancy of wild poinsettia (*Euphorbia heterophylla*) (Table 5). On the other hand, the chemical scarification method, performed with the help of sulfuric acid was efficient to overcome the dormancy of white mallow (*Sida glaziovii*) and alexandergrass (*Urochloa plantaginea*) (Salvador et al. 2007).

6.4 Seed Bank

The most common term to represent the amount of seeds and other propagation structures present in the soil or in plant remains is *seed bank*. However, the more correct term to represent the amount of propagation structures that come from seeds and also vegetative structures is *disasperm bank*. Other terms also used are *diaspore*, which is more closely related to sexual reproduction structures, and *propagules*, which is used for asexual reproduction structures.

Of all the disseminators, seeds are the most important weed reproduction structures. The weed seed bank is the pool of viable seeds present on the soil surface and also scattered along the profile. The seed bank consists of recently released seeds and older seeds that have persisted in the soil for several years (Menalled 2008). In the absence of seed reintroduction to the site, the persistence of weed infestation depends solely on the content of the seed bank and the natural longevity of the species.

Weed seeds can have several fates after being shed on the parent plant. Only some seeds from the bank, will emerge and produce a plant, most seeds will be decomposed, or ingested, before they even germinate (Menalled 2008). Many weed seeds are preyed upon, such as by beetles, crickets, and larger animals like rodents and birds. These animals can consume significant amounts of weed seeds, influencing the seed bank (Fig. 9).

The weed seed bank can be divided into two types: *transient* and *persistent* seed bank. The *transient* seed bank consists of seeds that can remain viable for a maximum of 1 year. On the other hand, the *persistent* seed bank contains seeds that do not germinate during the first year after they are produced and remain viable for more than a year. Non-emergence occurs mainly because many seeds are dormant, both primary and secondary (Braccini 2011).

Table 6 shows the longevity of 41 weed species in two soils over a 17-year period. With this, one can observe the variability of the plant species regarding the longevity of the seed bank. The species of barnyardgrass (*Echinochloa crus-galli*), bristly foxtail (*Setaria verticillata*), green foxtail (*Setaria viridis*), common lambsquarters (*Chenopodium album*), common sunflower (*Helianthus annuus*), tall waterhemp (*Amaranthus tuberculatus*), ivyleaf morning glory (*Ipomoea hederacea*), among others, showed significant emergence after being buried for 17 years. In contrast, species such as jimsonweed (*Datura stramonium*), shattercane (*Sorghum bicolor*) and prickly Russian thistle (*Salsola iberica*), the longevity of the seed bank did not exceed 6 years.

The soil type should not have so much influence in changing the longevity of weed species, because the longevity is more related to the characteristics of the species, environmental conditions and management.

Weed seed longevity in soil depends on the interaction of many factors, including population dormancy, environmental conditions such as moisture, temperature and light, and biological processes such as predation and allelopathy. Management practices also influence the longevity of the seed bank. Soil tillage generally increases seed longevity, as weed seeds generally remain viable longer if they are buried.

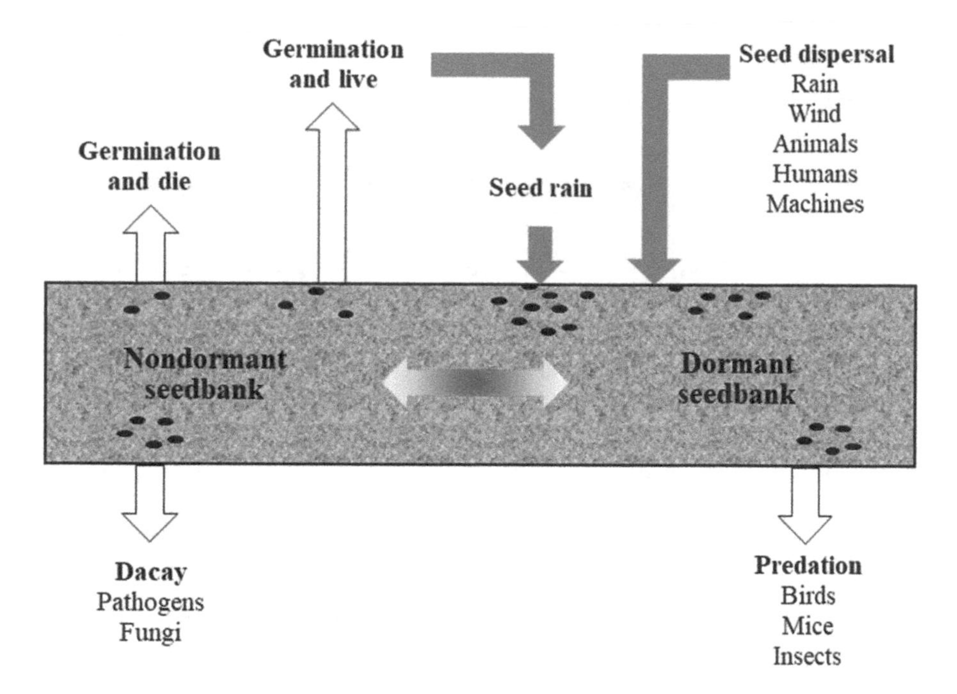

Fig. 9 Destination of weed seeds. Inputs to the seed bank are shown with blue arrows and losses with white arrows. (Source: Adapted from Menalled (2008))

No-till, on the other hand, reduces seed persistence by exposing seeds to predators and pathogens (Menalled 2008).

Weed seeds disperse vertically and horizontally in the soil profile. The cultivation systems are determining factors in the vertical distribution of the seeds in the soil profile. Under conventional tillage, most weed seeds are buried 10–15 cm below the soil surface. In minimum tillage, approximately 80–90% of weed seeds are distributed in the first 10 cm of soil (Hossain and Begum 2015). In no-till, most weed seeds remain at or near the soil surface.

In no-till systems, Clements et al. (1996) demonstrated that most weed seeds remained on the soil surface. However, with the use of a mouldboard plow most of the seeds moved to deeper layers, from 10 to 15 cm deep (Fig. 10). The management system associated with soil texture also influenced the depth of the seeds, in sandy soil the seeds reached the deeper layers more easily with the help of soil preparation equipment (Fig. 10).

The mechanisms of weed seed dispersal promote a change in the seed bank horizontally in the area. The seed bank, if well determined and considering the dormancy of the species, presents a more stable distribution than the emergent flora itself. Thus, the confection of maps of the seed bank can generate important information in the application of herbicides in a controlled manner at varying rates (Shiratsuchi et al. 2021).

Table 6 Seed germination of 41 weed species sown at 20 cm depth over 17 years on a clay soil in Lincoln, Nebraska, USA, from 1976 to 1992

	Years after sowing the seeds											
	0	1	2	3	4	5	6	7	8	9	12	17
Clay soil												
Annual Gramineae	Germination (%)											
Echinochloa crus-galli	17	4	4	19	16	35	20	8	3	8	3	1
Setaria verticillata	74	79	75	55	34	44	20	18	9	26	11	13
Bromus secalinus	20	16	10	7	8	14	0	1	1	5	1	0
Bromus tectorum	99	2	0	1	0	2	0	0	0	.	.	.
Setaria viridis	8	73	61	45	49	34	20	21	13	23	8	5
Bromus japonicus	95	5	1	1	0	0	0	0	.	0	.	.
Aegilops cylindrica	100	54	12	6	1	1	0	0	0	1	0	0
Digitaria sanguinalis	12	48	28	28	19	9	15	4	7	3	1	0
Cenchrus longispinus	25	16	14	4	2	4	2	1	0	0	0	0
Sorghum bicolor	99	43	4	2	0	0	0	0	0	0	0	1
Setaria glauca	94	73	48	36	10	13	7	6	7	22	2	0
Annual broadleaf												
Solanum rostratum	10	3	2	1	1	1	0	0	0	0	1	0
Xanthium strumarium	10	60	36	16	16	4	3	0	18	20	0	0
Chenopodium album	28	53	43	40	40	17	48	36	21	37	42	28
Helianthus annuus	15	1	2	2	1	15	2	1	1	1	5	3
Thlaspi arvense	81	72	3	36	26	43	4	10	52	11	13	2
Solanum sarrachoides	85	90	11	13	6	8	8	1	0	1	1	4
Datura stramonium	93	81	16	23	0	0	0	0	0	0	0	0
Kochia scoparia	100	0	2	0	1	0	11	0	1	1	1	1
Polygonum pensylvanicum	11	3	1	1	1	30	0	0	0	1	0	7

	Years after sowing the seeds											
	0	1	2	3	4	5	6	7	8	9	12	17
Tribulus terrestris	50	42	21	11	9	3	3	2	1	0	0	0
Amaranthus retroflexus	66	73	27	5	8	1	3	3	0	2	7	0
Salsola iberica	73	0	0	0	0	0	0	0	0	0	0	0
Amaranthus tuberculatus	40	38	10	7	12	10	14	3	2	7	6	3
Abutilon theophrasti	15	32	23	43	17	40	70	5	24	41	25	25
Broadleaf biennial												
Verbascum thapsus	98	91	87	83	86	73	92	62	63	74	85	72
Carduus nutans	44	6	6	34	40	33	19	33	30	.	.	.
Onopordum acanthium	79	68	51	45	60	45	50	39	47	.	.	.
Cirsium altissimum	16	7	2	8	14	33	9	16	26	.	.	.
Perennial broadleaf												
Cirsium arvense	60	47	39	44	40	35	31	29	28	34	14	9
Asclepias syriaca	33	36	21	34	28	28	26	5	6	13	0	0
Rumex crispus	76	83	73	88	88	89	87	87	91	86	83	77
Taraxacum officinale	2	3	6	3	1	5	1	0	0	1	0	0
Cirsiumflodmanii	2	0	1	4	5	27	4	5	9	2	3	1
Apocynum cannabinum	74	22	0	0	0	0	0	.	0	.	0	0
Cardaria draba	75	41	21	16	9	3	2	2	5	1	0	0
Solanum carolinense	0	12	2	2	1	2	2	1	1	1	0	1
Ipomoea hederacea	69	4	3	9	9	11	7	4	5	5	8	7
Amorpha canescens	5	2	0	1	0	0	0	0	0	.	.	0
Polygonum coccineum	5	3	2	1	1	20	2	0	1	0	1	14
Ambrosia grayi	0	6	5	7	1	4	1	2	4	.	0	1
Sand soil												

(continued)

Table 6 (continued)

	Years after sowing the seeds											
	0	1	2	3	4	5	6	7	8	9	12	17
Annual Gramineae	Germination (%)											
Echinochloa crus-galli	17	3	58	39	42	31	9	14	4	4	2	0
Setaria verticillata	74	73	33	34	38	22	22	26	33	6	10	0
Bromus secalinus	20	35	31	50	69	26	13	12	25	35	7	13
Bromus tectorum	99	1	0	0	0	0	0	0	0	1	.	.
Setaria viridis	8	51	59	45	11	26	6	18	20	2	0	0
Bromus japonicus	95	10	0	0	0	0	.	0	.	0	.	0
Aegilops cylindrica	100	72	55	37	29	2	0	0	0	0	0	0
Digitaria sanguinalis	12	79	45	42	43	12	1	12	2	0	0	0
Cenchrus longispinus	25	12	8	7	19	13	1	3	1	0	0	1
Sorghum bicolor	99	37	24	7	13	3	0	0	0	0	0	0
Setaria glauca	94	79	85	38	60	9	56	37	25	10	9	0
Annual broadleaf												
Solanum rostratum	10	2	2	2	3	3	1	2	2	0	1	2
Xanthium strumarium	10	60	59	51	65	33	37	41	15	21	0	1
Chenopodium album	28	49	35	23	44	31	14	6	11	16	16	7
Helianthus annuus	15	1	1	2	0	6	1	1	0	2	0	0
Thlaspi arvense	81	61	17	16	17	14	3	1	10	16	15	8
Solanum sarrachoides	85	94	88	79	47	70	82	59	80	49	4	65
Datura stramonium	93	93	93	94	96	89	88	82	92	78	95	90
Kochia scoparia	100	8	2	1	1	1	2	0	0	1	0	0
Polygonum pensylvanicum	11	20	12	7	6	7	0	2	7	1	0	0
Tribulus terrestris	50	43	38	29	23	15	3	15	6	1	3	2
Amaranthus retroflexus	66	69	38	40	40	37	9	2	6	5	7	1

	Years after sowing the seeds											
	0	1	2	3	4	5	6	7	8	9	12	17
Salsola iberica	73	0	1	1	0	0	0	0	0	0	0	0
Amaranthus tuberculatus	40	42	39	14	24	23	0	8	9	8	14	1
Abutilon theophrasti	15	35	27	35	53	50	60	40	57	36	29	35
Broadleaf biennial												
Verbascum thapsus	98	95	88	82	90	88	90	84	79	55	90	95
Carduus nutans	44	39	36	35	33	38	23	23	10	4	.	.
Onopordum acanthium	79	65	55	68	64	51	65	46	62	31	.	.
Cirsium altissimum	16	28	24	47	35	19	28	26	7	6	.	.
Perennial broadleaf												
Cirsium arvense	60	35	29	21	30	29	25	25	1	11	17	7
Asclepias syriaca	33	31	33	14	13	5	0	0	0	0	0	0
Rumex crispus	76	92	93	85	70	74	94	84	91	22	73	61
Taraxacum officinale	2	12	5	10	1	1	2	4	0	0	0	0
Cirsiumflodmanii	2	3	2	12	8	11	11	7	1	0	.	2
Apocynum cannabinum	74	52	13	0	1	0	1	.	0	0	0	0
Cardaria draba	75	37	28	8	30	13	4	1	2	0	0	0
Solanum carolinense	0	12	7	9	7	7	6	6	5	5	4	5
Ipomoea hederacea	69	5	10	6	10	9	6	4	6	3	6	3
Amorpha canescens	5	3	0	0	1	0	0	0	0	0	.	.
Polygonum coccineum	5	7	6	5	8	7	0	2	14	2	.	0
Ambrosia grayi	0	15	10	2	7	8	11	5	8	6	3	0

Source: Burnside et al. (1996)
(.) Missing Data

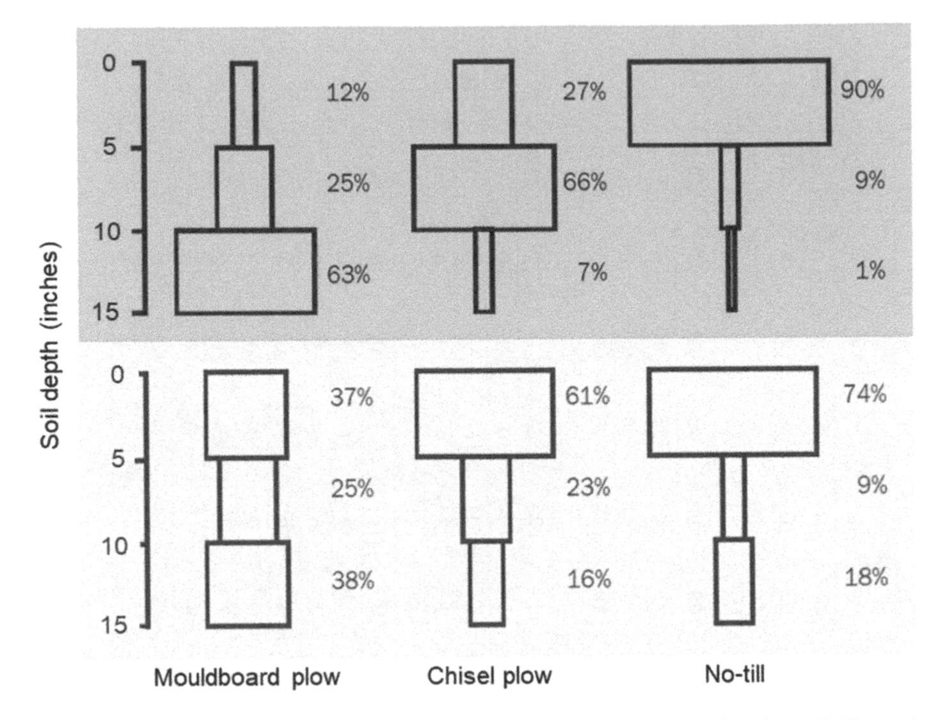

Fig. 10 Vertical seed distribution (%) of weeds in a sandy soil (Blue) and a clay soil (Orange). (Source: Adapted from Clements et al. (1996) and Menalled (2008))

The density of the seed bank in the soil can present high variability within the same area. The seed bank of alexandergrass (*Urochloa plantaginea*) varied in the range of 0–523 to 4438–4735 seeds m^{-2} and of benghal dayflower (*Commelina benghalensis*) from 0–261 to 1306–1392 seeds m^{-2} within the same crop plot (Fig. 11). Furthermore, the correlation of seed bank density as a function of evaluation time may be high, showing a high stability of the seed bank.

Weed seed banks can be greatly reduced by eliminating seed production for a few years, on the other hand, areas can be rapidly reinfested with seeds if weeds are not managed (Gulden and Shirtliffe 2009).

Reducing weed seeds entering the seed bank is the most efficient way to reduce the weed seed bank. Any method that reduces the weed population will also reduce the number of seeds deposited in the seed bank. There are several methods that increase the death of the seeds in it or encourage germination when the weeds can then be easily controlled (Hossain and Begum 2015). Among the methods to control the weed seed bank are the preventive method, use of herbicides, crop rotation, straw removal, soil preparation, mulching, and fertilization (Menalled 2008, Gulden and Shirtliffe 2009, Hossain and Begum 2015), as described below.

Preventive: The most effective approach to reducing weed seed banks is to not allow the production of new seed. In addition, care must be taken to prevent the

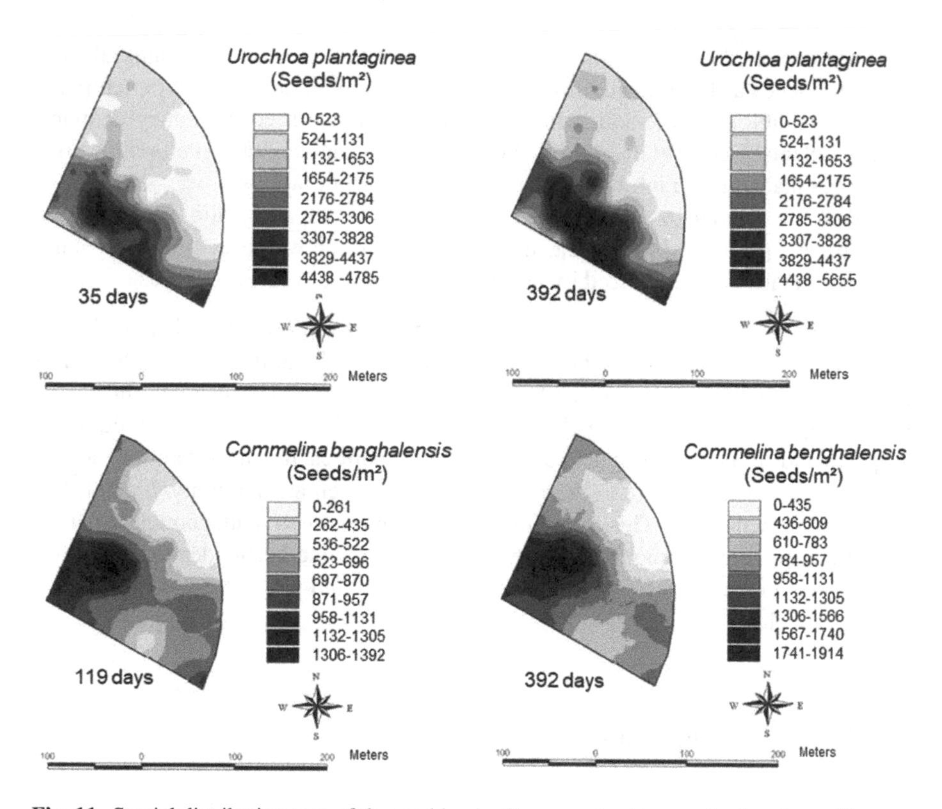

Fig. 11 Spatial distribution map of the seed bank of two weeds at two evaluation times. (Source: Shiratsuchi et al. (2021))

entry of new seed from other areas through irrigation, equipment (seeders and harvesters), and animals.

Herbicide use: The use of herbicides is still the most efficient weed control tool. Herbicides are very effective at reducing weed populations and at the same time reducing the number of seeds added to the soil seed bank.

Crop rotation: Rotation can cause a change in the species composition of the weed community. Knowing these changes can help change the composition of the seed bank and make it easier to employ other management methods.

Straw collection: Straw collection is an effective method to reduce entry into the weed seed bank. Weed seeds generally weigh less than crop seeds and therefore end up in the straw. Even for large weed seeds, straw collection can prevent over 90% of the number of weed seeds that would otherwise be added to the seed bank.

Soil preparation: The degree of soil inversion and the depth of soil preparation strongly affected the vertical distribution of weed seeds in the seed bank. Deeply buried seeds that remain intact may persist in the soil seed bank longer, but in the no-till system, most seeds remain on the soil surface and may be preyed upon by animals or become more prone to conditions that make it possible to initiate the germination process.

Mulch: Crop residues create microenvironments that provide cover for animals that use the seeds as food. In addition, residues have a moderating effect on temperature fluctuations in the soil, which in turn can impact seed dormancy of many weed plants. Recently incorporated crop residues may contain allelopathic substances that can also inhibit seed germination of some weeds.

Fertilization: Similar to crops, weeds can respond to the application of inorganic fertilizers, particularly nitrogen. In the long term, the weed seed bank of many weed species can be reduced by applying fertilizer at the correct time.

Understanding weed behavior, involving aspects of reproduction, dispersal, dormancy and seed bank, helps to better understand weed dynamics in the agricultural and non-agricultural environment. This information can help reduce the impacts of weed competition on crops and also help in deciding the best control techniques to adopt. In addition to better understanding these dynamics, it is important to classify and group weeds according to their characteristics, such as taxonomy, habitat, life cycle, and growth habit. This grouping is important in weed management, because some management techniques and the selectivity of some herbicides are based on specific weed characteristics, such as morphological and physiological differences between species.

7 Weeds Classification

To facilitate correct identification of the species, it is important to know some characteristics that allow grouping of weeds. In some cases, management techniques and the selectivity of some herbicides are based on specific weed characteristics, such as morphological and physiological differences between species.

The main forms of weed classification are based on taxonomy, habitat, life cycle and growth habit. In the following, the different weed species are grouped according to their characteristics:

7.1 Taxonomic Classification

The basis of modern classification is taxonomy, the identification, naming and grouping of plants according to their common characteristics. The goal of this classification is to correctly identify weed species so that the best method of weed community control can be chosen.

The taxonomic system accepted and used today classifies organisms into a hierarchy of categories: kingdom, class, order, family, genus, and species. Most weeds occur in the phylum Anthophyta (angiosperms). Angiosperms are subdivided into the classes Dicotyledons (eudicots or dicotyledons) and Monocotyledons (monocots). Regarding the order, there is no consensus for this type of weed classification.

Within orders. Are divided into families, which, like classes and orders, are composed of plants whose morphological similarities are greater than their differences (Radosevich et al. 2007).

The main weed families according to the number of species in Brazil are Poaceae and Asteraceae, with 40% of the total number of weeds. Next are Malvaceae, Fabaceae, and Amaranthaceae with 7, 6, and 5% of the total number of species. The other families have a smaller number of weed species (Table 7 and Fig. 12).

Table 7 shows the main weed species that occur in Brazil, divided according to their families, as well as their common name, life cycle, and reproduction method.

Many weeds show high similarity among species, which is the case for the genus Amaranthus. Amaranthus are present in much of the agricultural areas in Brazil, among the most common species can be highlighted: *Amaranthus deflexus* (perennial pigweed), *Amaranthus hybridus* (smooth pigweed), *Amaranthus lividus* (livid amaranth), *Amaranthus retroflexus* (redroot pigweed), *Amaranthus spinosus* (spiny amaranth), *Amaranthus viridis* (slender amaranth) (Carvalho et al. 2006), and more recently *Amaranthus palmeri* (palmer amaranth) (Fig. 13).

As the species of this genus vary in growth, development, and sensitivity when applied to the same herbicide, it is important to differentiate between the species (Carvalho et al. 2006, 2008). Some species have a more prostrate growth habit, with decumbent branches, especially *A. deflexus* and *A. lividus*, full of secondary branching. On the other hand, the species *A. hybridus, A. retroflexus* and *A. viridis* have more vertical growth, with emphasis on primary branching (Carvalho et al. 2008).

In the field, the identification of plants at the species level becomes more difficult due to the high phenotypic plasticity (Maluf 1994, Carvalho and Christoffoleti 2007), besides the existence of reports on the viability of hybridization, making identification more difficult. For the same species, differences are found in leaf length and width, as well as in the estimation of leaf area (Carvalho and Christoffoleti 2007).

The first step in the differentiation of plants can be done by the fruits, which are of two types: pixidia (dehiscent) or utricles (indehiscent). The dehiscence of the fruits is confirmed by projecting the inflorescence against the palm of the hand. In the case of dehiscent fruits numerous small, spherical, black colored seeds are released. The exception is made for *A. viridis*, of which Kissmann and Groth (1999) classify the fruit as utricle, that is, indehiscent; however, under practical conditions the seed is easily obtained, evidencing the high dehiscence of the fruit. Next, a simplified dichotomous key is proposed (Fig. 14), based on Kissmann and Groth (1999), Lorenzi (2000) and Moreira and Bragança (2010), Senna (2015) and Carvalho (2016).

7.2 Habitat Classification

Weeds can be classified according to where they grow. Most weeds are terrestrial, but some are restricted to the aquatic environment. There are plants that only infest a certain crop or crop system, plant communities, or growing condition. Therefore,

Table 7 Main weeds in cultivated areas in Brazil divided by family, scientific name, common name, life cycle, and reproduction method

Family	Scientific name	Common name	Life cycle	Reproduction method
Class Magnoliopsida (eudicotyledons)				
Amaranthaceae	*Alternanthera tenella* Colla.	Joyweed	Perennial	Seed
	Amaranthus deflexus L.	Perennial pigweed	Annual	Seed
	Amaranthus hybridus L.	Smooth pigweed	Annual	Seed
	Amaranthus lividus L.	Livid Amarant	Annual	Seed
	Amaranthus retroflexus L.	Redroot pigweed	Annual	Seed
	Amaranthus spinosus L.	Tumble pigweed	Annual	Seed
	Amaranthus viridis L.	Slender Amaranth	Annual	Seed
	Amaranthus palmeri	Palmer amaranth	Annual	Seed
	Chenopodium album L.	Common lambsquarters	Annual	Seed
	Chenopodium ambrosioides L.	Mexican tea	Annual	Seed
	Gomphrena celosioides Mart.	Prostrate globe-amaranth	Perennial	Seed
Asteraceae	*Acanthospermum australe* (Loefl.) Kuntze	Paraguayan starbur	Annual	Seed
	Acanthospermum hispidum DC.	Bristly starbur	Annual	Seed
	Achyrocline satureioides (Lam.) DC.	Marcela	Annual	Seed
	Ageratum conyzoides L.	Tropical whiteweed	Annual	Seed
	Ambrosia elatior L.	Ragweed	Annual	Seed
	Artemisia Verlotorum Lamott	Chinese wormwood	Perennial	Seed and rhizome
	Bidens pilosa L.	Hairy Beggarticks	Annual	Seed
	Bidens subalternans DC.	Greater Beggarticks	Annual	Seed
	Blainvillea acmella (L.) Philipson	Para cress flower	Annual	Seed
	Conyza bonariensis (L.) Cronquist	Hairy fleabane	Annual	Seed
	Conyza canadensis (L.) Cronquist	Horseweed	Annual	Seed
	Conyza sumatrensis	Sumatran fleabane	Annual	Seed

	Eclipta alba (L.) Hassk.	False daisy	Annual	Seed
	Emilia coccinea (Sims) G. Don	Tassel flower	Annual	Seed
	Emilia fosbergii Nicolson	Florida tasselflower	Annual	Seed
	Galinsoga parviflora Cav.	Gallant soldier	Annual	Seed
	Galinsoga quadriradiata Ruiz & Pav.	Shaggy soldier	Annual	Seed
	Gnaphalium spicatum Lam.	Spiked cudweed	Annual	Seed
	Hypochaeris chillensis (Kunth) Britton	Brazilian catsear	Annual or biennial	Seed
	Hypochaeris radicata L.	Hairy cat's ear	Perennial	Seed
	Jaegeria hirta (Lag.) Less.	Toothache plant	Annual	Seed
	Lourteigia ballotifolia (Kunth) R.M.King;H. Rob.	Purple flower	Annual	Seed
	Melampodium divaricatum (Rich. ex Pers.) DC.	Butter daisy	Annual	Seed
	Melampodium perfoliatum (Cav.) Kunth	Melampodium	Annual	Seed
	Parthenium hysterophorus L.	Ragweed parthenium	Annual	Seed
	Pluchea sagittalis	Wingstem camphorweed	Annual or perennial	Seed
	Porophyllum ruderale (Jacq.) Cass.	Yerba porosa	Annual	Seed
	Praxelis pauciflora (Kunth) R. M. King; H. Rob.	Praxelis	Annual	Seed
	Pterocaulon virgatum (L.) DC.	Wand blackroot	Annual	Seed
	Senecio brasiliensis (Spreng.) Less.	Hempleaf ragwort	Perennial	Seed
	Siegesbeckia orientalis L.	Indian weed	Annual	Seed
	Soliva pterosperma	Field burrweed	Annual	Seed
	Sonchus asper (L.) Hill	Spiny-leaved Sowthistle	Annual	Seed
	Sonchus oleraceus L.	Sowthistle	Annual	Seed
	Synedrellopsis grisebachii Hieron. & Kuntze	Straggler daisy	Perennial	Seed and stolon
	Tagetes minuta L.	Stinking roge	Annual	Seed

(continued)

Table 7 (continued)

Family	Scientific name	Common name	Life cycle	Reproduction method
	Taraxacum officinale (L.) Webber ex F. H. Wigg.	Dandelion	Annual	Seed
	Tridax procumbens L.	Coat buttons	Annual or biennial	Seed
	Xanthium strumarium L.	Common cocklebur	Annual	Seed
Boraginaceae	*Echium plantagineum* L.	Paterson's curse	Annual or biennial	Seed
	Heliotropium indicum L.	Indian turnsole	Annual	Seed
Brassicaceae	*Brassica rapa* L.	Birdsrape mustard	Annual	Seed
	Cleome affinis DC.	Spider flower	Annual	Seed
	Coronopus didymus (L.) Smith	Swine wartcress	Annual	Seed
	Lepidium virginicum L.	Virginia pepperweed	Annual	Seed
	Raphanus raphanistrum L.	Wild radish	Annual	Seed
	Raphanus sativus L.	Radish	Annual	Seed
	Rapistrum rugosum (L.) All.	Wild rape	Annual or biennial	Seed
	Sinapis arvensis L.	Wild mustard	Annual	Seed
Caryophyllaceae	*Silene gallica* L.	Common catchfly	Annual	Seed
	Spergula arvensis L.	Field spurry	Annual	Seed
	Stellaria media (L.) Vill.	Chickweed	Annual	Seed

Convolvulaceae	*Ipomoea cairica* (L.) Sweet	Five-fingered morning glory	Perennial	Seed
	Ipomoea hederifolia L.	Scarlet morning glory	Annual	Seed
	Ipomoea nil (L.) Roth.	Morning glory	Annual	Seed
	Ipomoea purpurea (L.) Roth.	Common morning glory	Annual	Seed
	Ipomoea quamoclit L.	Cypress vine	Annual	Seed
	Ipomoea triloba L.	Three-lobed morning glory	Annual	Seed
	Merremia aegyptia (L.) Urb.	Hairy wood rose	Annual	Seed
	Merremia cissoides (Lam.) Hall. f	Roadside woodrose	Annual	Seed
Cucurbitaceae	*Momordica charantia* L.	Bitter gourd	Annual	Seed
Euphorbiaceae	*Astraea lobata* (L.) Klotzsch	Lobed croton	Annual	Seed
	Chamaesyce hirta (L.) Millsp.	Snake weed	Annual	Seed
	Chamaesyce hyssopifolia (L.) Small.	Hyssopleaf sandmat	Annual	Seed
	Croton glandulosus L.	Tooth-leaved croton	Annual	Seed
	Croton lobatus L.	Lobed croton	Annual	Seed
	Euphorbia heterophylla L.	Wild poinsettia	Annual	Seed
	Ricinus communis L.	Castor oil plant	Perennial	Seed

(continued)

Table 7 (continued)

Family	Scientific name	Common name	Life cycle	Reproduction method
Fabaceae	*Aeschynomene denticulata* Rudd.	Jointvetch	Annual	Seed
	Aeschynomene rudis Benth	Sensitive jointvetch	Annual	Seed
	Calopogonium mucunoides Desv.	Calopo	Perennial	Seed
	Crotalaria incana L.	Shakeshake	Annual	Seed
	Crotalaria lanceolata E. Mey.	Lanceleaf rattlebox	Annual	Seed
	Crotalaria pallida Aiton	Rattlebox	Perennial	Seed
	Crotalaria spectabilis Roth.	Showy rattlebox	Annual	Seed
	Desmodium incanum DC.	Zarzabacoa comun	Perennial	Seed
	Desmodium tortuosum (Sw.) DC.	Florida beggarweed	Annual	Seed
	Indigofera hirsuta L.	Hairy indigo	Perennial	Seed
	Mimosa pudica L.	Sensitive plant	Perennial	Seed
	Senna obtusifolia (L.) H. S. Irwin & Barneby	Sicklepod	Perennial	Seed
	Senna occidentalis (L.) Link	Septicweed	Perennial	Seed
	Vigna unguiculata (L.) walp.	Cowpea	Annual	Seed
Lamiaceae	*Cantinoa americana* (Aublet.) Harley e J.F.B Pastore	Matubio	Annual	Seed
	Hyptis atrorubens Poit.	False ironwort	Perennial	Seed
	Hyptis suaveolens Poit.	Wild spikenard	Annual	Seed
	Leonotis nepetifolia (L.) R. Br.	Christmas candlestick	Annual	Seed
	Leonurus sibiricus L.	Honeyweed	Annual or biennial	Seed
	Marsypianthes chamaedrys (Vahl) Kuntze	Paracari	Annual	Seed
	Stachys arvensis L.	Staggerweed	Annual	Seed

Malvaceae	*Anoda cristata (L.) Schltdl*	Spurred anoda	Annual	Seed
	Malvastrum coromandelianum (L.) Garcke	Threelobe false mallow	Annual or biennial	Seed
	Sida acuta Burm. f.	Common wireweed	Perennial	Seed
	Sida ciliaris L.	Bracted sida	Annual	Seed
	Sida cordifolia L.	Flannel sida	Perennial	Seed
	Sida galheirensis Ulbr.	Musk mallow	Perennial	Seed
	Sida glaziovii K. Schum.	White mallow	Perennial	Seed
	Sida planicaulis Cav.	Flatstem sida	Perennial	Seed
	Sida rhombifolia L.	Arrowleaf sida	Annual or perennial	Seed
	Sida spinosa L.	Prickly sida	Perennial	Seed
	Sida urens L.	Bristly sida	Annual or biennial	Seed
	Sidastrum micranthum (St.-Hil.) Fryxell	Dainty sand mallow	Perennial	Seed
	Triumfetta rhomboidea Jacq.	Diamond burbark	Perennial	Seed
	Triumfetta semitriloba Jacq.	Burweed	Perennial	Seed
	Waltheria indica L.	Sleepy morning	Perennial	Seed
	Wissadula subpeltata	Flaxleaf fanpetals	Perennial	Seed
Onagraceae	*Ludwigia leptocarpa* (Nutt.) H. Hara	Anglestem primrose willow	Annual or perennial	Seed
	Ludwigia octovalvis (Jacq.) P. H. Raven	Mexican primrose willow	Perennial	Seed
Oxalidaceae	*Oxalis corniculata* L.	Yellow wood sorrel	Perennial	Seed and stolon
	Oxalis debilis Kunth	Large-flowered pink sorrel	Perennial	Seed, bulbs and stolon
	Oxalis latifolia Kunth	Common wood sorrel	Perennial	Seed and bulbs

(continued)

Table 7 (continued)

Family	Scientific name	Common name	Life cycle	Reproduction method
Papaveraceae	*Argemone mexicana* L.	Mexican prickly poppy	Annual	Seed
Phyllanthaceae	*Phyllanthus tenellus* Roxb	Mascarene island leaf flower	Annual	Seed
	Phyllanthus niruri L.	Seed-under-the-leaf	Annual	Seed
Plantaginaceae	*Plantago major* L.	Broad-leaved plantain	Perennial	Seed
	Plantago tomentosa Lam.	Tanchagem	Annual	Seed
Polygonaceae	*Polygonum convolvulus* L.	Black bindweed	Annual	Seed
	Polygonum hydropiperoides Michx.	Swamp smartweed	Perennial	Seed
	Polygonum persicaria L.	Lady's thumb	Annual or perennial	Seed
	Rumex crispus L.	Curly dock	Perennial	Seed
	Rumex obtusifolius L.	Broad-leaved dock	Perennial	Seed and rhizome
	Rumex acetosella L.	Sheep Sorrel	Perennial	Seed and rhizome
Rubiaceae	*Borreria capitata* (Ruiz & Pav.) DC.	Button weed	Annual	Seed
	Borreria latifolia (Aubl.) K. Schum.	Button weed	Annual	Seed
	Borreria verticillata (L.) G. Mey	Shrubby buttonweed	Perennial	Seed
	Richardia brasiliensis Gomes	White-eye	Annual	Seed
	Richardia grandiflora (Cham. & Schltdl.) Steud.	Largeflower mexican clover	Annual	Seed
	Richardia scabra L.	Rough mexican clover	Annual or perennial	Seed
	Spermacoce capitata Ruiz & Pav.	False buttonweed	Annual	Seed
	Spermacoce latifolia Aubl.	Broadleaf buttonweed	Annual	Seed
	Spermacoce verticillata L.	Shrubby false buttonweed	Perennial	Seed
Sapindaceae	*Cardiospermum halicacabum* L.	Balloon vine	Annual	Seed

Family	Scientific name	Common name	Life cycle	Reproduction method
Scrophulariaceae	*Buddleja stachyoides* Cham. & Schltdl.	Butterflybush	Perennial	Seed
	Verbascum virgatum Stockes	Twiggy mullein	Annual or biennial	Seed
Solanaceae	*Datura inoxia* Mill.	Downy thorn apple	Annual	Seed
	Datura stramonium L.	Jimsonweed	Annual	Seed
	Nicandra physalodes (L.) Gaertn.	Apple of Peru	Annual	Seed
	Physalis angulata L.	Cutleaf groundcherry	Annual	Seed
	Solanum americanum Mill.	American black nightshade	Annual	Seed
	Solanum granuloso-leprosum Dunal	Fumo bravo	Perennial	Seed
	Solanum mauritianum Scop.	Bugweed	Perennial	Seed
	Solanum paniculatum L.	Jurubeba	Perennial	Seed and rhizome
	Solanum sisymbriifolium Lam.	Sticky nightshade	Perennial	Seed
	Solanum viarum	Tropical soda apple	Annual	Seed
Class Liliopsida (monocots)				
Alismataceae	*Sagittaria montevidensis*	California arrowhead	Annual or perennial	Seed and rhizome
Commelinaceae	*Commelina benghalensis* L.	Benghal dayflower	Perennial	Rhizome, stolon and seed
	Commelina diffusa Burm. f.	Spreading dayflower	Annual or perennial	Stolon and seed

(continued)

Table 7 (continued)

Family	Scientific name	Common name	Life cycle	Reproduction method
Cyperaceae	*Cyperus difformis* L.	Smallflower umbrella sedge	Annual	Seed
	Cyperus esculentus L.	Yellow nutsedge	Perennial	Seed and tuber
	Cyperus iria L.	Rice flatsedge	Annual	Seed
	Cyperus odoratus L.	Fragrant flatsedge	Annual or perennial	Seed
	Cyperus rotundus L.	Purple nutsedge	Perennial	Seed, basal bulb, stolon and tuber
	Fimbristylis miliacea (L.) Vahl	Globe fringerush	Annual or perennial	Seed
Marantaceae	*Thalia geniculata* L.	Alligator flag	Perennial	Seed and rhizome
Molluginaceae	*Mollugo verticillata* L.	Green carpetweed	Annual	Seed
Poaceae	*Andropogon bicornis* L.	Broomsedge	Perennial	Seed and rhizome
	Andropogon leucostachyus Kunth	Beard grass	Perennial	Seed
	Andropogon gayanus	Gamba grass	Perennial	Seed
	Aristida longiseta Steud.	Red threeawn	Perennial	Seed
	Avena sativa L.	Common oat	Annual	Seed
	Urochloa brizantha (Hochst. ex. A. Rich.) Stapf	Palisade grass	Perennial	Seed and rhizome
	Urochloa decumbens Stapf.	Signal grass	Perennial	Seed, rhizome and stolon
	Bromus catharticus Vahl	Brome grass	Annual	Seed
	Cenchrus ciliaris L.	African foxtail grass	Perennial	Seed
	Cenchrus echinatus L.	Southern sandbur	Annual	Seed
	Chloris barbata Sw.	Swollen windmill grass	Annual or perennial	Seed
	Chloris gayana Kunth	Rhodes grass	Perennial	Seed and stolon

	Cynodon dactylon (L.) Pers.	Bermuda grass	Perennial	Seed, stolon and rhizome
	Digitaria ciliaris (Retz.) Koeler	Southern crabgrass	Annual	Seed
	Digitaria horizontalis Willd.	Jamaican crabgrass	Annual	Seed
	Digitaria insularis (L.) Fedde	Sourgrass	Perennial	Seed and rhizome
	Digitaria sanguinalis (L.) Scop.	Large crabgrass	Annual or perennial	Seed
	Echinochloa colona (L.) Link	Junglerice	Annual	Seed
	Echinochloa crus-galli (L.) P.Beauv.var.*crus-galli*	Barnyardgrass	Annual	Seed
	Echinochloa crus-pavonis (Kunth) Schult.	Gulf cockspur	Annual	Seed
	Eleusine indica (L.) Gaertn.	Goosegrass	Annual	Seed
	Eragrostis plana Nees	South African lovegrass	Perennial	Seed
	Eragrostis ciliaris (L.) R. Br.	Gophertail lovegrass	Annual	Seed
	Eragrostis pilosa (L.) P. Beauv.	India lovegrass	Annual	Seed
	Eustachys disticophylla (Lag.) Nees	Weeping finger grass	Perennial	Seed
	Leptochloa filiformis (Lam.) P. Beauv.	Red sprangletop	Perennial	Seed
	Lolium multiflorum Lam.	Italian ryegrass	Annual or biennial	Seed
	Megathyrsus maximus (Jacq.) B. K. Simon & S. W. L. Jacobs	Guinea grass	Perennial	Seed and rhizome
	Melinis repens (Willd.) Zizka	Natal redtop		Seed
	Oryza Sativa L.	Weedy rice	Annual	Seed
	Oryza Sativa var. nitrispina Portéres	Weedy rice	Annual	Seed
	Panicum dichotomiflorum Michx.	Fall panicgrass	Perennial	Seed and rhizome
	Panicum maximum Jacq.	Guinea grass	Perennial	Seed and rhizome
	Paspalum paniculatum L.	Russell river grass	Perennial	Seed and stolon

(continued)

Table 7 (continued)

Family	Scientific name	Common name	Life cycle	Reproduction method
	Pennisetum glaucum (L.) R. Br.	Pearl millet	Annual	Seed
	Pennisetum purpureum Schumach.	Elephant grass	Perennial	Seed, rhizome and stolon
	Pennisetum setosum (Sw.) Rich.	Fountain grass	Perennial	Seed and rhizome
	Poa annua L.	Annual bluegrass		
	Rhynchelitrum repens (Willd.) C. E. Hubb	Melinis repens	Annual or perennial	Seed
	Rottboellia cochinchinensis (Lour.) Clayton	Itchgrass	Annual or perennial	Seed
	Setaria parviflora (Poir.) Kerguélen	Knotroot foxtail	Perennial	Seed and rhizome
	Sorghum arundinaceum (Willd.) Stapf.	Wild sorghum	Annual or perennial	Seed
	Sorghum bicolor (L.) Moench	Shattercane	Annual	Seed
	Sorghum halepense (L.) Pers	Johnsongrass	Perennial	Seed and rhizome
	Sporobolus indicus (L.) R. Br.	Rat tail grass	Perennial	Seed
	Urochloa decumbens (Stapf) R. D. Webster	Signal grass	Perennial	Seed
	Urochloa mutica (Forssk.) T.Q. Nguyen	Para grass	Perennial	Seed, rhizome and stolon
	Urochloa plantaginea (Link) R. D. Webster	Alexandergrass	Annual	Seed
Pontederiaceae	*Eichhornia crassipes* (Mart.) Solms	Common water hyacinth	Perennial	Seed and rhizome
	Heteranthera reniformis Ruiz e Pav.	Kidneyleaf mudplantain	Perennial	Seed and rhizome
	Heteranthera limosa (SW.) Willd	Ducksalad	Perennial	Seed, rhizome and stolon
Portulacaceae	*Portulaca oleracea* L.	Common purslane	Annual	Seed
	Talinum paniculatum (Jacq.) Gaertn.	Fameflower	Annual	Seed

Source: Moreira and Bragança (2010, 2011) and Lorenzi (2014)

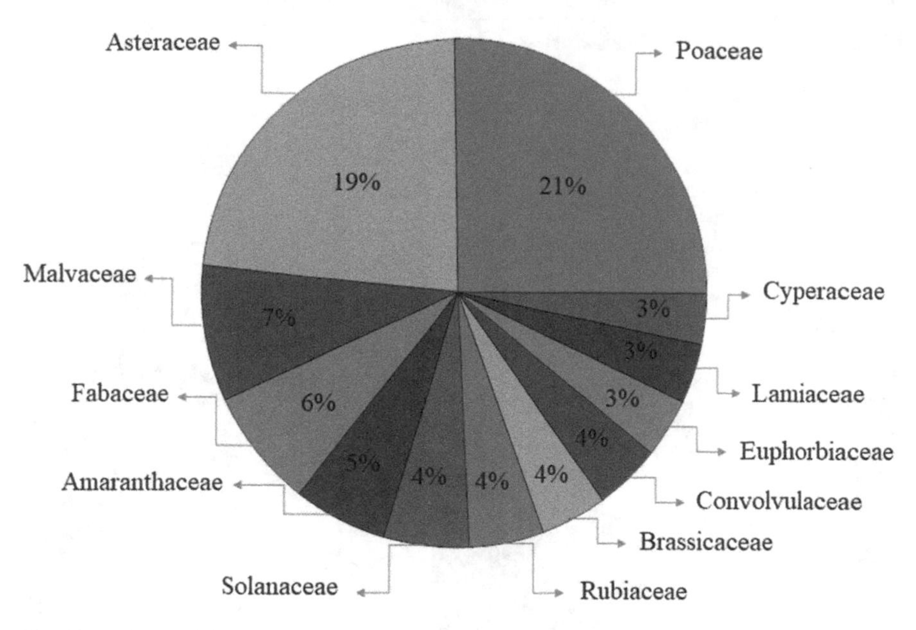

Fig. 12 Main weed families in Brazil according to the number of species, according to Table 2

it is common to find materials that carry the weeds that are usually found in specific environments, such as arable land, pasture, forest, and wilderness areas. In this sense it is important to classify weeds according to their habitat, which are divided into terrestrial, lowland, aquatic, indifferent environment and parasitic.

Terrestrial: Plants that live on the ground, some grow better in more fertile soil, such as amaranths (*Amaranthus* sp.) and common purslane (*Portulaca oleracea*). Others grow in low fertility soil, indicating poor soil, such as sidas (*Sida* spp.). There are also those that are indifferent to fertility, an example being the cyperaceaes (*Cyperus* spp.).

Lowland weeds: These are species that grow better in organic and humid soils. For example, *Cuphea carthaginensis* and *Alternanthera philoxeroides*.

Aquatic weeds: These are plants that live in aquatic environments, and can be divided into emergent (Bulrush—*Typha domingensis*), fixed floating (Tropical royalblue waterlily—*Nymphaea elegans*), free floating (Water lettuce—*Pistia stratiotes*), fixed submerged (Brazilian elodea—*Egeria densa*), and free submerged (Utricularia—*Utricularia* spp.) (Damo et al. 2020) (Fig. 15).

Indifferent environment weeds: Plants that live both in and out of water. For example, *Echinochloa* spp.

Parasitic weeds: Living on other plants and living off them. For example, dodder (*Cuscuta racemosa*).

Fig. 13 Features that differentiate *Amaranthus palmeri* from other *Amaranthus* species. Whitish spot in the "V" shape (**a**), spinescent structure (**b**), growth pattern with rosette appearance (**c**), presence of a small hair at the tip of the leaves (**d**), inflorescence ends long (**e**), petiole longer than the length of the leaf lamina (**f**), Dioecious plants (**g**, **h**). (Source: Legleiter and Johnson (2013) and Gazziero and Silva (2017) and Photo: Ednaldo A. Borgato)

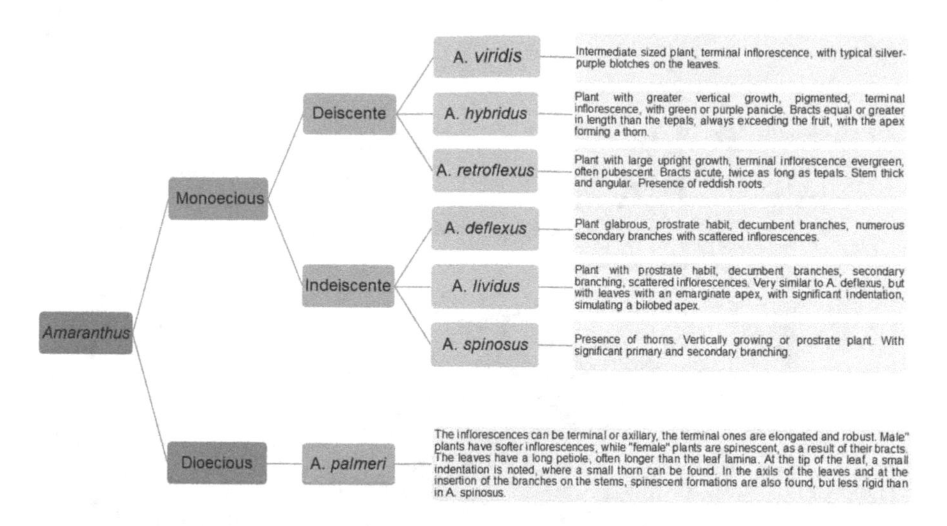

Fig. 14 Simplified dichotomous key, considering the main weed species of the genus *Amaranthus* found in the agricultural system in Brazil. (Source: Adapted from Carvalho (2016))

7.3 Life Cycle Classification

As for their life cycle or vegetative cycle, weeds are divided into *monocarpic* or *polycarpic*. *Monocarpic* include those species that flower and fruit only once at the end of the cycle; they can be annual or biennial. Annuals are those that complete their life cycle in up to 1 year, and biennial plants complete their life cycle in more than 1 year and up to 2 years. The *polycarpic* are the perennial species, which complete the life cycle in more than 2 years and may live a few or many years, flowering and fruiting annually.

In general, weeds are classified according to their life cycle into annuals, biennials, and perennials, as described below:

Annuals: it germinate, develop, flower, produce seeds, and die within a year (Fig. 16a). These can be *winter annuals* (which germinate in fall or winter, grow in spring and produce fruit, and die in midsummer) and *summer annuals* (which germinate in spring, grow in summer, and mature and die in fall) (Fig. 17). In certain regions of Brazil, especially in the South, where the seasons are well-defined, these facts are clearly observed. Annual plants are propagated by fruits and seeds. The best time of control for these species would be before the production of seeds. For example, Smooth pigweed (*A. hybridus*).

Biennials: These are plants whose complete development normally takes 2 years. In the first year, they germinate and grow, and in the second year they produce flowers, fruits, seeds and die. They must be controlled in the first year (Fig. 17). They can be annual in one region and biennial in another. For example, Honeyweed (*Leonurus sibiricus*).

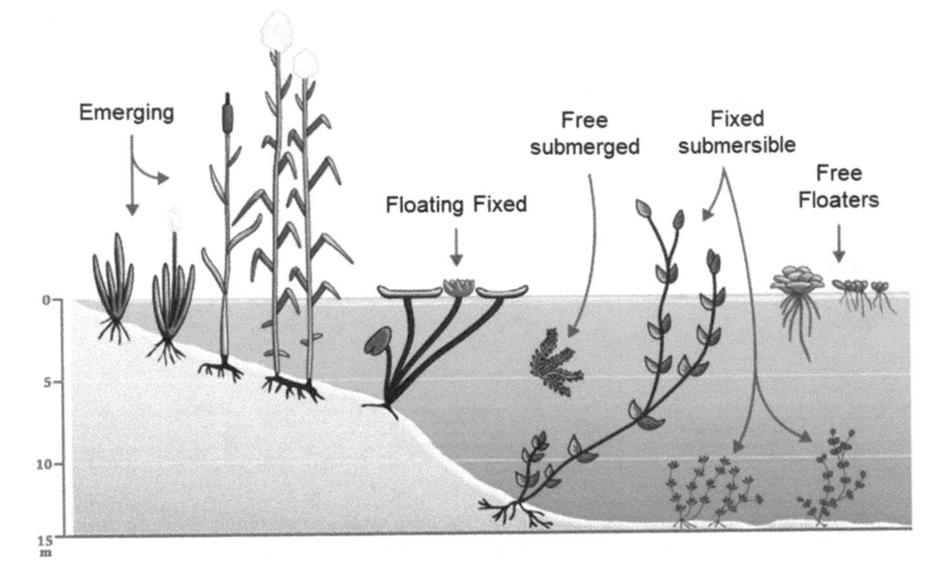

Fig. 15 Classification of aquatic weeds. (Source: Ferrarese et al. (2015) and Damo et al. (2020))

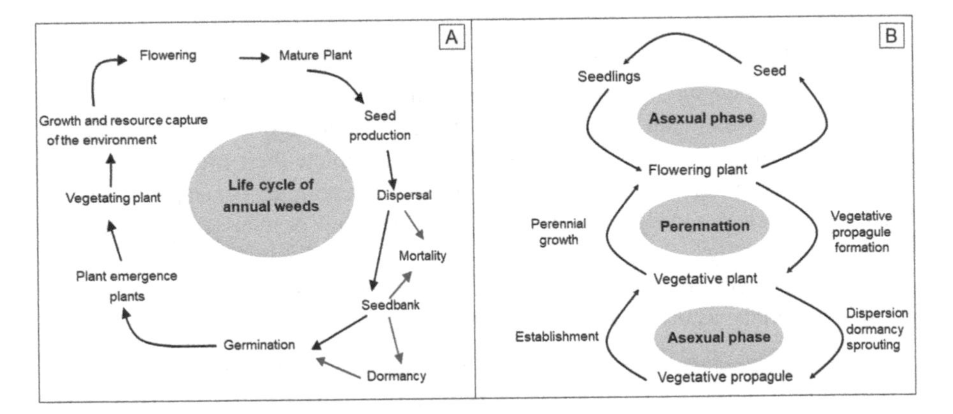

Fig. 16 Life cycle of annual (**a**) and perennial (**b**) weeds. (Source: Adapted from Grime (1979) and Radosevich et al. (2007))

Perennials: Perennial plants are those that live more than 2 years and are characterized by renewed growth year after year from the same root system (Fig. 16b). They are best controlled with the use of systemic herbicides, because the mechanical method of control causes them to multiply even more through their vegetative parts. For example, *Vernonia polyanthes*.

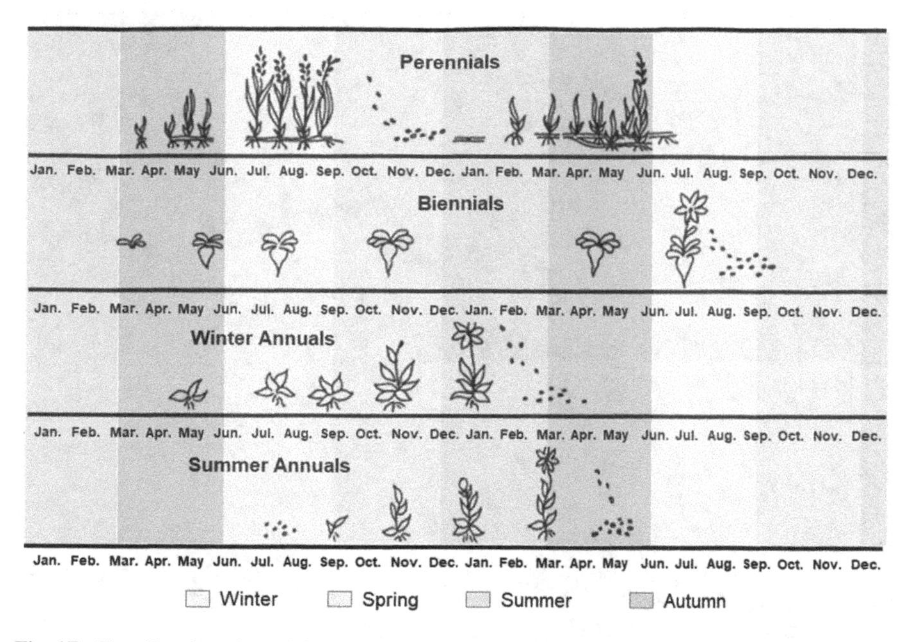

Fig. 17 Classification of weeds by vegetative cycle. (Source: Adapted from Bellinder et al. (1989))

7.4 Classification According to Growth Habit

The classification according to growth habit, divide weeds into herbaceous, subarbaceous, shrubs, trees, vines, epiphytes, hemiepiphytes, and parasites, as described below:

Herbaceous: Low herbaceous plants, smaller than 1.0 m, erect or prostrate and tender. For example, Gallant soldier (*Galinsoga parviflora).*

Subarbustive: They are 0.8–1.5 m tall, erect and woody-stemmed. For example, Sicklepod (*Senna obtusifolia*).

Shrubs: Height between 1.5 and 2.5 m, erect, woody and have branching from the base. For example, Wolf apple (*Solanum lycocarpum*).

Arboreal: Height above 2.5 m, woody plants, erect and also present well-defined branching, however above the stem base. For example, Lacre (*Vismia guianensis*).

Climbers: These are plants that benefit from other plants used as support for growth. They are divided into voluble, they climb by rolling. For example, Morning glory (*Ipomoea* spp.). And cirriferous, they attach themselves by means of tendrils. For example, Balloon vine (*Cardiospermum halicacabum*).

Epiphytes: These are plants that grow on others, without the use of photoassimilates from the host plant. For example, mosses, ferns and orchids.

Hemiephytes: They start their development as vines and, in later development, emit a radicular system. For example, *Caussopa schotii.*

Fig. 18 Plants *Orobanche* spp. (**a**, **b**) infesting tobacco crop and *Cuscuta racemosa* (**c**, **d**) in papaya plant. (Source: CIDASC (2018) and Correia (2018))

Parasites: Plants that grow on another, benefiting from the photoassimilates of the parasitized plant species. They can be divided into aerial part parasites. For example, Dodder (*Cuscuta* spp.). And root system parasites. For example, Broomrape (*Orobanche* spp.) (Fig. 18).

8 Concluding Remarks

Weeds are plants that become undesirable at certain times and places related to human activity. Plant species become weeds because they are competitive and adaptable, able to exploit habitats naturally disturbed or altered by human activity. However, many weeds have useful characteristics when present, favoring certain agricultural practices at certain times and conditions.

The high production of seeds and propagules, associated with factors involving dormancy, longevity and viability of the seed bank, influence the infestation capacity and survival of weed species. Therefore, knowing these characteristics is important for weed management, in order to avoid damage and maintain the balance of the agroecosystem.

Being able to classify and group weed species according to their taxonomy, the habitat where these plants live, plant life cycle, and growth habit is also important, as many management methods need this information to be positioned correctly to increase weed control efficiency.

Weeds will always be present in man's agricultural activities. In this sense, it is necessary to learn to live with their presence, always trying to get the most out of what weeds have to contribute and manage them efficiently so that they do not generate losses in agricultural production systems.

References

Baker HG (1974) The evolution of weeds. Annu Rev Ecol Syst 5(1):1–24. https://doi.org/10.1146/annurev.es.05.110174.000245

Bararpour MT, Oliver LR (1998) Effect of tillage and interference on common cocklebur (*Xanthium strumarium*) and sicklepod (*Senna obtusifolia*) population, seed production, and seedbank. Weed Sci 46(4):424–431. https://doi.org/10.1017/S0043174500090846

Barrat-Segretain MH (1996) Strategies of reproduction, dispersion, and competition in river plants: a review. Vegetatio 123(1):13–37. https://doi.org/10.1007/BF00044885

Baskin JM, Baskin CC (2004) A classification system for seed dormancy. Seed Sci Res 14(1):1–16. https://doi.org/10.1079/SSR2003150

Batlla D, Benech-Arnold RL (2014) Weed seed germination and the light environment: implications for weed management. Weed Biol Manag 14(2):77–87. https://doi.org/10.1111/wbm.12039

Bellinder RR, Kline RA, Warholic DT (1989) Weed control for the home vegetable garden. Bulletin informative. Cornell University no 216, pp 1–32

Benech-Arnold RL, Sánchez RA, Forcella F, Kruk BC, Ghersa CM (2000) Environmental control of dormancy in weed seed banks in soil. Field Crops Res 67(2):105–122. https://doi.org/10.1016/S0378-4290(00)00087-3

Benvenuti S (2007) Weed seed movement and dispersal strategies in the agricultural environment. Weed Biol Manag 7(3):141–157. https://doi.org/10.1111/j.1445-6664.2007.00249.x

Bhowmik PC, Bekech MM (1993) Horseweed (*Conyza canadensis*) seed production, emergence, and distribution in no-tillage and conventional-tillage corn (*Zea mays*). Agron Trends Agric Sci 1:67–71

Blackshaw RE, Lemerle D, Mailer R, Young KR (2002) Influence of wild radish on yield and quality of canola. Weed Sci 50(3):344–349. https://doi.org/10.1614/0043-1745(2002)050[0344:IOWROY]2.0.CO;2

Bostock SJ, Benton RA (1979) The reproductive strategies of five perennial Compositae. J Ecol 67:91–107. https://doi.org/10.2307/2259339

Braccini A (2011) Bancos de sementes e mecanismos de dormência em sementes de plantas daninhas. In: Oliveira RS Jr, Constantin J, Inoue HM (eds) Biologia e Manejo de Plantas Daninhas. Omnipax, Curitiba, pp 37–66

Brighenti AM, Oliveira MF (2011) Biologia de Plantas Daninhas. In: Oliveira RS Jr, Constantin J, Inoue HM (eds) Biologia e Manejo de Plantas Daninhas. Omnipax, Curitiba, pp 1–36

Burnside OC, Wilson RG, Weisberg S, Hubbard KG (1996) Seed longevity of 41 weed species buried 17 years in eastern and western Nebraska. Weed Sci 44(1):74–86. https://doi.org/10.1017/S0043174500093589

Cardoso VJM (2009) Conceito e classificação da dormência em sementes. Oecologia Brasiliensis 13(4):619–631. https://doi.org/10.4257/oeco.2009.1304.06

Carneiro RG, Mônaco APA, Moritz KC, Nakamura KC, Scherer A (2006) Identificação de *Meloidogyne mayaguensis* em goiabeira e em plantas invasoras, em solo argiloso, no Estado do Paraná. Nematologia Brasileira 30(2):293–298. https://doi.org/10.1590/S1982-56762009000400009

Carvalh SJP, López-Ovejero RF, Nicolai M, Christoffoleti PJ (2005) Crescimento, desenvolvimento e produção de sementes da planta daninha capim-branco (*Chloris polydactyla*). Planta Daninha 23(4):603–609. https://doi.org/10.1590/S0100-83582005000400007

Carvalho LB (2013) Plantas Daninhas. Editado pelo autor, Lages, p 82

Carvalho SJP (2016) Identificação de espécies de plantas daninhas do gênero *Amaranthus*. HRAC-BR. DIALOG. https://nomato.files.wordpress.com/2020/03/aula-2_leitura-bc3a1sica.pdf. Accessed 8 March 2021

Carvalho SJP, Christoffoleti PJ (2007) Estimativa da área foliar de cinco espécies do gênero *Amaranthus* usando dimensões lineares do limbo foliar. Planta Daninha 25(2):317–324. https://doi.org/10.1590/S0100-83582007000200011

Carvalho SJP, Buissa JAR, Nicolai M, López-Ovejero RF, Christoffoleti PJ (2006) Suscetibilidade diferencial de plantas daninhas do gênero *Amaranthus* aos herbicidas trifloxysulfuron-sodium e chlorimuron-ethyl. Planta Daninha 24(3):541–548. https://doi.org/10.1590/S0100-83582006000300017

Carvalho SJP, López-Ovejero RF, Christoffoleti PJ (2008) Crescimento e desenvolvimento de cinco espécies de plantas daninhas do gênero *Amaranthus*. Bragantia 67(2):317–326. https://doi.org/10.1590/S0006-87052008000200007

Chandrasena N (2014) Living with weeds-a new paradigm. Indian j Weed sci 46(1):96–110

Chauhan BS (2013) Shade reduces growth and seed production of *Echinochloa colona*, *Echinochloa crus-galli*, and *Echinochloa glabrescens*. Crop Prot 43:241–245. https://doi.org/10.1016/j.cropro.2012.10.009

Chauhan BS, Abugho S (2013) Effects of water regime, nitrogen fertilization, and rice plant density on growth and reproduction of lowland weed *Echinochloa crus-galli*. Crop Prot 54:142–147. https://doi.org/10.1016/j.cropro.2013.08.005

Chauhan BS, Johnson DE (2010) Implications of narrow crop row spacing and delayed *Echinochloa colona* and *Echinochloa crus-galli* emergence for weed growth and crop yield loss in aerobic rice. Field Crops Res 117(2–3):177–182. https://doi.org/10.1016/j.fcr.2010.02.014

Christoffoleti PJ, López-Ovejero R (2003) Principais aspectos da resistência de plantas daninhas ao herbicida glyphosate. Planta Daninha 21(3):507–515. https://doi.org/10.1590/S0100-83582003000300020

CIDASC - Companhia Integrada de Desenvolvimento Agrícola de Santa Catarina (2018) Procedimentos para realização do levantamento de detecção da praga *Orobanche* spp., planta parasita, em cultivos de tabaco no Estado de Santa Catarina. DIALOG. http://www.cidasc.sc.gov.br/defesasanitariavegetal/files/2018/10/IS-011-2018-DEDEV-Inspe%C3%A7%C3%A3o-OROBANCHE.pdf. Accessed 3 March 2021

Clements DR, Benoit DL, Swanton CJ (1996) Tillage effects on weed seed return and seedbank composition. Weed Sci 44(2):314–322. https://doi.org/10.1017/S0043174500093942

Correia NM (2018) Saiba como retirar a parasita *cuscuta* do mamoeiro. Globo Rural. DIALOG https://revistagloborural.globo.com/vida-nafazenda/noticia/2018/10/saiba-como-retirar-parasita-cuscuta-do-mamoeiro.html. Accessed 14 March 2021

Costa ML, Radünz Neto J, Lazzari R, Losekann ME, Sutili FJ, Brum ÂZ, Veiverberg CA, Grzeczinski JA (2008) Juvenis de carpa capim alimentados com capim teosinto e suplementados com diferentes taxas de arraçoamento. Ciência Rural 38(2):492–497. https://doi.org/10.1590/S0103-84782008000200031

Cousens R, Mortimer M (1995) Dynamics of weed populations. Cambridge University Press, New York, p 348

Crowley RH, Buchanan GA (1982) Variations in seed production and the response to pests of morningglory (*Ipomoea*) species and smallflower morningglory (*Jacquemontia tamnifolia*). Weed Sci 30(2):187–190. https://doi.org/10.1017/S0043174500062305

Cury JP, Santos JB, Silva EB, Byrro ECM, Braga RR, Carvalho FP, Valadão Silva D (2012) Acúmulo e partição de nutrientes de cultivares de milho em competição com plantas daninhas. Planta Daninha 30(2):287–296. https://doi.org/10.1590/S0100-83582012000200007

D'oliveira OS, Brighenti AM, Oliveira VM, Miranda JEC (2018) Plantas tóxicas em pastagens: cafezinho (*Palicourea marcgravii* St. Hill, Família Rubiaceae). Comunicado Técnico Embrapa 85, Juiz de Fora, 8p

Damo L, Mendes KF, Freitas FCL (2020) Plantas daninhas aquáticas: da classificação ao controle. Boletim informativo. Universidade Federal de Viçosa, MIPD, Viçosa, no. 5, pp 1–16. DIALOG.https://www.researchgate.net/publication/344455645_PLANTAS_DANINHAS_AQUATICAS_DA_CLASSIFICACAO_AO_CONTROLE. Accessed 18 March 2021

De Wet JM, Harlan JR (1975) Weeds and domesticates: evolution in the man-made habitat. Econ Bot 29(2):99–108. https://doi.org/10.1007/BF02863309

Ellstrand NC, Heredia SM, Leak-Garcia JA, Heraty JM, Burger JC, Yao L, Nohzadeh-Malakshah S, Ridley CE (2010) Crops gone wild: evolution of weeds and invasives from domesticated ancestors. Evol Appl 3(5–6):494–504. https://doi.org/10.1111/j.1752-4571.2010.00140.x

Eslami SV (2011) Comparative germination and emergence ecology of two populations of common lambsquarters (*Chenopodium album*) from Iran and Denmark. Weed Sci 59(1):90–97. https://doi.org/10.1614/WS-D-10-00059.1

Ferrarese MD, Xavier RA, Cantorodow TS (2015) As plantas aquáticas e a saúde da água, 1 rd edn. Fundação Mo'Ã, Itaara, p 27

Ferreira ACB, Costa AGF, Lamas FM, Bogiani JC, Sofiatti V (2017) Plantas daninhas. In: In: Araujo A E, Sofiatti V. (ed) Cultura do Algodão no Cerrado, 2nd edn. Embrapa, Brasília

Ferreira SD, Exteckoetter V, Gibbert AM, Barbosa JA, Costa NV (2018) Biological cycle of susceptible and glyphosate-resistant sourgrass biotypes in two growth periods. Planta Daninha 36:1–9. https://doi.org/10.1590/S0100-83582018360100077

Fleck NG, Rizzardi MA, Agostinetto D, Vidal RA (2003) Produção de sementes por picão-preto e guanxuma em função de densidades das plantas daninhas e da época de semeadura da soja. Planta Daninha 21(2):191–202. https://doi.org/10.1590/S0100-83582003000200004

Gazziero DLP, Silva AF (2017) Caracterização e manejo de *Amaranthus palmeri*. In: Documentos-Embrapa Soja, no. 384, 39p. DIALOG. https://ainfo.cnptia.embrapa.br/digital/bitstream/item/159778/1/Doc-384-OL.pdf. Accessed 12 Feb 2021

Gomes FG Jr, Christoffoleti PJ (2008) Biologia e manejo de plantas daninhas em áreas de plantio direto. Planta Daninha 26(4):789–798. https://doi.org/10.1590/S0100-83582008000400010

Grime JP (1979) Plant strategies and vegetation processes. Wiley, New York, p 222

Grundy AC, Mead A, Burston S, Overs T (2004) Seed production of *Chenopodium album* in competition with field vegetables. Weed Res 44(4):271–281. https://doi.org/10.1111/j.1365-3180.2004.00399.x

Gulden RH, Shirtliffe SJ (2009) Weed seed banks: biology and management. Prairie Soils Crops J 2:46–52

Guo L, Qiu J, Li LF, Lu B, Olsen K, Fan L (2018) Genomic clues for crop–weed interactions and evolution. Trends Plant Sci 23(12):1102–1115. https://doi.org/10.1016/j.tplants.2018.09.009

Hassanpour-Bourkheili S, Heravi M, Gherekhloo J, Alcántara-De-La-Cruz R, De Prado R (2020) Fitness cost of imazamox resistance in wild poinsettia (*Euphorbia heterophylla* L.). Agronomy 10(12):1859. https://doi.org/10.3390/agronomy10121859

Hejcman M, Křišťálová V, Červená K, Hrdličková J, Pavlů V (2012) Effect of nitrogen, phosphorus and potassium availability on mother plant size, seed production and germination ability of *Rumex crispus*. Weed Res 52(3):260–268. https://doi.org/10.1111/j.1365-3180.2012.00914.x

Horowitz M (1973) Spatial growth of *Sorghum halepense* (L.) Pers. Weed Res 13(2):200–208. https://doi.org/10.1111/j.1365-3180.1973.tb01264.x

Hosamani MM, Shivaraj B, Kurdikeri CB (1971) Seed production potentialities of common weeds of Dharwar. PANS Pest Articles News Summ 17(2):237–239. https://doi.org/10.1080/09670877109413355

Hossain MM, Begum M (2015) Soil weed seed bank: importance and management for sustainable crop production-a review. J Bangladesh Agric Univ 13(2):221–228. https://doi.org/10.3329/jbau.v13i2.28783

Kaspary TE, Lamego FP, Cutti L, Aguiar ACDM, Rigon CAG, Basso CJ (2017) Growth, phenology, and seed viability between glyphosate-resistant and glyphosate-susceptible hary fleabane. Bragantia 76(1):92–101. https://doi.org/10.1590/1678-4499.542

Kissmann KG, Groth D (1999) Plantas infestantes e nocivas, vol 2, 2nd edn. BASF, São Paulo, p 978

Lacey JR, Wallander R, Olson-Rutz K (1992) Recovery, germinability, and viability of leafy spurge (*Euphorbia esula*) seeds ingested by sheep and goats. Weed Technol 6(3):599–602. https://doi.org/10.1017/S0890037X00035867

Legleiter TR, Johnson B (2013) Palmer amaranth biology, identification, and management. Purdue Extension, West Lafayette

López-Ovejero RF, Soares DJ, Oliveira NC, Kawaguchi IT, Berger GU, Carvalho SJPD, Christoffoleti PJ (2016) Interferência e controle de milho voluntário tolerante ao glifosato na cultura da soja. Pesq Agropecu Bras 51(4):340–347. https://doi.org/10.1590/S0100-204X2016000400006

Lorenzi H (2000) Plantas daninhas do Brasil: terrestres, aquáticas, parasitas e tóxicas, 3rd edn. Instituto Plantarum, Nova Odessa, p 608

Lorenzi H (2014) Manual de identificação e controle de plantas daninhas, 7th edn. Instituto Plantarum de Estudos da Flora Ltda, Nova Odessa, p 379

Maluf AM (1994) Plasticidade fenotípica em *Amaranthus hybridus* L. (Amaranthaceae). Acta Bot Bras 8(2):213–218. https://doi.org/10.1590/S0102-33061994000200006

Marchezan E (1994) Arroz vermelho: caracterização, prejuízos e controle. Ciênc Rural 24(2):423–429. https://doi.org/10.1590/S0103-84781994000200036

Martins JD, Restle J, Barreto IL (2000) Produção animal em capim papuã (*Brachiaria plantaginea* (Link) Hitchc) submetido a níveis de nitrogênio. Ciênc Rural 30(5):887–892. https://doi.org/10.1590/S0103-84782000000500025

Meda AR, Pavan MA, Miyazawa M, Cassiolato ME (2002) Plantas invasoras para melhorar a eficiência da calagem na correção da acidez subsuperficial do solo. Rev Bras Ciênc Só 26(3):647–654. https://doi.org/10.1590/S0100-06832002000300009

Menalled F (2008) Weed seedbank dynamics & integrated management of agricultural weeds. In: Agriculture and natural resources (weeds). Montana State University, Bozeman, pp 200–211

Mobli A, Yadav R, Chauhan BS (2020) Enhanced weed-crop competition effects on growth and seed production of herbicide-resistant and herbicide-susceptible annual sowthistle (*Sonchus oleraceus*). Weed Biol Manag 20(2):38–46. https://doi.org/10.1111/wbm.12197

Modro AFH, Marchini LC, Moreti ACCC (2011) Origem botânica de cargas de pólen de colmeias de abelhas africanizadas em Piracicaba, SP. Ciênc Rural 41(11):1944–1951. https://doi.org/10.1590/S0103-84782011005000137

Monaco TJ, Weller SC, Ashton FM (2002) Weed science: principles and practices, 4th edn. Wiley, New York, 671p

Mondo VHV, Carvalho SJPD, Dias ACR, Marcos Filho J (2010) Efeitos da luz e temperatura na germinação de sementes de quatro espécies de plantas daninhas do gênero *Digitaria*. Revista Brasileira de Sementes 32(1):131–137. https://doi.org/10.1590/S0101-31222010000100015

Moreira HJC, Bragança HBN (2010) Manual de identificação de plantas infestantes (Cultivos de Verão). Campinas, FMC Agricultural Products, 642p

Moreira HJC, Bragança HBN (2011) Manual de identificação de plantas infestantes (Hortifrúti). FMC Agricultural Products, Campinas, 1017p

Musik TJ (1970) Weed biology and control. McGraw-Hill, New York, 273p

Nikolaeva MG (1977) Factors controlling the seed dormancy pattern. In: Khan AA (ed) The physiology and biochemistry of feed dormancy and germination. North-Holland, Amsterdam, pp 51–74

Pancho JV (1964) Seed sizes and production capabilities of common weed species in the rice fields of the Philippines. Phil Agric Sci 48:307–316

Piasecki C, Mazon AS, Monge A, Cavalcante JA, Agostinetto D, Vargas L (2019) Glyphosate applied at the early reproductive stage impairs seed production of glyphosate-resistant hairy fleabane. Planta Daninha 37:e019196815. https://doi.org/10.1590/S0100-83582019370100104

Pitelli RA (2015) O termo planta-daninha. Planta Daninha 33(3):622–623. https://doi.org/10.1590/S0100-83582015000300025

Pontes VB Jr, Mendes KF, Silva AA, Reis MR (2021) Cebola. In: Mendes KF (ed) Atualidades no Manejo de Plantas Daninhas em Hortaliças Tuberosas. Brazil Publishing, Curitiba, pp 151–184

Popinigis F (1985) Fisiologia da semente, 2nd edn. Brasília, AGIPLAN, p 289

Qasem JR (2013) Herbicide resistant weeds: the technology and weed management. In: Price AJ, Kelton JA (eds) Herbicides-current research and case studies in use. IntechOpen, London, pp 445–471

Radosevich SR, Holt JS, Ghersa CM (2007) Ecology of weeds and invasive plants: relationship to agriculture and natural resource management, 3rd edn. Wiley, Chishester, p 473

Reeves TG, Code GR, Piggin CM (1981) Seed production and longevity, seasonal emergence and phenology of wild radish (*Raphanus raphanistrum* L.). Aust J Exp Agric 21(112):524–530. https://doi.org/10.1071/EA9810524

Saeed A, Hussain A, Khan MI, Arif M, Maqbool MM, Mehmood H, Iqbal M, Alkahtani J, Elshikh MS (2020) The influence of environmental factors on seed germination of *Xanthium strumarium* L.: implications for management. PLoS One 15(10):e0241601. https://doi.org/10.1371/journal.pone.0241601

Sahu TR (1984) Less known uses of weeds as medicinal plants. Anc Sci Life 3(4):245

Salvador FL, Victoria Filho R, Alves ASR, Simoni F, San Martin HAM (2007) Efeito da luz e da quebra de dormência na germinação de sementes de espécies de plantas daninhas. Planta Daninha 25(2):303–308. https://doi.org/10.1590/S0100-83582007000200009

Santos JB, Procópio SO, Silva AA, Costa LC (2002) Produção e características qualitativas de sementes de plantas daninhas. Planta Daninha 20(2):237–241. https://doi.org/10.1590/S0100-83582002000200010

Santos JBD, Procópio SDO, Pires FR, Silva AAD, Santos EAD (2006) Fitorremediação de solo contaminado com trifloxysulfuron-sodium por diferentes densidades populacionais de feijão-de-porco (*Canavalia ensiformis* (L) DC.). Cienc Agrotecnol 30(3):444–449. https://doi.org/10.1590/S1413-70542006000300009

Scheren MA, Palagi CA, Jurach J, Richart A, Contiero RL (2013) Germinação de Sementes de *Euphorbia heterophylla* e *Brachiaria plataginea* a Profundidades Variadas em Latossolo Vermelho. Acta Iguazu 2(2):49–57. https://doi.org/10.48075/actaiguaz.v2i2.8357

Schutte BJ (2017) Measuring interference from midseason tall morningglory (*Ipomoea purpurea*) to develop a model for teaching weed seedbank effects on Chile pepper. Weed Technol 31(1):155–164. https://doi.org/10.1017/wet.2016.19

Sellers BA, Smeda RJ, Johnson WG, Kendig JA, Ellersieck MR (2003) Comparative growth of six *Amaranthus* species in Missouri. Weed Sci 51(3):329–333. https://doi.org/10.1614/0043-1745(2003)051[0329:CGOSAS]2.0.CO;2

Senna LR (2015) Identificação de espécies de plantas daninhas do gênero *Amaranthus* L. (Amaranthaceae Juss.) no Brasil. In: Inoue MH, Oliveirja RS Jr, Mendes KF, Constantin J (eds) Manejo de *Amaranthus*. RiMa Editora, São Carlos, pp 1–20

Shen-Miller J, Mudgett MB, Schopf JW, Clarke S, Berger R (1995) Exceptional seed longevity and robust growth: ancient sacred lotus from China. Am J Bot 82(11):1367–1380. https://doi.org/10.2307/2445863

Shields EJ, Dauer JT, Vangessel MJ, Neumann G (2006) Horseweed (*Conyza canadensis*) seed collected in the planetary boundary layer. Weed Sci 54(6):1063–1067. https://doi.org/10.1614/WS-06-097R1.1

Shiratsuchi LS, Fontes JRA, Silva RR (2021) Metodologia de determinação do banco de sementes de plantas daninhas para confecção de mapas de distribuição especial. EMBRAPA Cerrados. DIALOG. file:///C:/Users/T.i/Desktop/Cap.%20livro,%20AA/Biologia%20de%20plantas%20daninhas/banco%20de%20seeds/MAPA%20distribui%C3%A7%C3%A3o.pdf. Accessed 11 March 2021

Shivakumar KV, Devendra R, Muniswamappa MV, Halesh GK, Mahadevamurthy M (2014) Weed seed production potentials in *Bidens pilosa* L. in plantation crops in hill zone of Karnataka. Int J Res Appl 2:11–18

Silva AA, Ferreira FA, Ferreira LR, Santos JB (2007) Biologia De Plantas Daninhas. In: Silva AA, Silva JF (eds) Tópicos em Manejo de Plantas Daninhas. Editora UFV, Viçosa, pp 17–61

Takano HK, Oliveira RS Jr, Constantin J, Braz GBP, Padovese JC (2016) Growth, development and seed production of goosegrass. Planta Daninha 34(2):249–258. https://doi.org/10.1590/S0100-83582016340200006

Tanaka RH, Cardoso LR, Martins D, Marcondes DAS, Mustafá AL (2002) Ocorrência de plantas aquáticas nos reservatórios da Companhia Energética de São Paulo. Planta Daninha 20(spe):101–111. https://doi.org/10.1590/S0100-83582002000400012

Thomaz SM (2002) Fatores ecológicos associados à colonização e ao desenvolvimento de macrófitas aquáticas e desafios de manejo. Planta Daninha 20(spe):21–33. https://doi.org/10.1590/S0100-83582002000400003

Vargas L, Roman ES, Rizzardi MA, Silva VC (2005) Alteração das características biológicas dos biótipos de azevém (*Lolium multiflorum*) ocasionada pela resistência ao herbicida glyphosate. Planta Daninha 23(1):153–160. https://doi.org/10.1590/S0100-83582005000100018

Vidal RA, Kalsing A, Goulart IDR, Lamego FP, Christoffoleti PJ (2007) Impacto da temperatura, irradiância e profundidade das sementes na emergência e germinação de *Conyza bonariensis* e *Conyza canadensis* resistentes ao glyphosate. Planta Daninha 25(2):309–315. https://doi.org/10.1590/S0100-83582007000200010

Vivian R, Silva AA, Gimenes M Jr, Fagan EB, Ruiz ST, Labonia V (2008) Dormência em sementes de plantas daninhas como mecanismo de sobrevivência: breve revisão. Planta Daninha 26(3):695–706. https://doi.org/10.1590/S0100-83582008000300026

Walker E, Oliver LR (2008) Weed seed production as influenced by glyphosate applications at flowering across a weed complex. Weed Technol 22(2):318–325. https://doi.org/10.1614/WT-07-118.1

Wu H, Walker S, Rollin MJ, Tan DKY, Robinson G, Werth J (2007) Germination, persistence, and emergence of flaxleaf fleabane (*Conyza bonariensis* [L.] Cronquist). Weed Biol Manag 7(3):192–199. https://doi.org/10.1111/j.1445-6664.2007.00256.x

Yamashita OM, Guimarães SC (2010) Germinação das sementes de *Conyza canadensis* e *Conyza bonariensis* em função da disponibilidade hídrica no substrato. Planta Daninha 28(2):309–317. https://doi.org/10.1590/S0100-83582010000200010

Weed Competition and Interference in Crops

Elisa Maria Gomes da Silva, Adalin Cezar Moraes de Aguiar, Kassio Ferreira Mendes, and Antonio Alberto da Silva

Abstract In agricultural areas, the limitation of resources such as water, light, and nutrients allows competition between crops and weeds to occur. In most cases, the crop of interest is damaged, resulting in reduced growth, development, and yield losses. The intensity of competition and the capacity of the weeds to interfere depends on the species present in the area and the density at which they are found, the competitive capacity of the cultivar and the density at which it is planted, and environmental conditions. The effects of competition through weed interference can be understood by studies capable of providing important information to assist in weed management. Thus, this chapter discusses competition for resources and the factors influencing the degree of interference between the plant community, and uses experimental models to study competition and interference between weeds and crops.

Keywords Competitive ability · Competition factors · Plant community · Interference period · Competition studies

1 Introduction

In a plant community the interactions that occur between species can be positive, negative, or neutral. In agricultural areas the negative aspects of interactions are usually observed. Thus, the crop of interest is harmed when in competition with other species, which as a consequence interferes with the reduction of growth, development, and productivity of these crops.

E. M. G. da Silva (✉) · A. C. M. de Aguiar · K. F. Mendes · A. A. da Silva
Department of Agronomy, Federal University of Viçosa, Viçosa, Minas Gerais, Brazil
e-mail: kfmendes@ufv.br; aasilva@ufv.br

© The Author(s), under exclusive license to Springer Nature Switzerland AG 2022
K. F. Mendes, A. Alberto da Silva (eds.), *Applied Weed and Herbicide Science*,
https://doi.org/10.1007/978-3-031-01938-8_2

Competition is the best known form of direct weed interference in agricultural crops. The resources normally subject to competition between plants are essential mineral nutrients, water, light, CO_2, as well as competition for space and the release of substances with allelopathic effects (Pitelli 1985). The resources of the environment susceptible to competition between weeds and cultivated plants can be seen in Fig. 1.

The intensity of competition and the ability of weeds to interfere depends on community factors such as the species present and the density at which they are found. The species cultivated also has an influence, due to the competitive capacity of the cultivar and planting density, as well as the environmental conditions. These conditions are influenced by the type of management adopted in the conduction of the crop, which can influence the availability of resources (Silva et al. 2007; Pitelli 2014).

The effects of competition by weed interference can be understood by studies capable of providing important information to assist in weed management. Studies such as the economic damage level (EDL) and the determination of critical periods of interference, assist farmers in making decisions about the need for control and when to enter control measures, in order to avoid losses in crop yield (Swanton et al. 2015).

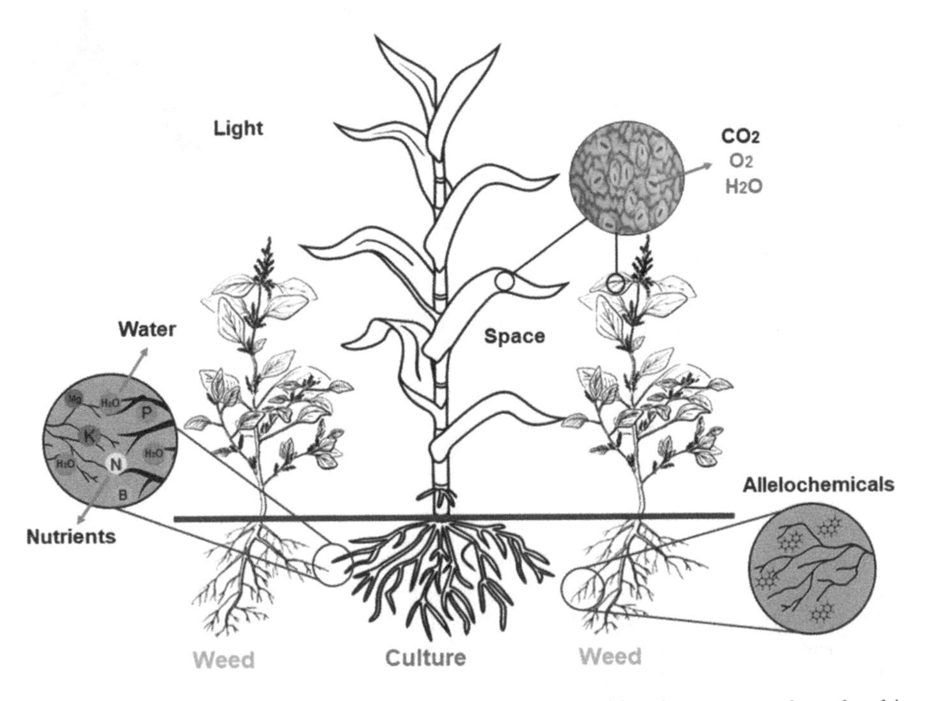

Fig. 1 Resources in the environment susceptible to competition between weeds and cultivated plants

Based on the importance of knowing the interactions between crops and weeds, this chapter highlights competition for resources and the factors that influence the degree of interference between the plant community, as well as the use of experimental models to study competition and interference between weeds and crops.

2 Competition

The presence of natural resources, such as water, light, nutrients, and CO_2, in adequate amounts are necessary for plant growth and development. Under limited conditions of these resources, competition can occur between plants of the same or different species. In this sense, competition can be defined as a biotic relationship in which different individuals use the same resource of the environment, whose availability and provision are not sufficient to meet the demand of the species (Pitelli 2014).

In agricultural systems, the selection of more productive biotypes was accompanied by a decrease in the competitive potential of cultivars, altering the ability of these plants to acquire resources when limited, impairing their development. On the other hand, weeds have characteristics that provide their survival in environments subject to various types of interferences and limitations to growth and development, attributes that favor the competitive ability of weeds in relation to crops in agricultural environments (Brighenti and Oliveira 2011).

The effect of weed competition on crops is driven by *three main variables*. *The first* and most important variable is the timing of weed emergence relative to the crop of interest. Weeds that emerge before the crop are more competitive and result in greater yield losses. The *second variable* is weed density, and there is a close relationship between weed density and the intensity and duration of interference. *The third variable* is in relation to the species of weeds predominant in the area, which differ in their ability to compete with crops (Swanton et al. 2015).

When living together, weeds almost always outperform the growth parameters of cultivated plants. This occurs due to characteristics that make weeds behave superior to crops as described below (Silva et al. 2007).

(a) Life cycle similar to that of the crop.
(b) Rapid initial development of the roots and/or aboveground parts.
(c) Phenotypic and population plasticity.
(d) Uniform germination in time and space (presence of dormancy).
(e) Production and release of allelopathic substances in the soil.
(f) Production of a high number of propagules per plant.
(g) Capacity of dispersion of seeds and reproductive structures.
(h) Adaptation to the environmental conditions.

One of the basic principles of competition is based on the fact that the first plants that appear in the environment tend to dominate the others. Therefore, it is important to provide conditions for the crop to establish itself before the weed community emerges. Techniques that promote crop growth and development such as soil preparation, planting time, planting depth, germination percentage and seed vigor, plant spacing and density are essential in reducing the competitive ability and interference potential of weeds with the crop of interest.

3 Competing Environmental Factors

In agricultural areas, most often the crop and the weeds grow together. Since both have their own demands for resources and most of the time these are not available in sufficient quantities, competition is established. In addition to this factor, the limited surface and soil space provided by weeds can also affect the growth and development of crop plants. The environmental factors that determine plant growth and influence competition can be divided into **resources** and **conditions** (Radosevich et al. 1997). **Resources** are consumable factors such as water, light, nutrients, and CO_2. **Conditions** are factors not directly consumable, such as pH, soil density, and temperature, which exert influence on the use of resources by plants, and thus influence competition.

The main resources that are the target of competition between weeds and crops are described below.

3.1 Competition for Water

Plants need a large amount of water for many processes, such as photosynthesis and cell turgescence. Thus, a reduction in the availability of this resource influences the development and transport of minerals and photo-assimilates in the plant, which can lead to a reduction in growth and productivity.

Mechanisms such as resistance to cavitation (formation of air bubbles during the transport of liquids), physiological characteristics for the removal of water in the soil, stomatic regulation, and the ability of the roots to adjust osmotically (Craine and Dybzinski 2013; Pitelli 2014) are developed by plants to respond and adapt to the stresses promoted by the environment in which water availability is reduced. These adaptations can influence the competitive ability, both in the acquisition of the water present in the soil, and the ability to explore larger volumes of soil for the growth and development of their roots.

Certain weed species are able to use less water per unit of dry matter produced, i.e. they present a higher water use efficiency. Therefore, it is expected that plants with low water requirements are more productive during periods of limited water availability compared to plants with high water requirements, and therefore more

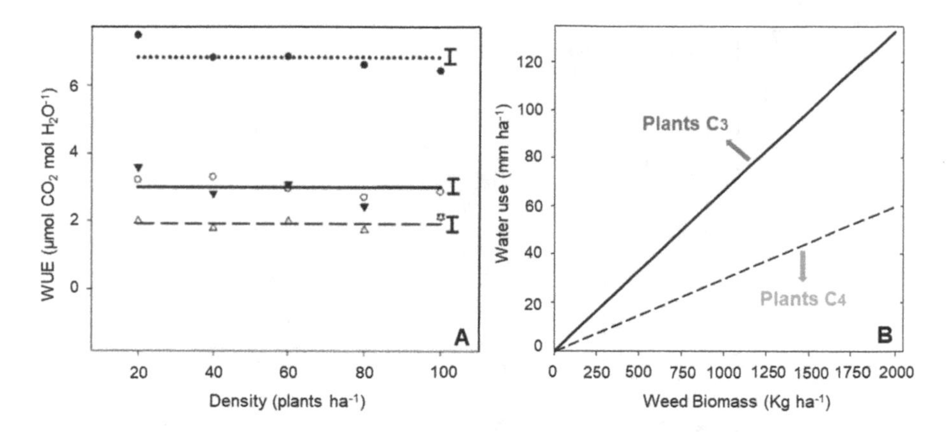

Fig. 2 Water use efficiency (WUE) of *Urochloa plantaginea* (●), *Bidens pilosa* (○), *Ipomoea indivisa* (▼), and *Conyza bonariensis* (Δ) (**a**). Water use by C_3 and C_4 weeds (**b**). (Source: Adapted from Norris (1996); Thiel et al. (2018))

competitive (Silva et al. 2007). A study conducted with four weed species *Urochloa plantaginea*, *Bidens pilosa*, *Ipomoea indivisa,* and *Conyza bonariensis*, showed that barnyard grass was the species with the highest water use efficiency (Fig. 2a). In general, weeds with C_4 metabolism, such as barnyard grass, show higher water use efficiency compared to C_3 plants. C_4 plants besides the enzyme Rubisco, have the enzyme PEP-carboxylase, which assists in carbon fixation. Thus, these species fix more CO_2 per unit of water lost, representing a higher efficiency in the use of the water present in the soil. This characteristic benefits C_4 plants, especially under water deficit conditions, but not under conditions of high soil water availability (Fig. 2b).

Some weed species can be more competitive under water deficit conditions. The presence of *Walternaria indica* did not affect the growth of sunflower plants, regardless of water availability, in contrast, the *Amaranthus spinosus* reduced the growth of sunflower plants under conditions of lower water availability (Fig. 3) (Soares et al. 2019). This shows that competition between plants depends on water availability, the species that are coexisting, such as crops and weeds.

The ability to extract water from the soil is under conditions of low availability is an important characteristic of weeds. The *B. pilosa* plants were able to extract water from the soil at tensions three times higher than those reached by soybean and bean crops, as well as other weeds, such as the *Desmodium tortuosum* and *Euphorbia heterophylla* (Fig. 4). The high survivability of the *B. pilosa* with little water in the soil may be related to the fact that this species presents at the beginning of its development the ability to drain a large part of photo-assimilates for the production of roots, which promote, in later stages of development, and assist in the greater exploration of the soil in search of water (Procópio et al. 2002, 2004).

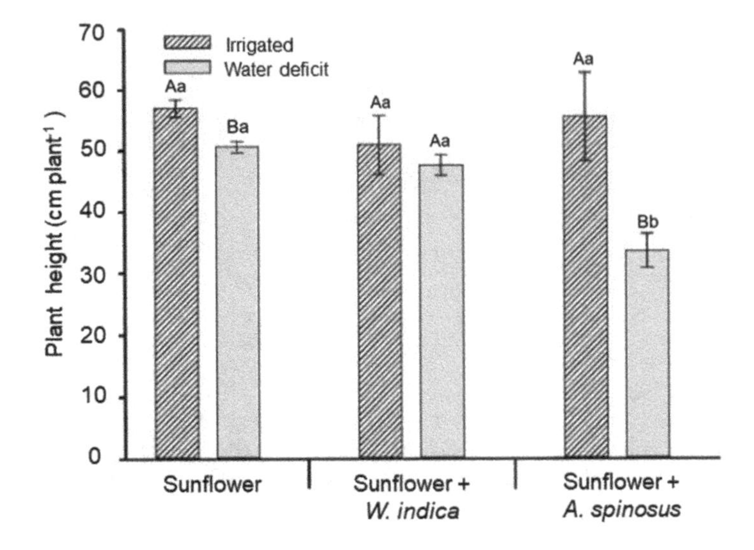

Fig. 3 Sunflower plant height as a function of water regimes and competition with *Waltheria indica* and *Amaranthus spinosus*. Means followed by the same lower case letter (between competition arrangements within each water regime) and capital letters (between water regimes within each competition arrangement) do not differ compared by Tukey's test. (Source: Adapted Soares et al. (2019))

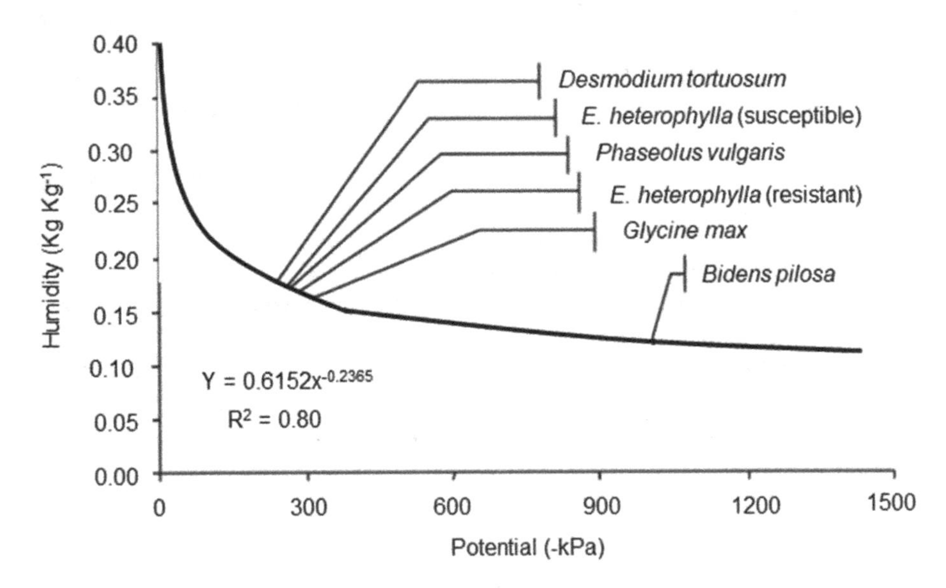

Fig. 4 Water potential in the soil, cultivated with different plant species, at the permanent wilting point. (Source: Adapted from Procópio et al. (2004))

3.2 Competing for Nutrients

The nutrients necessary for plant development are often present in scarce quantities in the soil, which makes fertilization necessary. However, this practice can also favor weeds present in the area, decreasing the benefits of fertilization (Little et al. 2021).

Weeds, in general, have a great capacity to extract from the soil the nutrients essential for their growth and development and, as a result, exert high competition with crops for these nutrients (Silva et al. 2007). Factors such as exploitation capacity of the root system, ability to absorb nutrients and efficiency of utilization are essential for a weed in competition (Silva et al. 2007).

The ability to compete for soil nutrients depends on the weed and crop species, as well as the amount and nutrient available. The presence of weeds, such as *Amaranthus viridis*, *B. pilosa,* and *Ipomoea grandifolia* with corn plants did not affect the levels of nitrogen (N), phosphorus (P), and calcium (Ca) accumulated in the aerial part of the crop, and did not affect the accumulation of N and Ca in the root system (Fig. 5). In contrast, the presence of *I. grandifolia* reduced the contents of accumulated P in the roots of corn plants, and the *A. viridis* reduced the content

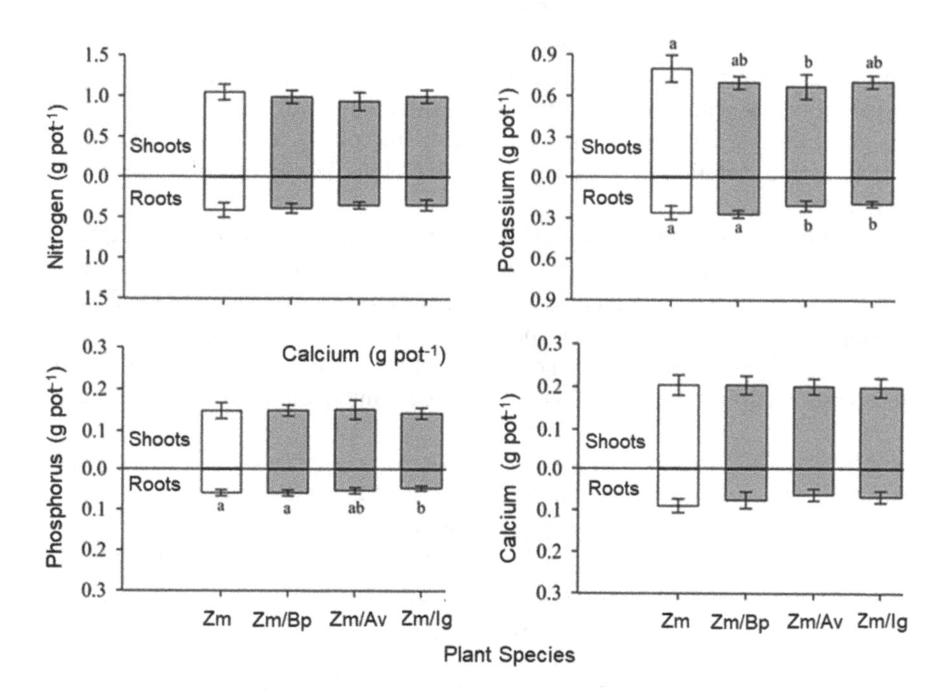

Fig. 5 Nitrogen, potassium, phosphorus, and calcium contents in the dry matter of the aerial part and root of corn (*Zea mays*) in association corn (Zm), *Bidens pilosa* (Bp), *Amaranthus viridis* (Av), and *Ipomoea grandifolia* (Ig) cultivated up to 60 days in monoculture or interspecific competition. Different letters indicate differences between nutrient levels in the associations based on Tukey's test. (Source: Adapted from Matos et al. (2019))

of potassium (K) accumulated in both the root and the aerial part of these plants (Matos et al. 2019).

The presence of weeds can affect the accumulation of nutrients in perennial plants, such as coffee seedlings. The evaluation of the coexistence periods and accumulation of nutrients of different weeds and coffee, Ronchi et al. (2003). These authors found that the weed species, even at low densities, caused decreases in the relative content of nutrients in coffee plants. Species such as *Commelina diffusa* caused the greatest reduction in the relative content of nutrients in coffee plants, and *Sida rhombifolia* in the least ability to reduce the accumulation of nutrients in coffee plants (Table 1). The authors also stated that the degree of nutrient interference varied considerably with the species and also with weed density.

3.3 Competition for Light

Light is a constant resource, but liable to competition when its availability is reduced due to the shading that plants can perform on each other within a community. Therefore, the speed and intensity of interception of solar radiation by plants are determining factors in the ability of competition (Pitelli 2014).

The direct effect of the amount of light that falls on plants influences photosynthesis and ontogenesis, characterizing this resource as one of the most important factors affecting plant growth. However, the effects of light are not only related to the magnitude of the photon flux that reaches the plants, but also to the quality, direction, and duration of irradiance (Smith 2000; Jiao et al. 2007; Merotto Jr et al. 2009).

Regarding photosynthetic pathways, C_4 plants are considered more efficient under conditions of relatively high photon flux density, temperature, and limited water, availability, i.e. in tropical and mostly subtropical environments. On the other hand, plants possessing only the C_3 pathway are more favored under relatively temperate conditions, with lower temperatures and photon flux density, and a less limited water supply (Fig. 6). Under crop-weed interaction conditions, climatic

Table 1 Relative content of macro and micronutrients (% of control treatment) in the aerial part of coffee plants, grown in pots, under the interference of different weed species

		Macronutrients and micronutrients (%)											
Weed	PTC[a]	N	P	K	Ca	Mg	S	Cu	Zn	B	Mn	Fe	Na
Bidens pilosa	77	59	72	67	67	74	97	106	66	76	59	54	69
Commelina difusa	180	30	42	37	45	48	69	69	37	54	19	41	35
Leunurus sibiricus	82	35	33	38	36	40	41	66	37	41	30	57	39
Nicandra physaloides	68	37	62	68	72	76	86	114	69	101	50	107	68
Richardia brasiliensis	148	49	61	57	53	50	67	43	51	63	57	61	59
Sida rhombifolia	133	97	83	105	90	88	98	93	77	138	102	80	106

Source: Adapted from Ronchi et al. (2003)

[a] Total period of coexistence of the weed with the coffee seedling in the pot

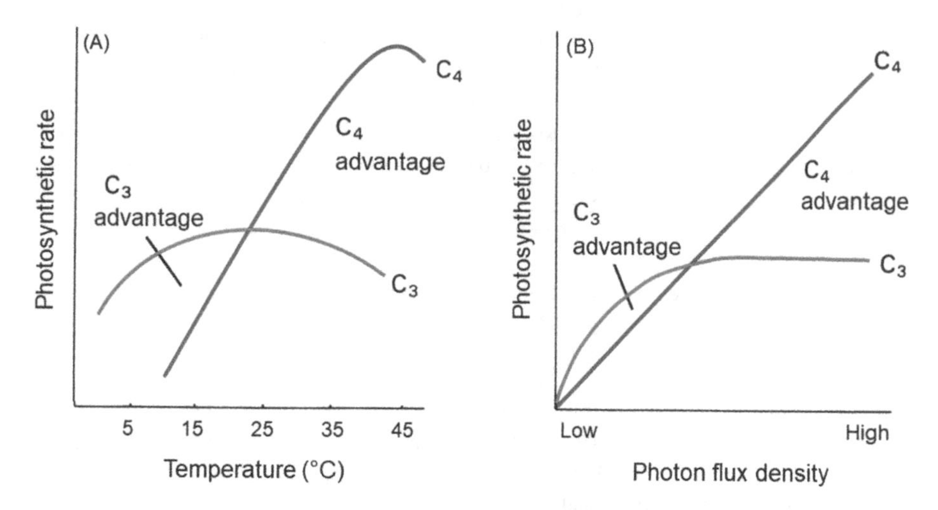

Fig. 6 Expected rates of photosynthesis by C_3 and C_4 plants in relation to varying temperature (**a**) and varying photosynthetic photon flux density (**b**). (Source: Adapted from Cobb (1992); Cobb and Reade (2011))

conditions are important in predicting low or severe competition (Cobb and Reade 2011).

The magnitude of competition for light can be influenced by the competing species, mainly by the photosynthetic route of these species, whether they are C_3, C_4, or the crassulacean acid mechanism (CAM). Plants with C_3 type photosynthetic routes present only the Calvin cycle, responsible for CO_2 fixation, so that the first stable product of photosynthesis is a three-carbon compound (3-phosphoglyceric acid). The enzyme responsible for the primary carboxylation of CO_2 from the air is ribulose 1-5 biphosphate carboxylase-oxygenase (Rubisco), which has both carboxylase and oxygenase activities. This enzyme has a low affinity for CO_2 and catalyzes the production of 3-phosphoglyceric acid, and also of glycolate, the initial substrate of respiration. As a consequence of the action of this enzyme, C_3 plants photorespire intensely, present low affinity for CO_2, and have high CO_2 compensation point, low light saturation point, low water use efficiency, and lower biomass production rate, when compared to plants with C_4-type metabolism (Table 2).

Plants C_4 have two enzymes responsible for CO_2 fixation. These plants have no detectable photorespiration, so they do not lost the carbon that is fixed during photosynthetic carbon assimilation. The primary carboxylating enzyme is PEP-carboxylase, located in the cells of the leaf mesophyll, which carboxylates the CO_2 absorbed from the air via stomata into phosphoenolpyruvic acid, forming oxaloacetic acid (OAA). The OAA is converted into malate or aspartate, depending on the plant species, and then, by diffusion, is transported to the vascular sheath cells of the leaves, where these products are decarboxylated, releasing CO_2 and pyruvic acid into the medium. This released CO_2 is again fixed, now by the enzyme ribulose 1,5 diphosphate carboxylase, occurring the Calvin cycle; the pyruvic acid, by diffusion, returns to the mesophyll cells, where it is phosphorylated, consuming two ATPs, regenerating the enzyme PEP-carboxylase and starting the cycle again.

Table 2 Differential characteristics between plants with C_3 and C_4 photosynthetic pathways

Feature	Photosynthesis C_3	Photosynthesis C_4
Photorespiration	Present: 25–30% of photosynthesis	Present: Not measurable
CO_2 compensation point	High: 50–150 ppm of CO_2	Bass: 0 to 10 ppm of CO_2
Primary carboxylative enzyme	RuDP-carboxylase	PEP-carboxylase
Effect of oxygen	Inhibition	No effect
Relationship CO_2:ATP:NADPH	1:3:2	1:5:2
Photosynthesis x light intensity	Saturation at 1/3 maximum brightness	Does not saturate with increasing brightness
Optimum temperature for photosynthesis	About 25 °C	About 35 °C
Net photosynthesis rate	15–35 mg CO_2 dm^{-2} h^{-1}	40–80 mg CO_2 dm^{-2} h^{-1}
Transpiratory coefficient	450–1000 g H_2O/g dry biomass	150–350 g H_2O/g dry biomass

Source: Adapted from Ferri (1985) and Silva et al. (2007)

In this sense, C_4 plants are mostly more efficient in CO_2 fixation than C_3 plants. However, C_4 plants, for presenting two carboxylative systems, require more energy for the production of photo-assimilates, because they need to recover two enzymes to carry out photosynthesis. As all this energy comes from light, in low light conditions, these plants start to lose the competition with the C_3 plants (Fig. 7). However, the enzyme (PEP-carboxylase) responsible for primary carboxylation in C_4 plants presents some characteristics, such as: high affinity for CO_2, acts specifically as a carboxylase, optimum activity at higher temperatures, and does not saturate at high light intensity. Due to these and other characteristics (Table 2), the C_4 species tend to be superior when in competition with C_3 plants, when they develop under conditions of high temperatures, high light intensity, and even temporary water deficit.

Since most agricultural crops in the tropics and subtropics (cotton, soybeans, corn, rice, sugarcane, beans, cassava) are grown in the months of the year that coincide with periods of high light intensity and temperature, C_4 weeds tend to exert greater competition with crops. Currently, problematic weeds in Brazil, such as *Digitaria insularis*, *Eleusine indica*, *Echinochloa* spp. and other species such as *Amaranthus* spp. and *Cyperus* spp. mostly present C_4 photosynthetic pathway (Table 3).

Due to the importance of light in development, plants, in order to try to avoid competition for light, have developed complex systems of information acquisition that allow them to adjust their phenotype due to light signals that indicate present or future competition with neighboring plants for access to solar radiation (Ballaré 2009). This information is collected and processed by various photoreceptors, such as phytochromes, cryptochromes, and phototropins. Differential absorption by chlorophyll reduces red light (R) relative to far-red radiation (FR), and this reduction provides information about the proximity of neighboring plants. R: FR ratios between 1 and 1.2 are indicative of direct sunlight, while lower values indicate some degree of shading or the proximity of potential competitors (Fig. 8) (Gundel et al. 2014).

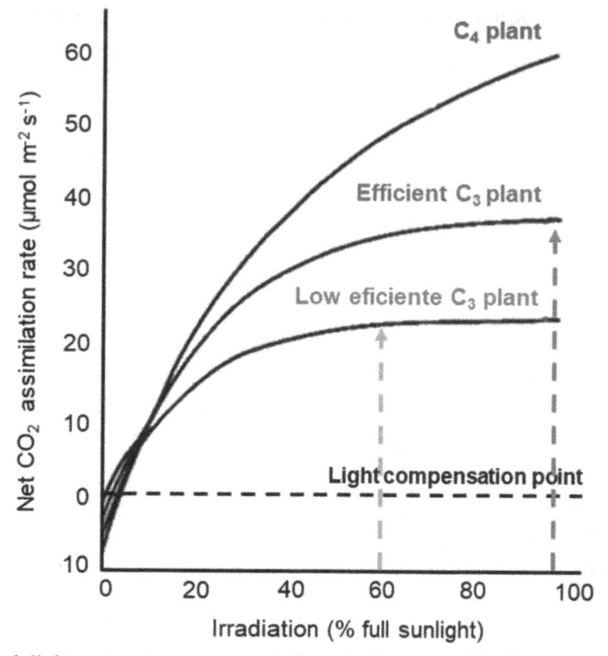

Fig. 7 Idealized light saturation curves of C_3 and C_4 plants. The light compensation point (intersection with the dotted black line), photosynthetic efficiency (slope under light-limited conditions), and light saturation point (dotted blue arrows) are higher for efficient C_3 plants compared to non-efficient C_3 plants. C_4 plants often do not reach the clearing point in full sunlight. (Source: Koyro et al. (2014))

The plants make some adjustments in their growth, to avoid competition for light, such as stem elongation, increase in height and leaf area, reduction of leaf thickness, leaf reorientation, changes in chlorophyll concentration, senescence, reduction of branching, and accelerated flowering. It is important to emphasize that these changes in most cases cause a high energy expenditure by the plant, and may interfere with the accumulation of biomass and grain production.

In a study of the interaction between corn and *Amaranthus retroflexus*, Liu et al. (2009) grew corn plants in isolation and together, but provided enough water, light and nutrients for both species to prevent competition. Maize plants living together with giant caruru, even with resource availability, showed a reduction in dry matter compared to monoculture. This reduction in dry matter without competition may then be related to an energy expenditure of plants with the objective of promoting an escape from neighboring plants to avoid future shading (Fig. 9) (Liu et al. 2009).

3.4 Competition for CO_2

Currently, the global average atmospheric CO_2 concentration is 415.13 ppm and is steadily increasing (ESRL 2021). As already discussed, both crops and weeds can be classified according to their photosynthetic pathways, C_3 or C_4. This

Table 3 Photosynthetic pathway of different weed species of economic importance in Brazil

Scientific name	Photosynthetic pathway
Alternanthera tenella	C_3
Chenopodium album	C_3
Bidens pilosa	C_3
Conyza bonariensis	C_3
Galinsoga parviflora	C_3
Sonchus oleraceus	C_3
Xanthium strumarium	C_3
Raphanus sativus	C_3
Ipomoea purpurea	C_3
Euphorbia heterophylla	C_3
Senna obtusifolia	C_3
Sida rhombifolia	C_3
Richardia brasiliensis	C_3
Solanum americanum	C_3
Spermacoce verticillata	C_3
Commelina benghalensis	C_3
Amaranthus retroflexus	C_4
Cyperus rotundus	C_4
Cenchrus echinatus	C_4
Chloris gayana	C_4
Cynodon dactylon	C_4
Digitaria horizontalis	C_4
Digitaria insularis	C_4
Echinochloa crus-galli	C_4
Eleusine indica	C_4
Lolium multiflorum	C_4
Rottboellia cochinchinensis	C_4
Sorghum halepense	C_4
Urochloa plantaginea	C_4
Portulaca oleracea	C_4

differentiation between species can alter the efficiency of CO_2 capture from the air, and these characteristics can interfere with the competitive ability between crops and weeds when they live together, mainly by increasing biomass production.

The current amount of CO_2 in the atmosphere is low enough to saturate the Rubisco enzyme, which drives photosynthesis in C_3 plants. In contrast, C_4-type plants tend to respond less to elevated CO_2 levels because they have an innate concentration mechanism capable of increasing the CO_2 level at the Rubisco site of action to 2000 ppm (Fig. 10) (Korres et al. 2016). Therefore, with this possibility of future increases in atmospheric CO_2 concentrations, it will likely favor both weed and C_3 crop species.

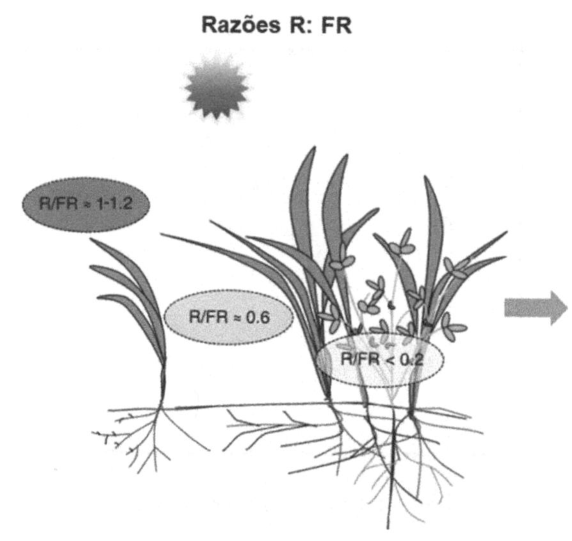

Fig. 8 Shade avoidance syndrome responses triggered by the red: far red (R: FR) ratio (**a**), and schematic representation of the effects of phytochrome perceived informational signals on shade escape (**b**). (Source: Adapted from Gundel et al. (2014))

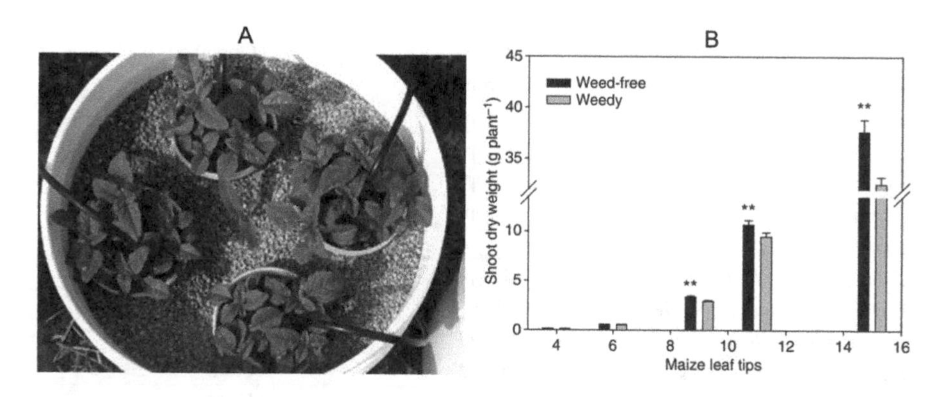

Fig. 9 Maize at the 3-leaf tip stage growing in the bucket surrounded by *Amaranthus retroflexus* (**a**) (the central corn nutrient solution supply tube was removed for photographic purposes) and aboveground dry matter (**b**) of maize plants sampled in the absence and coexistence with *Amaranthus retroflexus* at different times. (Source: Adapted from Liu et al. (2009))

Some of the world's most problematic weed species are of the C_4 type and are found in C_3 crops. An example of this, is ricegrass competing with cultivated rice, in such cases, a positive response of the C_3 crop to increased CO_2 can occur, which can make the weed less competitive. In Table 4, some associations of C_3 and C_4 metabolism plants are described, and their competitive responses under conditions to increased atmospheric CO_2.

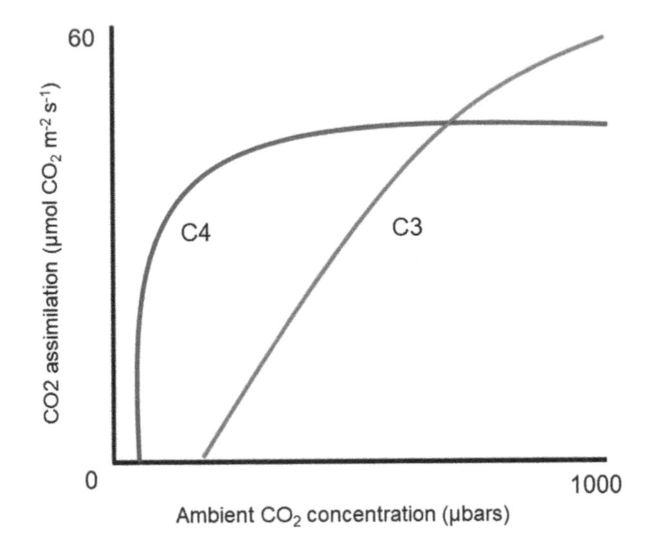

Fig. 10 Response of CO_2 assimilation in C_3 vs. C_4 plants to increases in CO_2 concentration. (Source: Adapted from Korres et al. (2016))

Table 4 Competition between crops and weeds as a function of high atmospheric CO_2 concentration

Weed and crop	Favored specie	Environment
Weed C_4 vs. crop C_3		
Sorghum halepense vs. Festuca	Crop	Greenhouse
Sorghum halepense vs. Soybean	Crop	Growth chamber
Amaranthus retroflexus vs. Soybean	Crop	Field
Echinochloa glabrescens vs. Rice	Crop	Greenhouse
Paspalum dilatatum vs. Grasses	Crop	Growth chamber
Gramíneas vs. *Medicago sativa*	Crop	Field
Weed C_3 vs. crop C_4		
Chenopodium album vs. Beet	Crop	Growth chamber
Taraxacum officinale vs. *Medicago sativa*	Weed	Field
Plantago lanceolata vs. Pasture	Weed	Growth chamber
Taraxacum e *Plantago* vs. Pasture	Weed	Field
Cirsium arvensis vs. Soybean	Weed	Field
Chenopodium album vs. Soybean	Weed	Field
Weed C_4 vs. crop C_4		
Amaranthus retroflexus vs. Sorghum	Weed	Field
Weed C_3 vs. crop C_4		
Xanthium strumarium vs. Sorghum	Weed	Greenhouse
Abutilon theophrasti vs. Sorghum	Weed	Field

Source: Adapted from Brunce and Ziska (2000); Korres et al. (2016)

Increased atmospheric CO_2 concentration can influence the interactions that occur between crops and weeds. Previously competitive crops may be more susceptible to this change in CO_2 and therefore need more monitoring and control of the weeds present, while other cultivars may need less control.

4 Factors Affecting Weed-Crop Interference

Interference can be defined as the adverse effect that one plant can have on the growth and development of other plants in its vicinity. In the case of weeds, this interference occurs when the presence of these species near the crop interferes in some way with human interests (Hijano et al. 2021). Another important issue in the interference of weeds in crops is the degree of interference, which is the intensity of the effects caused by the same, and will be influenced by the weed community, the characteristics related to the crop, the time and extent of the coexistence period.

The degree of weed interference with agricultural crops depends on three main factors involving the weed community, the crop, and the environment in which they are present. In addition, the coexistence period, i.e. the time at which weeds emerge in relation to the crop and also the length of time they coexist, can influence the intensity of weed interference with the crops of interest.

The main factors involving the weed community are the plant species present in the area, their density and distribution. The crop-related factors follow the same principles as weeds, the main factors involving the cultivar used, plant density, and also spacing. Environmental factors such as soil characteristics, climate, and management influence the degree of weed interference in crops, as some species may be more adaptable to the environmental conditions. The period of coexistence is also an important factor, that is, depending on the time at which weeds emerge in relation to the crop and the period they stay together, will influence a greater or lesser interference (Fig. 11) (Silva et al. 2007; Pitelli 2014).

4.1 Culture-Related Factors

Cultivated species vary in their ability to compete and withstand the interference imposed by weeds. Within the same crop, the agricultural market has several cultivars with adaptations for different agricultural aptitudes, such as grain production, biomass, and quality of the final product. These adaptations promoted by genetic improvement can change the competitive ability of these cultivars when in coexistence with weeds (Fig. 12). Rice is a crop with constant modifications caused by breeding. Sharma et al. (2013) devised a concept of a new plant type to further increase the yield potential of rice, the modifications included reduced number of tillers, increased number of grains per panicle, and increased stem stiffness. However, many modifications, such as reduced size, tillering, and increased reproductive structures, can reduce the competitive ability of these cultivars with weeds, resulting in increased interference.

A number of cultivar characteristics assist in the crops' ability to compete with weeds. These characteristics include fast germination and emergence, plant height, rapid biomass accumulation and ground cover, leaf characteristics such as high leaf area index, leaf length, and leaf angle, and canopy structure such as high light-intercepting capacity, lateral branching, and tillering ability.

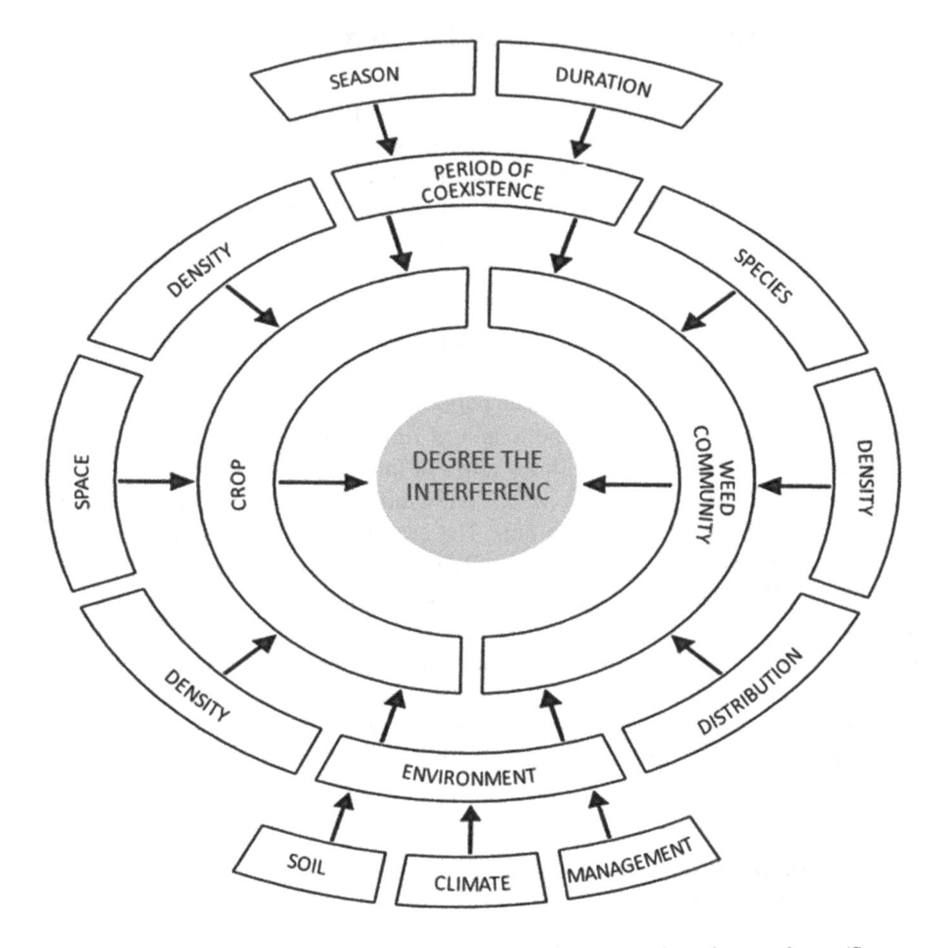

Fig. 11 Factors affecting the degree of interference between weeds and crop plants. (Source: Adapted from Pitelli (1985))

Competitive ability studies are effective in demonstrating the ability of some cultivars to withstand weed interference. In studies of competitive ability of canola cultivars (*Brassica napus* var. *Oleifera*) with *Raphanus sativus*, Durigon et al. (2019) observed that turnip presents greater competitive ability than canola, moreover, the canola cultivar Hyola 571CL, had the accumulation of dry matter less affected in the presence of turnip than the canola cultivar Hyola 555TT. This fact shows that the choice of cultivars with greater ability to withstand weed competition prevents yield losses (Fig. 13).

Another factor that influences the degree of interference is the arrangement of plants, which is altered as a function of the spacing between rows when sowing and the density of plants per area. Crops with an adequate plant arrangement tend to grow and cover the soil in a shorter period of time, besides avoiding interspecific competition. Crops grown in narrow rows may present better light interception,

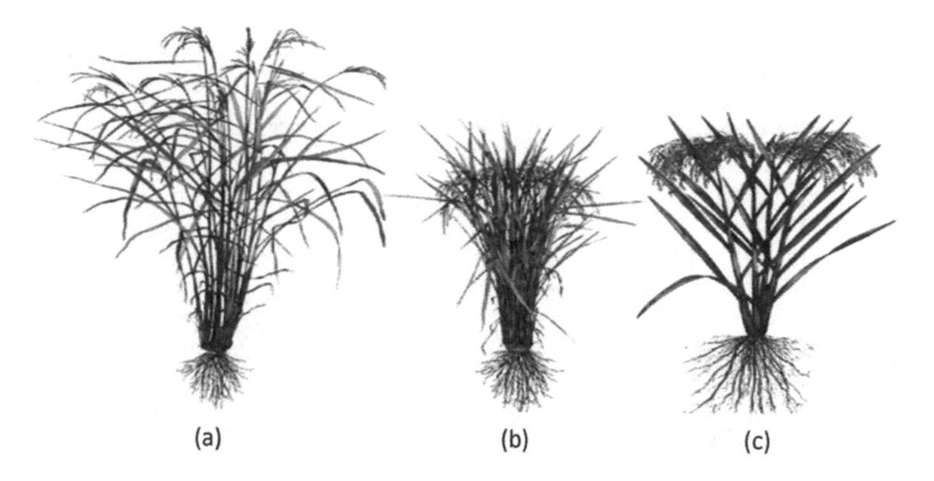

Fig. 12 Suggested ideotype changes for continuous improvement of rice productivity. Traditional plant type (**a**); semi-dwarf plant type (current varieties) (**b**); and new plant type (**c**). (Source: Adapted from Sharma et al. (2013))

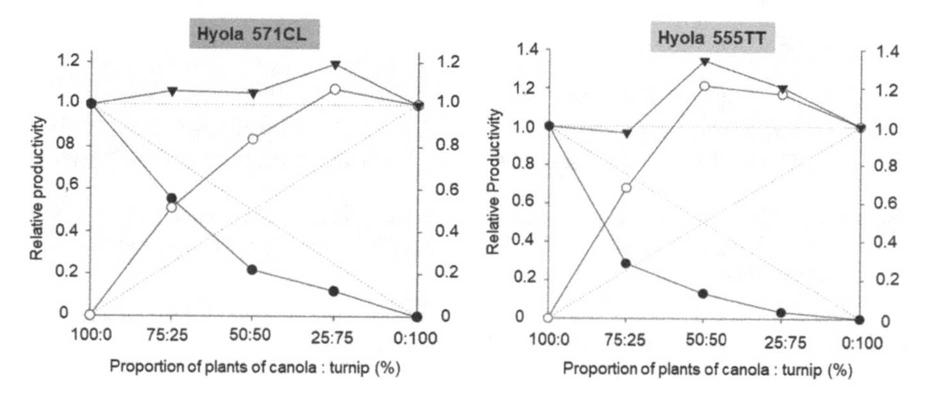

Fig. 13 Relative productivity of canola plants of *Hyola hybrid* 571CL and Hyola 555TT (●) and turnip (○), and total relative productivity (▼), of aerial part dry matter according to the proportions between species. (Source: Adapted from Durigon et al. (2019))

resulting in efficient development of leaf area and early canopy closure, which modifies the patterns of emergence and growth of weeds, reducing their competitiveness (Knezevic et al. 1999).

When evaluating the effect of row spacing and timing of weed control in cotton, Tursun et al. (2016) observed an increase in weed dry matter production with increasing cotton row spacing (50, 70, and 90 cm) (Fig. 14). This wider row spacing causes the crop to take longer to close the canopy, which allows more light to reach the soil in a more intense way. The greater amount of light encourages the emergence of photoblastic positive weeds and favors the growth and accumulation of biomass of the plants, favoring the competition of plants with the crop.

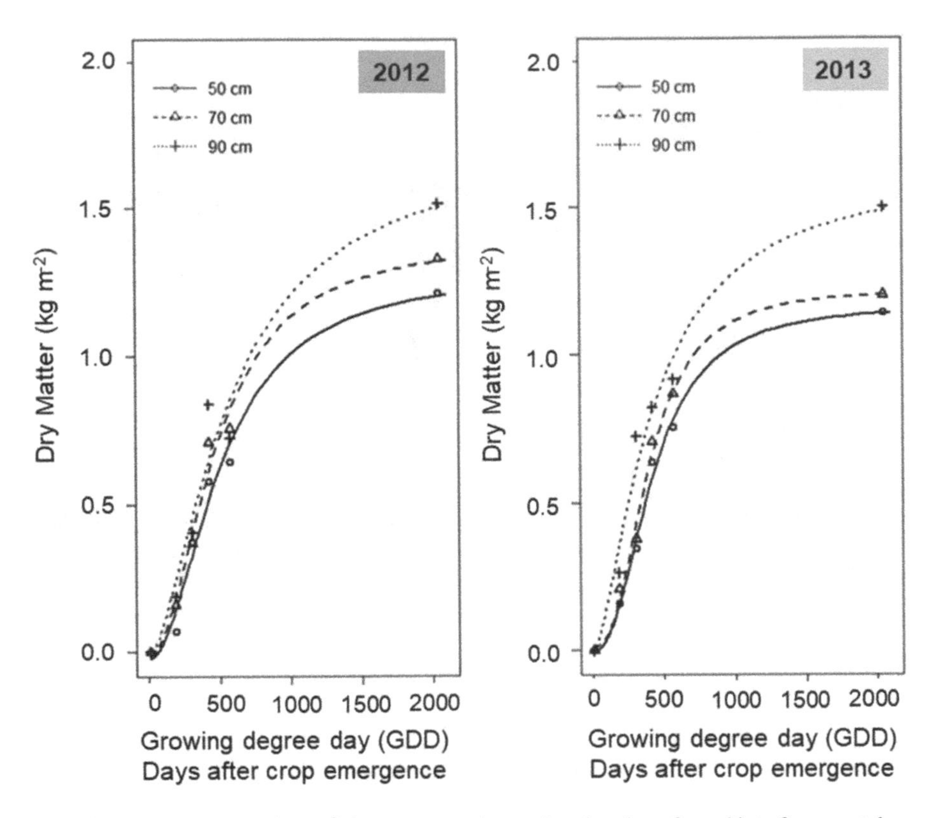

Fig. 14 Weed dry matter (kg m^{-2}) in response to increasing duration of weed interference at three row spacing (50, 70, and 90 cm) of cotton in the years 2012 and 2013. (Source: Adapted from Tursun et al. (2016))

4.2 Factors Related to the Weed Community

There is a constant competition between weeds and crop plants for resources in the environment. The intensity of this competition depends on the composition of the weed community present in the area. Factors such as species, density, and distribution influence the ability of these plants to interfere with the crop of interest, i.e., the degree of interference.

Characteristics that vary among species, such as the ability to produce propagules, speed of emergence and initial growth, efficiency of resource use, photosynthetic metabolism and also the phenotypic plasticity to respond to adverse environmental conditions, influence the speed of growth of weeds in relation to crops (Pitelli 2014; Hijano et al. 2021). These growth characteristics combined with the size of the plants and the ability to cover the ground influence the competitive ability and the degree of interference.

Weed community and density are important factors in crop interference and yield loss. Adeux et al. (2019) when evaluating wheat yield loss in coexistence with six

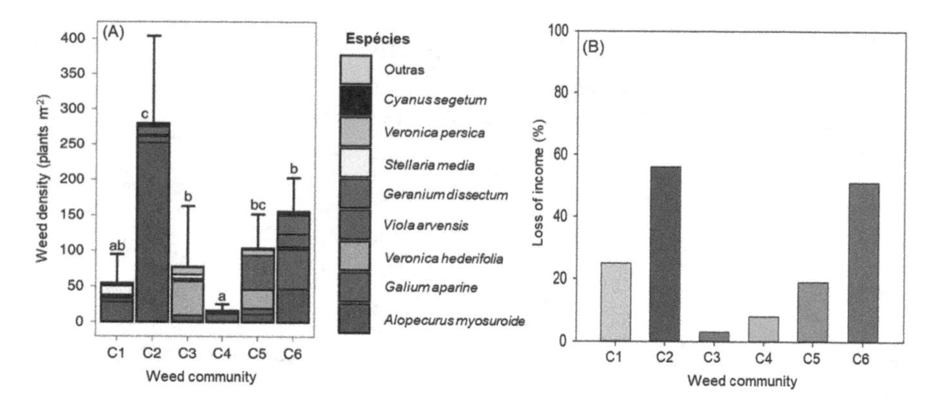

Fig. 15 Average density and composition of the observed distinct weed communities (denoted C1 to C6), equal letters show no significant differences in terms of total weed density by least squares test at $p < 0.05$ (**a**) and yield loss of wheat (*Triticum aestivum*) coexisting with the plant communities (**b**). (Source: Adapted from Adeux et al. (2019))

distinct plant communities, observed that losses ranged from 3 to 56%, and weed density and weed community composition were instrumental in increasing crop yield losses (Fig. 15a, b).

Plants of the same genus may vary as to the degree of interference. Bensch et al. (2003) studying the interference of *Amaranthus* species on soybean, observed that *Amaranthus palmeri*, followed by *Amaranthus rudis* and *Amaranthus retroflexus* have the greatest capacity to reduce soybean yield. The yield results showed the estimated maximum yield loss of 86.9%, 44.9%, and 62.8% when in coexistence with *A. palmeri*, *A. rudis,* and *A. retroflexus*, respectively (Fig. 16).

4.3 Environmental Factors

Environmental factors such as climate, soil, and management influence the emergence and growth of crops and weeds in the environment. This influence can favor or disadvantage one or both species, changing the degree of interference between them. Thus, climatic and edaphic changes determine changes in the community's balance, influencing the competitive balance. Weeds have several mechanisms that confer dormancy, so the emergence of plants is responsive to environmental conditions such as temperature, humidity, and light. These species characteristics and environmental conditions influence the flow of weed emergence, and this flow is a determining factor in the degree of interference, because it influences the density of plants and the time of emergence in relation to the crop (Fig. 17) (Mortensen et al. 2021).

The timing of emergence of weed species is a key determinant of weed control. Many weeds have only one emergence flush during the year, for example, in

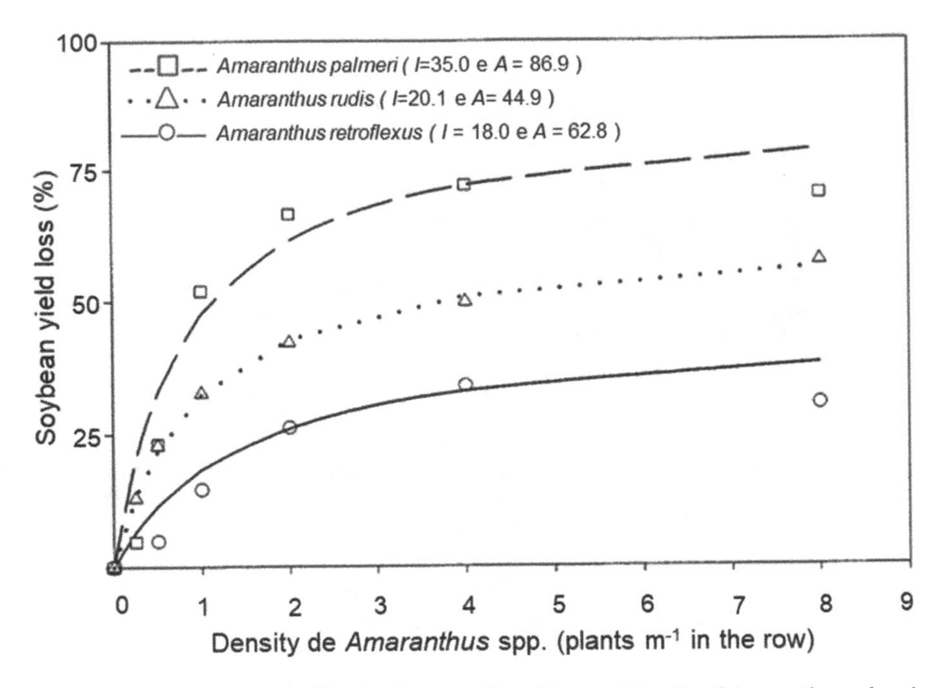

Fig. 16 Estimated yield loss (%) of soybean as affected by weed density of *Amaranthus palmeri*, *Amaranthus rudis,* and *Amaranthus retroflexus*. (Source: Adapted from Bensch et al. (2003))

temperate climate conditions such as *Ambrosia artemisiifoli*, *Polygonum pensylvanicum*, *Amaranthus retroflexus,* and *Setaria* spp. (Fig. 17), in the case of these plants, the introduction of control in a strategic way helps in the management, avoiding reinfestation in the rest of the year.

On the other hand, plants that have a longer period of emergence varying throughout the seasons, such as *Chenopodium album*, *Galinsoga ciliata*, *Capsela bursa-pastoris,* and *Thlaspi arvense*, make it difficult to adopt control strategies, because they will possibly need control methods that allow them to avoid competition for a longer period of time.

The management or cultural practices adopted can also influence the degree of interference (Pitelli 2014). When evaluating the effect of the cultivation system (conventional—SPC and direct—SPD), Cunha et al. (2015) observed that in the bell pepper crop, SPD reduced the density and accumulation of dry matter of weeds compared to SPC. The presence of straw in SPD when well managed provides a physical barrier that prevents the passage of light, reducing seed germination of photoblastic positive weeds and preventing the growth of species that cannot pass through the mulch.

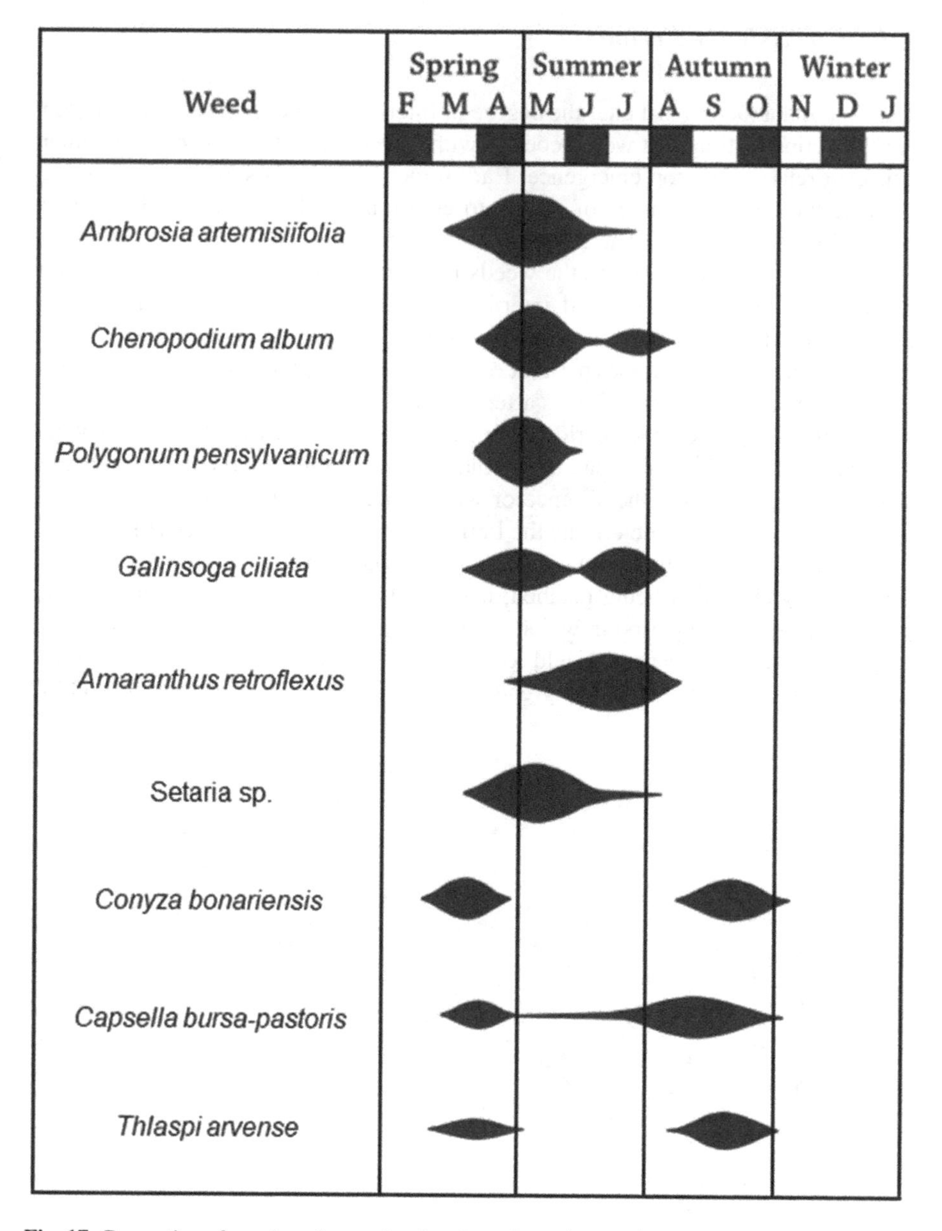

Fig. 17 Proportion of weed seeds germinating throughout the year in central Pennsylvania, letters represent months. Larger bar width represents greater germination. (Source: Mortensen et al. (2021))

4.4 Coexistence Period

Two important factors that alter the degree of interference between weeds and crops are the period of time that weeds coexist with the crop and the time of weed emergence in relation to crop emergence. Early-emerging weeds show greater growth because they gain advantage of access to environmental resources and are more competitive (Agostinetto et al. 2004).

Establishing the crop before the weeds is important in overcoming competition by reducing the degree of weed interference. The emergence of volunteer corn together with the bean crop was more competitive compared to corn that emerged 7 days after the bean. The losses per unit of volunteer corn plant reduced from 27.1 to 11.1% when plants emerged 7 days after beans (Fig. 18) (Silva et al. 2019).

The effect of coexistence periods is well represented in Critical Period of Weed Interference (CPWI) studies, and these studies highlight the period that crops exclusively need to remain in the absence of weeds to avoid yield losses. To define the CPWI, it is necessary to determine the Period Previous to Interference (PPI), which is the period in which the weeds can remain in coexistence with the crop, without acceptable yield levels being reached; and the Total Period of Weed Interference (TPWI), which is the period in which the plants can emerge and coexist within the crop and not cause acceptable yield losses. These studies show that the longer the weeds coexist with the crop, the greater the yield losses. Furthermore, the TPWI

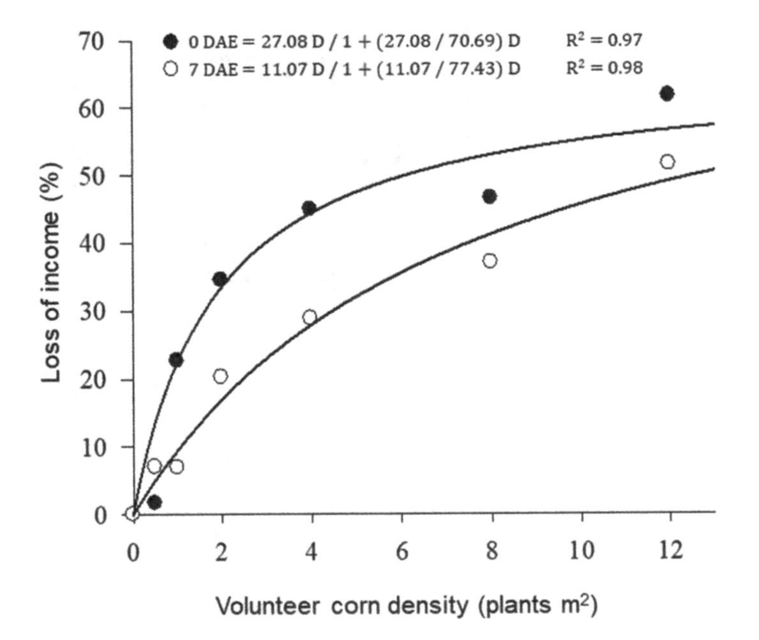

Fig. 18 Loss of bean yield as a function of time of emergence and density of the volunteer at 0 days after emergence (DAE), bean and volunteer corn plants emerged together, and 7 DAE in which corn emerged 7 days after bean emergence. (Source: Adapted from Silva et al. (2019))

studies show that weeds can coexist with crops provided they are not present at critical times, and especially when the crop is already well established. These studies will be discussed in more detail in Sect. 5.5 of this chapter.

5 Experimental Methods for Studying Weed Competition and Interference in Crops

Studies of competition between weeds and crops are able to provide important information that assists in weed management. Studies such as EDL and CPWI help farmers to make a decision on the need for control to avoid yield losses. Based on the importance of knowing the interactions between crops and weeds, the experimental models most commonly used in studies of competition and interference between weeds and crops are described below.

5.1 Additive Experiment

The additive experiment is most commonly used to study the effect of competition between weeds and crops. The crop density is held constant while the weed density is altered (Fig. 19a). It is a relevant study for most agricultural situations where the crop occurs at a fairly uniform density and weed density varies between areas, over years, or according to management practices (Swanton et al. 2015).

In this type of study, the percent crop yield loss with increasing weed density is calculated by dividing the yield at each weed density by the weed-free yield. The data are most commonly fitted to a nonlinear regression model (rectangular hyperbola) (Fig. 19b) (Cousens 1985).

$$\mathrm{YL} = \frac{Id}{1+\left(\frac{I}{A}\right)d}$$

The YL is the percent yield loss, d is the weed density, I is the percent yield loss per unit of weeds when d approaches zero, and A is the maximum asymptote of yield loss when d approaches infinity (Cousens 1985; Radosevich et al. 2007; Swanton et al. 2015).

The main advantage of this method is the ease of establishing the study under field conditions. This study is most suitable for determining the potential for crop yield loss in the presence of weeds. In addition, the study is useful for evaluating the effect of relative emergence times of competing weeds and crops, and is also appropriate in determining economic thresholds of weeds. The disadvantage of this study is related to the simultaneous changes in population and total weed density, making

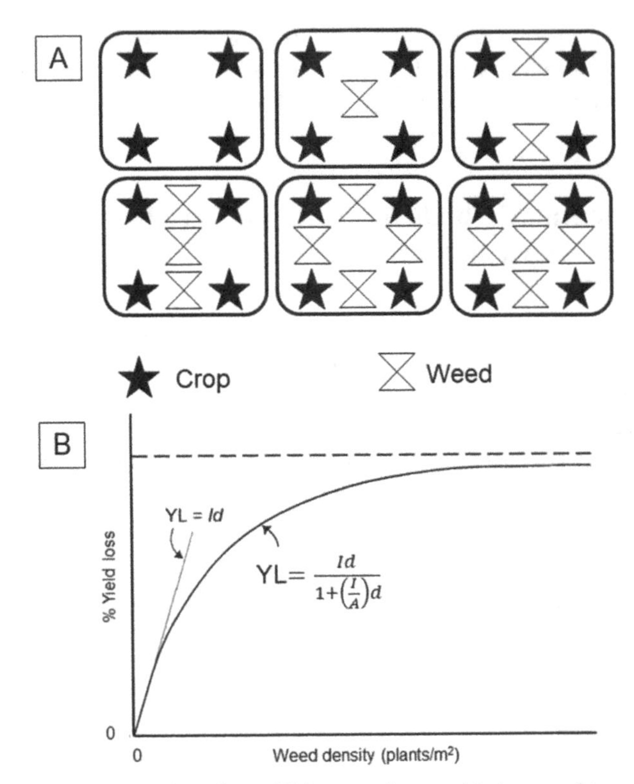

Fig. 19 Schematic representation of an additive experiment with the use of two plant species, in which the crop density is maintained and the weed density is increased (**a**) and the nonlinear regression model (rectangular hyperbola) that makes it possible to relate the yield loss to the weed density, through the parameters *A* and *I* (**b**). (Source: Adapted from Cousens (1985), Radosevich et al. (2007), Swanton et al. (2015))

it difficult to determine specific plant interactions and to obtain information about competitive processes. The results also depend on local soil and environmental conditions and therefore caution should be exercised when extrapolating the results to wide geographic areas (Cousens 1985; Blackshaw 1993; Weaver and Ivany 1998; Swanton et al. 2015).

Using the *I* parameter from the previous equation, it is possible to calculate the economic thresholds or EDL, with the help of the equation below, adapted from Lindquist and Kropff (1996); Agostinetto et al. (2017).

$$\text{EDL} = \frac{\text{Wc}}{\text{Yg}^{*} P^{*} \left(\frac{I}{100}\right) * \left(\frac{H}{100}\right)}$$

where EDL is the Economic Damage Level (plants m^{-2}), Wc is the weed control cost (herbicide and application, in real ha^{-1}), Yg is the crop grain yield (kg ha^{-1}), *P* is the price paid for the crop (R$ kg^{-1}), *I* is the crop yield loss (%) per unit of competing plant, when the density level approaches zero and *H* is the control efficiency

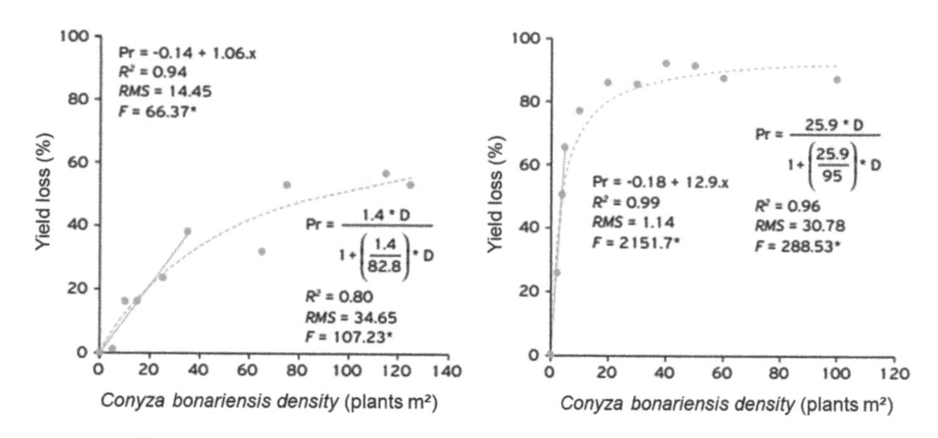

Fig. 20 Soybean yield loss in the cultivars BRS Estância and BMX Turbo due to increased density of *Conyza bonariensis*. (Source: Adapted from Agostinetto et al. (2017))

with the use of herbicide (%). For EDL calculation purposes, three values are normally estimated (low, medium and high) according to yield potential of the crop, price paid for the product, control cost and efficiency of the applied herbicides.

When studying the yield loss of soybean (cultivars BRS Estância and BMX Turbo) and the interference EDL of glyphosate-resistant *Conyza* spp., Agostinetto et al. (2017) observed different values of I in the cultivars. In BRS Estancia, the I (yield loss per unit) was 1.4%, and the A (maximum yield loss) was 82.8% (Fig. 20). In contrast, in the BMX Turbo cultivar, the I and A parameters were estimated to be 25.9 and 95.0%, respectively.

In calculating the EDL, the same authors estimated three values of soybean yield potential (2000, 3500, and 5000 kg ha^{-1}), soybean prices (R$ 40, 60, and 80 bags^{-1}), control cost (R$ 50, 100, 150), and efficiency of herbicides in the control of the horseweed (80, 90, and 100%).

In general, increases in crop yield, price received for the product, and herbicide efficiency reduce EDL, indicating that control measures should be adopted when weed density is low. However, when the cost of control is high, an increase in EDL occurs, indicating that it is more economical to control weeds when the density is higher (Fig. 21) (Agostinetto et al. 2017).

5.2 *Replacement Series*

The study was developed to overcome some of the criticisms of the additive experiment, often used to determine which species under study is the most competitive, as well as to gain information about plant-to-plant interactions (Radosevich 1987). The overall plant density is held constant while the proportion of species is varied. A common approach is to grow plant species in a uniform spatial arrangement in various mixing ratios, such as 25:75, 50:50, and 75:25, along with monoculture of each species (Fig. 22a, b).

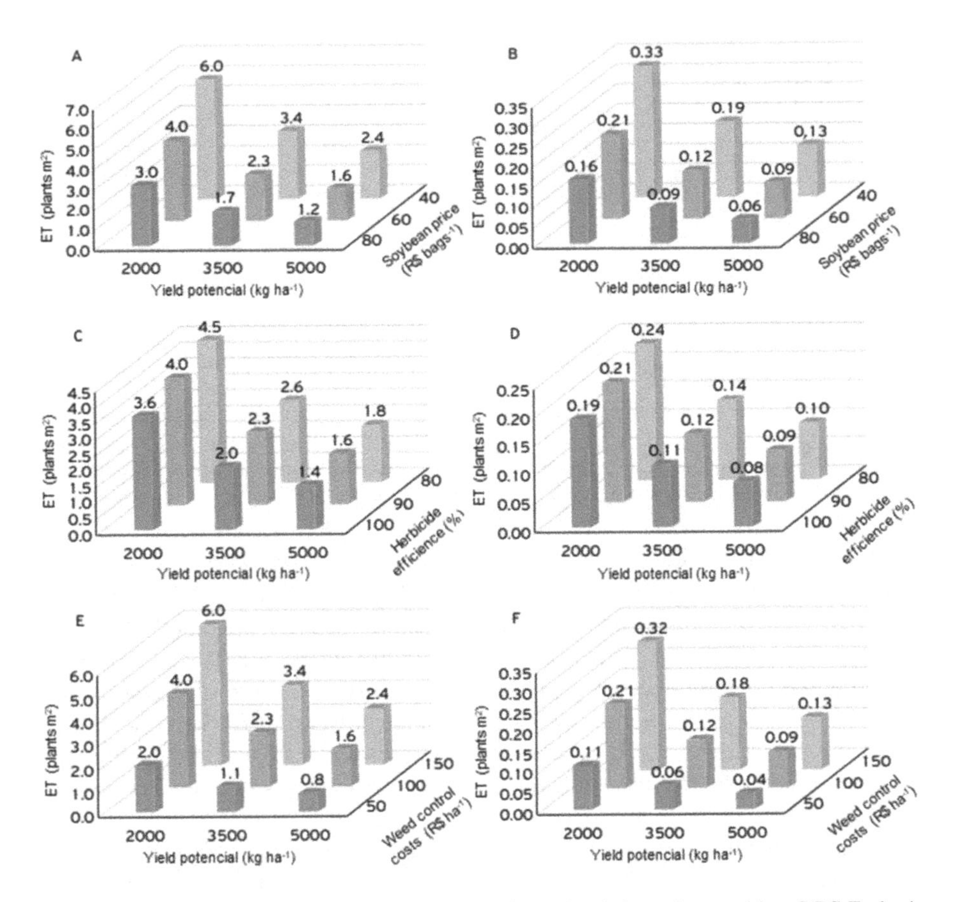

Fig. 21 Economic damage level (EDL) of *Conyza bonariensis* in soybean cultivar BRS Estância (**a**, **c**, and **e**) and BMX Turbo (**b**, **d**, and **f**) as a function of yield potential, soybean price, herbicide efficiency, and weed control cost. (Source: Agostinetto et al. (2017))

The results of substitution series studies are often presented graphically with diagrams, in which the absolute or relative yield of each of the two species is plotted against their proportion in the mixture (Fig. 23a, b).

In the interaction of a crop species and a weed when grown in a replacement scheme there are four possible outcomes. When the relative productivity (RP) of both species A and B (RP_A and RP_B) behave like straight lines, the ability of each species to interfere with the other is equivalent or the two species are located so far apart that no interaction occurs between them. When one response curve is concave and the other is convex: one species is more aggressive than the other. When both response curves are concave and the total yield of the two species in a mixed stand is lower than their respective yields in a pure stand, it is a case of mutual antagonism. When both species have convex response curves and the total yield of the two species in a mixed stand is greater than that of their respective yields in a pure stand, and it is considered a case of symbiosis (Swanton et al. 2015).

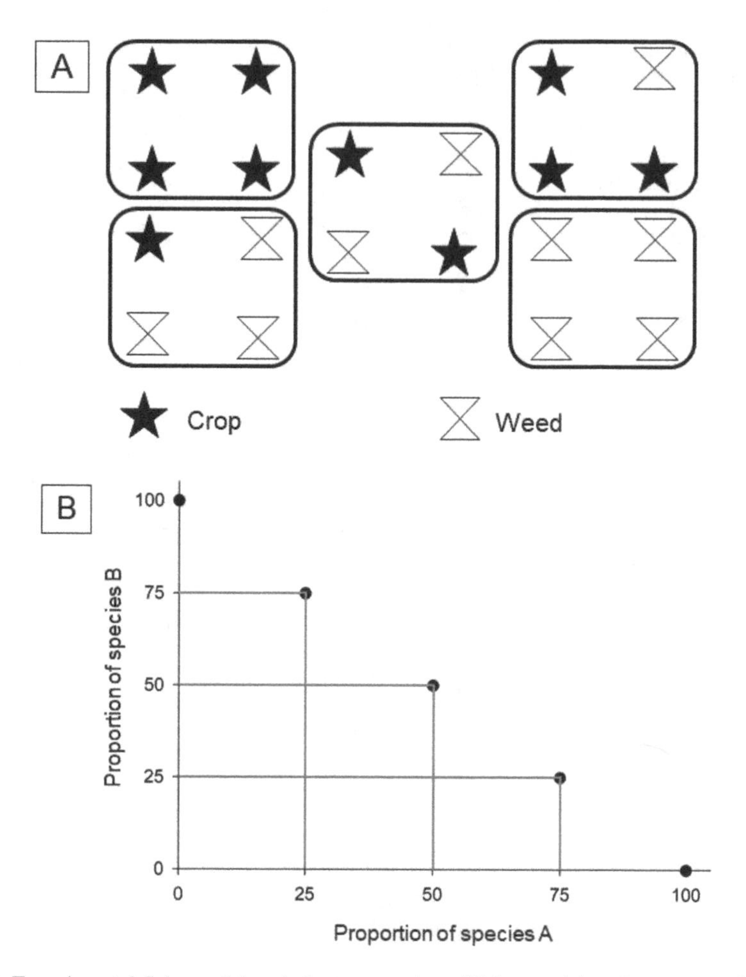

Fig. 22 Experimental Scheme (**a**) and plant proportions (**b**) in a serial replacement study using two plant species. (Source: Adapted from Rejmánek et al. (1989), Cousens (1991), Swanton et al. (2015))

Regarding the total relative productivity (TRP), when it behaves as a straight line, it means that competition occurred for the same resource(s). If the value is greater than 1 (convex line), no competition occurs, because the supply of resources exceeds demand, or because the species have different demands for the resources in the environment. When less than 1 (concave line), it means that antagonism occurred, with mutual prejudice to the growth of both species (Fig. 23b) (Radosevich et al. 2007; Bastiani et al. 2016).

At the 50/50 ratio of species, the relative competitiveness (RC), relative clustering coefficients (*K*), and competitiveness (*C*) indices can be calculated. The CR represents the comparative growth of species A over B. The K indicates the relative dominance of one species over the other. And C determines which species is more aggressive.

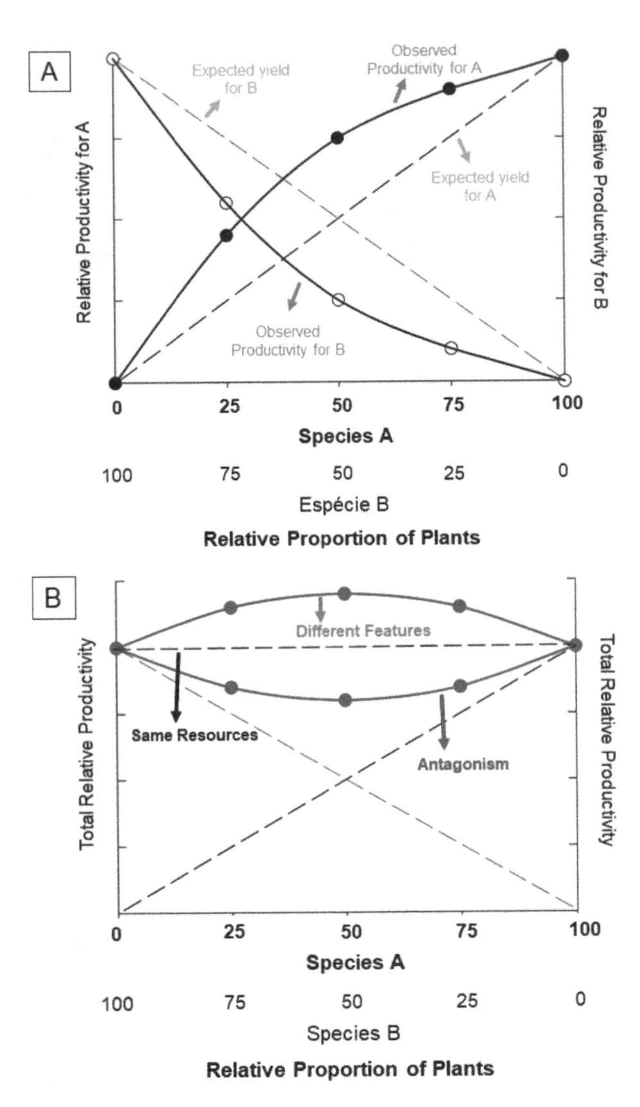

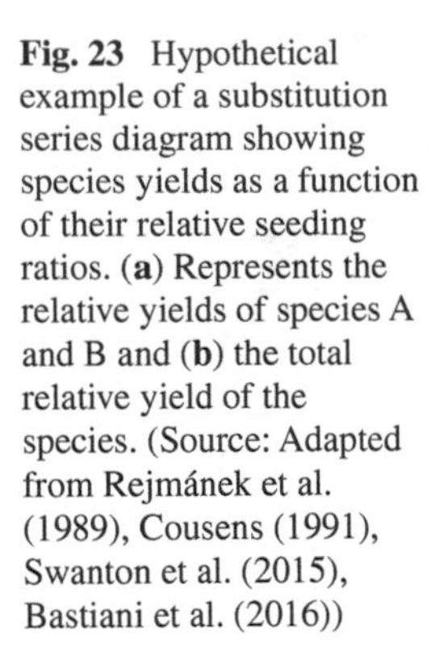
Fig. 23 Hypothetical example of a substitution series diagram showing species yields as a function of their relative seeding ratios. (**a**) Represents the relative yields of species A and B and (**b**) the total relative yield of the species. (Source: Adapted from Rejmánek et al. (1989), Cousens (1991), Swanton et al. (2015), Bastiani et al. (2016))

Calculations of these indices are obtained using the following equations:

$RC = RP_A/RP_B$
$K_A = RP_A/(1 - RP_A)$
$K_B = RP_B/(1 - RP_B)$
$C = RP_A - RP_B$

In this sense, species A is more competitive than B when RC > 1; KA > KB and $C > 0$; on the other hand, species B will be more competitive than A when CR < 1; KA < KB and $C < 0$ (Cousens 1991; Cousens and O'Neill 1993; Hoffman and Buhler 2002; Bianchi et al. 2006a, b; Bastiani et al. 2016).

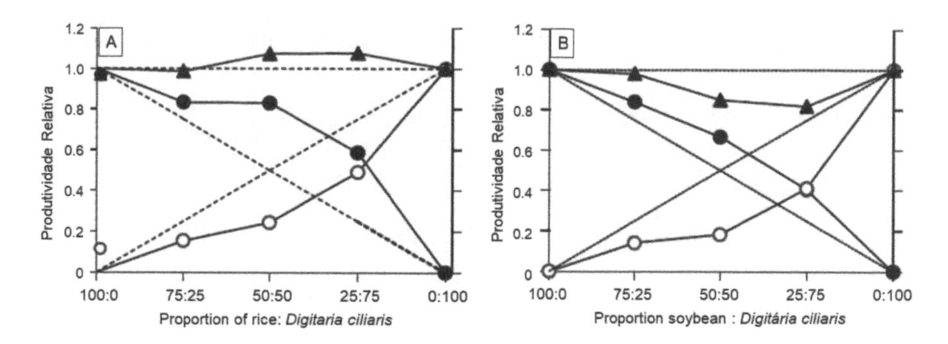

Fig. 24 Relative productivity of irrigated rice (*Oryza sativa*) (**a**), soybean (*Glycine max*) (**b**) (●), and *Digitaria ciliaris* (○) and total relative productivity (▼), regarding the accumulation of aboveground dry matter, due to the proportions among the species. The dashed lines refer to the hypothetical relative productivity when there is no interference of one species over the other. (Source: Adapted from Agostinetto et al. (2013))

The advantage of this study is its good suitability for ranking the competitive ability of a species under a given set of conditions appropriate for examining the mechanisms of competition between crops and weeds. Yield rates can be used to describe patterns of resource use and relative aggressiveness among species. As for the disadvantages, the study may not be representative of most field situations, where crops occur at a constant density. In addition, experimental results may depend on the overall density of the selected plant and the specific conditions of increasing resource use (Swanton et al. 2015).

In studies of competition of irrigated rice and soybean with millet grass (*Digitaria ciliaris*), Agostinetto et al. (2013) observed a higher competitive ability of irrigated rice and also soybean (convex line) with *D. ciliaris* (concave line) (Fig. 24a, b). The TRP data showed that irrigated rice and *D. ciliaris* compete for the same resources in the environment, due to the fact that the TRP line of the plants were similar to the hypothetical line. The TRP of soybean: *D. ciliaris* shows an antagonistic effect of the two species, however, the reduction of TRP is related to the low dry matter accumulation in the presence of soybean.

The RC, *K*, and *C* values reinforce that *D. ciliaris* shows less competitive ability per individual than irrigated rice and soybean plants, when the crops are in the same proportion of plants with the weed (Table 5).

5.3 Neighborhood Studies

Most competition studies focus on assessing crop yields, however, this parameter can be influenced by the proximity of individuals and local variation in the environment (microsites). The performance of a species in neighborhood studies is recorded as a function of the number, biomass, cover, aggregation, or distance of its neighbors.

Table 5 Competitiveness indices between irrigated rice or soybean and *Digitaria ciliaris*, expressed by relative competitiveness (CR) and relative clustering (*K*) and competitiveness (*C*) coefficients

Irrigated rice: *Digitaria ciliaris*			
CR	*K* rice	*K D. ciliaris*	*C*
3.47	5.01	0.32	0.59
Soybeans: *Digitaria ciliaris*			
CR	*K* soybean	*K D. ciliaris*	*C*
3.68	2.13	0.22	0.48

Source: Adapted from Agostinetto et al. (2013)

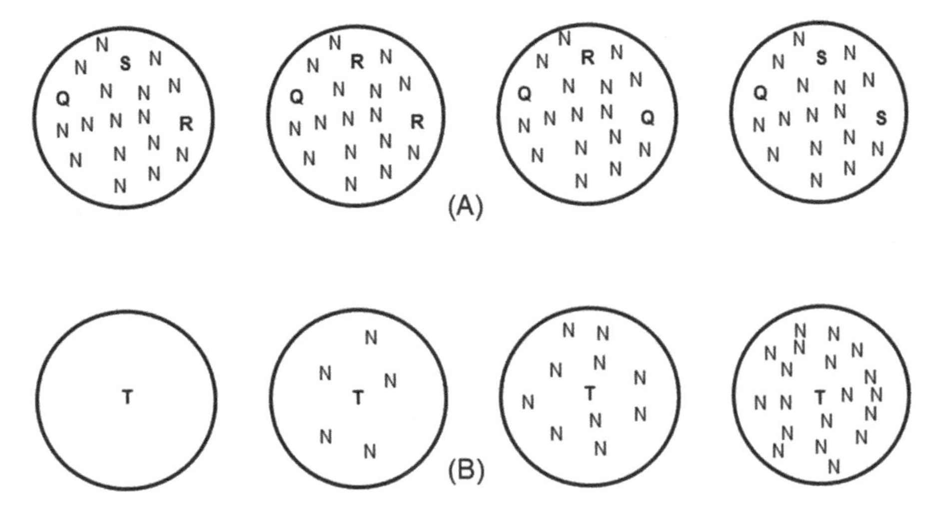

Fig. 25 Experimental design to evaluate the competitive effects of a neighboring species (N) on a target species (T). R, Q, and S represent individuals that do not belong to the neighboring species selected for study (**a**). Only four steps of the neighboring density gradient (after treatment) are shown (**b**), but experiments should include a wider range of densities to accurately estimate the slope of the regression equation. (Source: Adapted from Goldberg and Werner (1983))

A common experimental approach in which the performance of a single target species is evaluated over a range of densities of a neighboring species is described by Goldberg and Werner (1983). The target species is grown alone or is surrounded by individuals of neighboring species. The spatial arrangement of individual plants may also vary between the target and neighboring species (Fig. 25a, b).

The effect of neighboring species on target species is defined as the slope of the regression of performance (e.g., growth rate, survival, or reproduction) of target individuals on the quantity (e.g., density, biomass, or leaf area) of neighboring species (Radosevich 1987). The relationship is expressed by means of the mathematical model:

$$P(T) = Y + X_N \left[A(N) \right]$$

where $P(T)$ is the performance of the target individual, X_N is the slope of the regression equation, and $A(N)$ is the density of neighbors. A refinement of the neighborhood method is to measure the distance of the target individuals' neighbors, so that the diminishing effects of more distant neighbors can be incorporated into the regression equations (Goldberg and Werner 1983). In general, if the slope of the curve is not significantly different from 0, no interaction is present. However, if the slope is greater than 0 it indicates beneficial effects of neighbors; slopes less than 0 indicate competitive effects.

An additional approach can be taken, according to the distance of the target's neighbors within each plot, so that the decreasing competitive effects of more distant neighbors can be incorporated into the regression equations. Thus, the term for quantity of neighbors [$A(N)$] can be replaced by, for example, $\sum 1(A_i/d_i^2)$, where A_i is the biomass of an individual plant of the neighboring species at a distance d from the target (or the biomass of plants within concentric rings of mean distance d from the target) (Weiner 1982).

The slope of the regression can be varied so that the competitive effects of several neighboring species can be compared statistically directly, and these neighboring species can be ranked in terms of the magnitude of their effect on a given target species (Fig. 26).

5.4 *Splitting the Competition*

Studies of competition for available resources below and above the soil surface are mostly carried out under controlled conditions, because of the ease of separation compared to field conditions. The plant row method is the most common technique used for partitioning studies, developed by McPhee and Aarssen (2001). The method consists of partitioning the space below and above and the soil surface, a partition set up, separates the competition below the surface and the other that occurs above the soil surface. The method generates four competition situations, namely: no competition, competition only for soil resources, only for solar radiation, and for both resource sources (total competition) (Fig. 27a, b).

The competition in the area can be isolated with the use of partitions covered with aluminum film, to reflect solar radiation and avoid competition for the light resource (Bianchi et al. 2006a, b).

In studies of competition of soybean with volunteer corn, Caratti et al. (2016) observed a reduction in dry matter accumulation of soybean when in competition with volunteer corn competing for light, soil, and for both resources. However, corn dry matter was not negatively affected in the presence of soybean, moreover, when the competition involved soil resources an increase in dry matter accumulation could be observed (Fig. 28).

Fig. 26 Possible relationships between the performance of individuals of the target species and the abundance of neighboring species. Curve *a* corresponds to the equation provided in the text (linear relationship). Curve *b* represents a relationship in which competitive effects are minimal at low neighboring density, but increasingly severe as density increases. Curve *c* represents a quadratic function with a peak at some intermediate neighboring density, indicating beneficial effects at low density, but negative effects at high densities. Curve *d* represents a negative exponential function, indicating decreasing competitive effects by quantity as density increases. (Source: Adapted from Goldberg and Werner (1983))

5.5 Critical Period of Weed Interference

Critical interference period studies are widely used to study the interference between weeds and crops. This period is defined as the time (growth stages) in the crop cycle when weeds must be controlled to avoid unacceptable yield losses.

In critical period studies, crops are kept weed-free for varying time intervals after planting or emergence, and after this period, weeds are allowed to grow for the rest of the growing season (Fig. 29). The resulting data are compared with those from a complementary study in which weeds are allowed to grow for varying time intervals after planting or crop emergence, with the remainder of the growing season free of weeds.

The critical period is determined using two sets of experimental treatments (Swanton et al. 2015). In the first set of treatments, weeds are grown next to the crop for increasing periods of time, with the aim of determining the maximum period that a crop can tolerate infestations before yield losses occur, defined as PPI

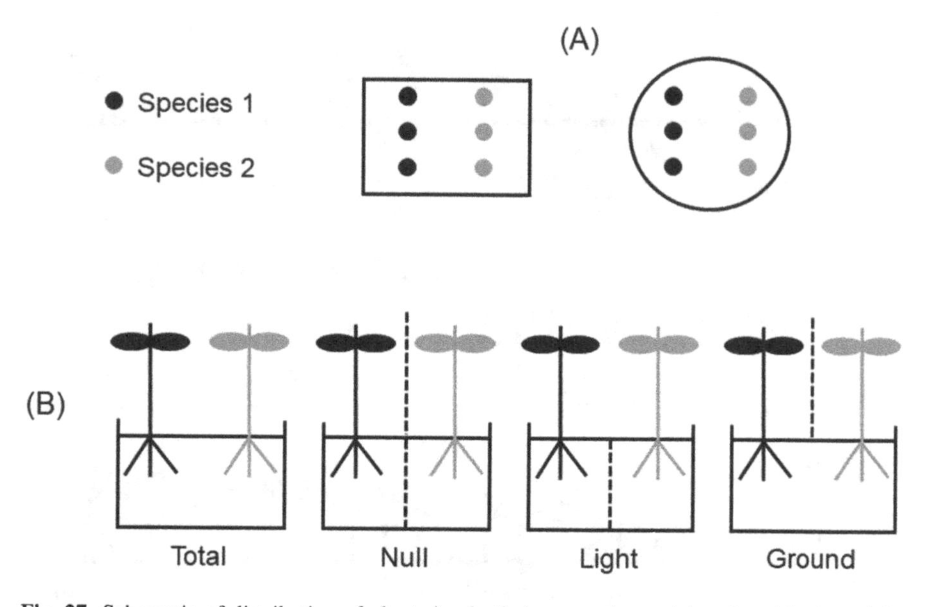

Fig. 27 Schematic of distribution of plants in plastic boxes and pots (**a**) and positioning of the partition (dashed line) in the pots (**b**), for dividing competition for resources. (Source: Adapted from Mcphee and Aarssen (2001), Bianchi et al. (2006a, b), Caratti et al. (2016))

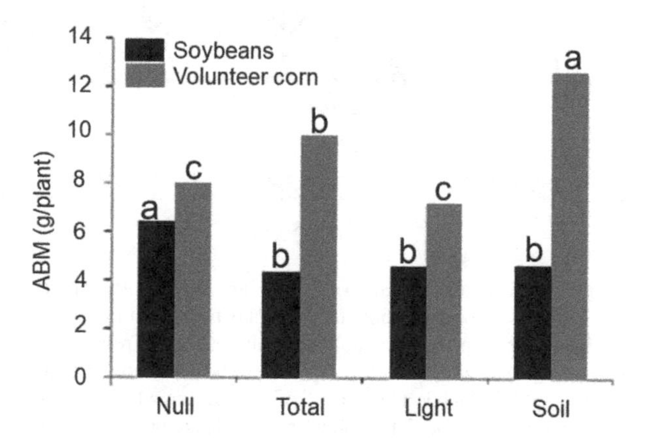

Fig. 28 Aboveground dry matter (ABM) at 42 days after emergence (DAE): of soybean and corn plants competing in the following competition conditions: no competition (Null), soil resources + solar radiation (Total), solar radiation (Light), and soil resources (Soil) means followed by different letters differ according to the *t*-test ($p \leq 0.05$). (Source: Adapted from Caratti et al. (2016))

(Fig. 30a). For this, mechanical weed control measures or the use of herbicides can be used to remove the weeds, depending on the treatment.

In the second set of treatments, the crop is kept free of weeds for increasing periods of time, with the goal of determining the period when the crop should be kept free of weeds to avoid yield loss, defined as TPWI (Fig. 30b).

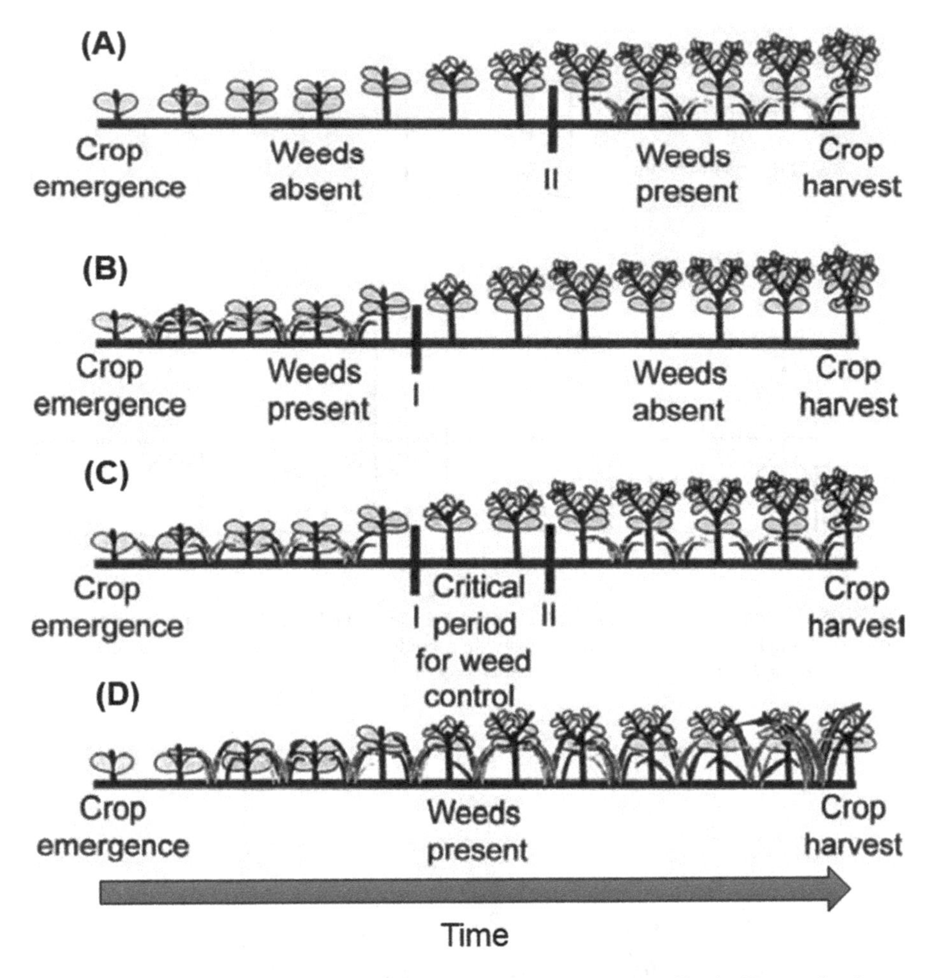

Fig. 29 Critical period of weed interference. (**a**) If weeds are absent until point II, crop dominance is established and no yield losses occur, although weeds may be present later. (**b**) If weeds are present for a period of time after crop emergence, but are absent for the rest of the growing season, yield losses do not occur. (**c**) The combination of the results of (**a**) and (**b**) indicates the critical period between points I and II, which is a "window" of time during which weeds must be removed or suppressed to avoid yield losses. (**d**) Situation where weeds are present throughout the growing season and the results of crop yield loss. (Source: Adapted from Radosevich et al. (1997 and 2007))

In both sets of treatments, the intervals that crops remain in the presence or absence of weeds can be implemented as a specified number of days or weeks after crop emergence, or at specific crop growth stages (two, four, six leaves, and so on). Weed removal at these specified time intervals can be accomplished by either manual weeding or the use of herbicides.

The CPWI is the combination of these two periods PPI and TPWI, that is, it is the period in which the presence of the weed influences the crop yield (Fig. 31).

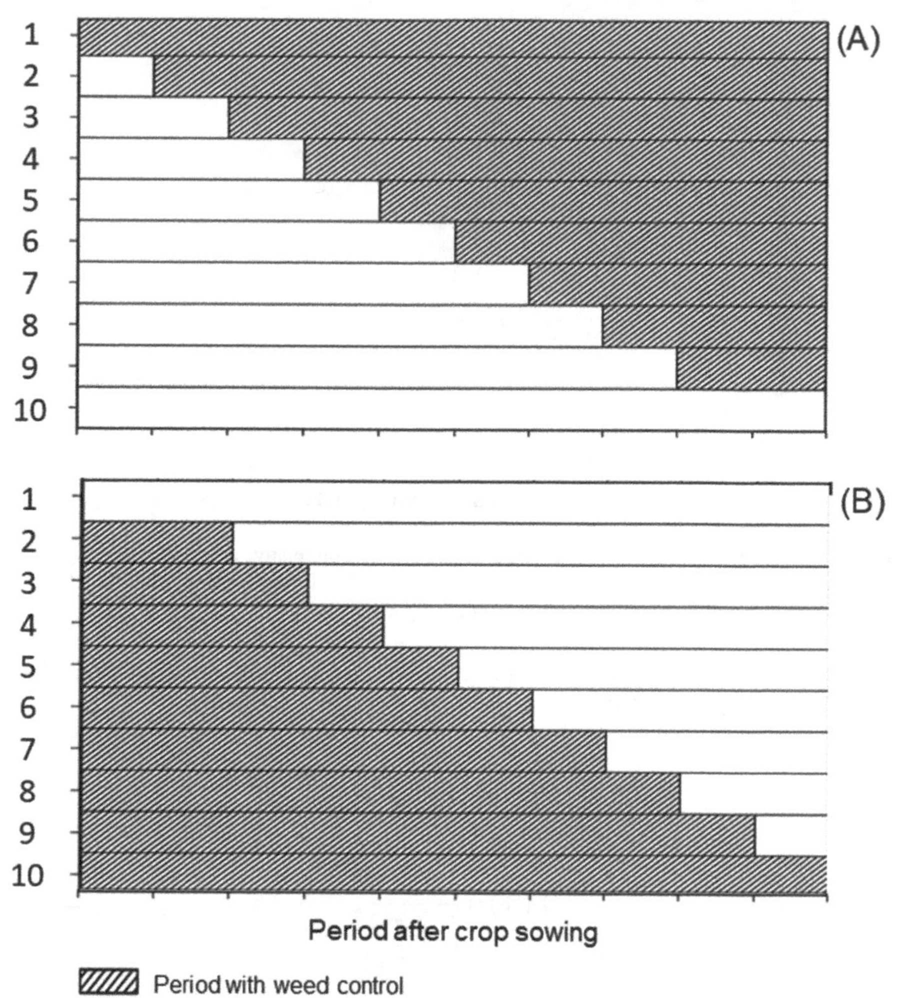

Fig. 30 Diagram of graphical representation of the treatments (with and without control, A and B), for determining the interference periods. (Source: Adapted from Dawson (1970), Swanton et al. (2015))

Acceptable yield loss values vary from 1 to 10%, according to the added value of the product.

Several models are described to calculate the critical interference periods of weeds (Knezevic et al. 2002; Agostinetto et al. 2008; Chauhan and Johnson 2011; Stagnari and Pisante 2011; Odero and Wright 2013; Souza et al. 2020). Chauhan and Johnson (2011) described some models commonly used to calculate interference periods. The Gompertz equation is used to model the effect of the duration of weed control periods, on relative grain yield, while the logistic equation is used to

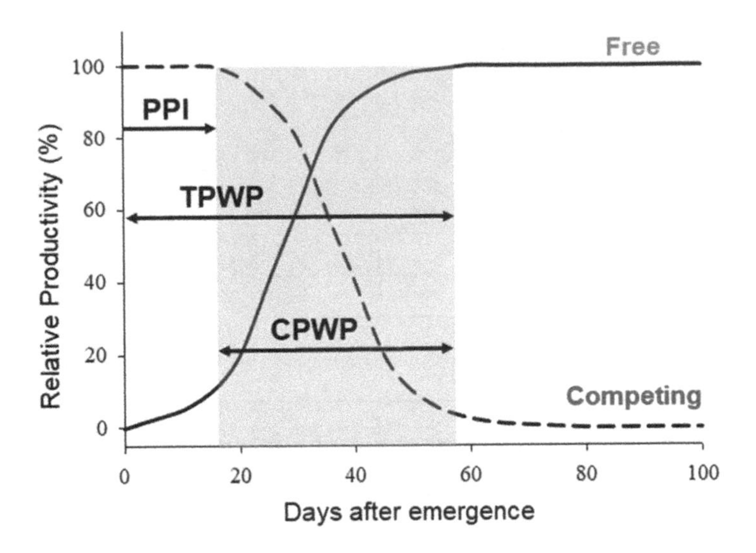

Fig. 31 Influence of timing of weed emergence and removal on relative crop yield and magnitude of interference periods. Period Previous to Interference (PPI), Total Period of Weed Interference (TPWI), and Critical Period of Weed Interference (CPWI)

model the effect of the duration of weed interference, on relative grain yield (Chauhan and Johnson 2011).

$$Y(\%) = A^{*} \exp - \left(\exp \left(\frac{-(T - T0)}{B} \right) \right)$$

$$Y(\%) = \frac{A}{1 + \left(\frac{T}{T0} \right)^{B}}$$

In the Gompertz equation *Y* is the yield (%) of the weed-free control, *A* is the asymptote, *B* is the exponential rate, *T* is the number of days after sowing, and *T0* is the inflection point. In the logistic equation, *Y* is the yield (%) of weed-free control, *A* is the lower asymptote, *B* is the measure of the rate of reduction in yield, *T* is the number of days after sowing, and *T0* is the duration of weed coexistence in days after sowing, when there is a reduction of half the maximum yield that would occur at A.

Although critical periods of interference provide valuable information to growers, it should be noted that the scientific community is currently placing more emphasis on controlling all weed seed production as a means of preventing infestation in future crops, in addition to some weeds interfering with harvest operations, and/or quality of the final product.

The main advantage of this study is the ease in establishing it under field conditions, where the crop is present at constant density. As for disadvantages, the study is usually carried out with only one weed species or density, so it is difficult to extend the results to situations where several weed species occur.

The interference period is an important study to evaluate the importance of cultural control in weed management, several factors such as fertilization, growing seasons, irrigation systems, row spacing, cultivars, weed density, alter the interference periods of weeds in crops, demonstrating the importance of cultural management (Ahmadvand et al. 2009; Chauhan and Johnson 2011; Silva et al. 2013; Singh et al. 2014; Souza et al. 2020).

Phosphate (P) fertilization was able to alter the critical period of weed interference in lettuce. Odero and Wright (2013) reported that the CPWI of weeds was estimated between 15 and 48, 16 and 40, and 20 and 36 days after transplanting due to the amount of P applied when planting lettuce seedlings (93, 196, and 293 kg P ha^{-1}, respectively) (Fig. 32). The results showed that in soils with inadequate levels of P fertilization, there is a need for more intensive weed management practices to achieve acceptable yields.

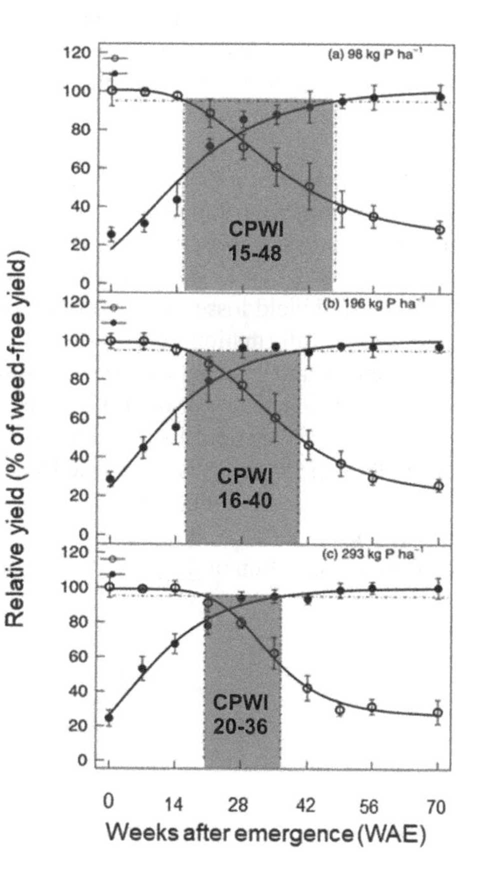

Fig. 32 Period of weed interference on lettuce productivity as a result of phosphate fertilization [(**a**) 98, (**b**) 196, and (**c**) 293 kg P ha^{-1}]. Duration of weed interference (○) and duration of weed-free period (●) on the relative productivity of lettuce. (Source: Adapted from Odero and Wright (2013))

6 Concluding Remarks

As crops grow, environmental factors such as water, light, nutrients, and CO_2 can become limited. This situation is aggravated by the presence of other plants of no economic interest growing in the same space. These plants are considered weeds and the effect they cause is called competition, which depending on the intensity is capable of interfering negatively with the crop.

Simple strategies can minimize the effects of competition such as choosing the most competitive cultivars with fast growth and good ground covering capacity; clean sowing; adequate planting time; adequate plant density and spacing. These strategies, allow crops to excel over weeds, minimizing the effects of competition for resources.

Simple strategies can minimize the effects of competition such as choosing the most competitive cultivars with fast growth and good ground covering capacity; clean sowing; adequate planting time; adequate plant density and spacing. These strategies, allow crops to excel over weeds, minimizing the effects of competition for resources.

In order to prevent that coexistence brings damage, it is not enough to observe not only the factors related to the crop, but also the behavior of weeds. Information regarding weed emergence time, population dynamics, and morphological and physiological characteristics of the species helps to control weeds and assists in crop decisions that maximize the competitive potential of crops, which reduces competition.

The effects of competition can be minimized if these interactions are further studied. Competition and interference studies between weeds and crops can provide information to determine the competitive ability of weeds. It also helps to determine whether weed control measures are necessary and when to implement control practices to avoid yield losses and reduced quality of the final product. The studies can also provide information on the effect of cultural practices, such as crop rotation, plant density and spacing, fertilization, sowing time, consortia, and choice of cultivars, which are capable of improving the overall competitiveness of the crop.

Several trials are described to study weed–crop interactions. The choice of study depends on the hypothesis to be tested and the resources available. Care in conducting the study is important, such as determining the variables to be analyzed, the methodology adopted, and the selection of appropriate statistical analyses. The appropriate selection of the experimental design and procedures will ensure the success of the study.

References

Adeux G, Vieren E, Carlesi S et al (2019) Mitigating crop yield losses through weed diversity. Nat Sustain 11(2):1018–1026. https://doi.org/10.1038/s41893-019-0415-y

Agostinetto D, Fleck NG, Rizzardi MA (2004) Perdas de rendimento de grãos na cultura de arroz irrigado em função da população de plantas e da época relativa de emergência de arroz-vermelho ou de seu genótipo simulador de infestação de arroz-vermelho. Planta Daninha 22(2):175–183. https://doi.org/10.1590/S0100-83582004000200002

Agostinetto D, Rigoli RP, Schaedler CE et al (2008) Período crítico de competição de plantas daninhas com a cultura do trigo. Planta Daninha 26(2):271–278. https://doi.org/10.1590/S0100-83582008000200003

Agostinetto D, Fontana LC, Vargas L et al (2013) Habilidade competitiva relativa de milhã em convivência com arroz irrigado e soja. Pesq Agropecu Bras 48(10):1315–1322. https://doi.org/10.1590/S0100-204X2013001000002

Agostinetto D, Silva DRO, Vargas L (2017) Soybean yield loss and economic thresholds due to glyphosate resistant hairy fleabane interference. Arq do Inst Biológico 84:e0022017. https://doi.org/10.1590/1808-1657000022017

Ahmadvand G, Mondani F, Golzardi F (2009) Effect of crop plant density on critical period of weed competition in potato. Sci Hortic 121(3):249–254. https://doi.org/10.1016/j.scienta.2009.02.008

Ballaré CL (2009) Illuminated behaviour: phytochrome as a key regulator of light foraging and plant anti-herbivore defence. Plant Cell Environ 32(6):713–725. https://doi.org/10.1111/j.1365-3040.2009.01958.x

Bastiani MO, Lamego FP, Agostinetto D et al (2016) Relative competitiveness of soybean cultivars with barnyardgrass. Bragantia 75(4):435–445. https://doi.org/10.1590/1678-4499.412

Bensch CN, Horak MJ, Peterson D (2003) Interference of redroot pigweed (*Amaranthus retroflexus*), Palmer amaranth (*A. palmeri*), and common waterhemp (*A. rudis*) in soybean. Weed Sci 51(1):37–43. https://doi.org/10.1614/0043-1745(2003)051[0037:IORPAR]2.0.CO;2

Bianchi MA, Fleck NG, Dillenburg LR (2006a) Partição da competição por recursos do solo e radiação solar entre cultivares de soja e genótipos concorrentes. Planta Daninha 24(4):629–639. https://doi.org/10.1590/S0100-83582006000400003

Bianchi MA, Fleck NG, Lamego FP (2006b) Proporção entre plantas de soja e plantas competidoras e as relações de interferência mútua. Ciênc Rural 36(5):1380–1387. https://doi.org/10.1590/S0103-84782006000500006

Blackshaw RE (1993) Downy brome (*Bromus tectorum*) density and relative time of emergence affects interference in winter wheat (*Triticum aestivum*). Weed Sci 41(4):551–556. https://doi.org/10.1017/S004317450007630X

Brighenti AM, Oliveira MF (2011) Biologia de Plantas Daninhas. In: Oliveira RS Jr, Constantin J, Inoue HM (eds) Biologia e Manejo de Plantas Daninhas. Omnipax, Curitiba, pp 1–36

Brunce JA, Ziska LH (2000) Crop ecosystems responses to climatic change. Crop/weed interactions. In: Reddy KR, Hodges HF (eds) Climate change and global crop productivity. Wallingford, CAB International, pp 333–348

Caratti FC, Lamego FP, Silva JDG et al (2016) Partitioning of competition for resources between soybean and corn as competitor plant. Planta Daninha 34(4):657–666. https://doi.org/10.1590/S0100-83582016340400005

Chauhan BS, Johnson DE (2011) Row spacing and weed control timing affect yield of aerobic rice. Field Crops Res 121(2):226–231. https://doi.org/10.1016/j.fcr.2010.12.008

Cobb AH (1992) Auxin-type herbicides. In: Cobb AH (ed) Herbicides and plant physiology. Chapman and Hall, London, pp 82–125

Cobb AH, Reade JPH (2011) An introduction to weed biology. In: Cobb AH, Reade JP (eds) Herbicides and plant physiology, 2nd edn. Wiley-Blackwell, Hoboken, pp 1–26. https://doi.org/10.1111/j.1744-7348.1985.tb01567.x

Cousens R (1985) A simple model relating yield loss to weed density. Ann Appl Biol 107(2):239–252. https://doi.org/10.1111/j.1744-7348.1985.tb01567.x
Counsens R (1991) Aspects of the design and interpretation of competition (interference) experiments. Weed Technol 5(3):664–673. https://doi.org/10.1017/S0890037X00027524
Cousens R, O'neill M (1993) Density dependence of replacement series experiments. Oikos 66(2):347–352. https://doi.org/10.2307/3544824
Craine JM, Dybzinski R (2013) Mechanisms of plant competition for nutrients, water and light. Funct Ecol 27(4):833–840. https://doi.org/10.1111/1365-2435.12081
Cunha JLXL, de Freitas FCL, Coelho MH, da Silva MGO, de Mesquita HC, Silva KDS (2015) Períodos de interferência de plantas daninhas na cultura do pimentão nos sistemas de plantio direto e convencional. REVISTA AGRO@MBIENTE ON-LINE 9(2):175. https://doi.org/10.18227/1982-8470ragro.v9i2.2187
Dawson JH (1970) Time and duration of weed infestations in relation to weed-crop competition. Paper presented at the international symposium on the Southern Weed Science Society. SWSS, Birmingham
Durigon MR, Mariani F, Cechin J et al (2019) Competitive ability of canola hybrids resistant and susceptible to herbicides. Planta Daninha 37:e019180593. https://doi.org/10.1590/S0100-83582019370100133
ESRL – Earth System Research Laboratories. Global monthly mean CO_2. DIALOG. https://www.esrl.noaa.gov/gmd/ccgg/trends/global.html. Accessed 22 Abr de 2021
Ferri MG (1985) Fisiologia vegetal. EPU, São Paulo, 362p
Goldberg DE, Werner PA (1983) Equivalence of competitors in plant communities: a null hypothesis and a field experimental approach. Am J Bot 70(7):1098–1104. https://doi.org/10.2307/2442821
Gundel PE, Pierik R, Mommer L et al (2014) Competing neighbors: light perception and root function. Oecologia 176(1):1–10. https://doi.org/10.1007/s00442-014-2983-x
Hijano N, Orzari I, Colombo WL, Nepomuceno MP, Alves PLCA (2021) Interferência: Conhecer para usá-la a nosso favor. In: Barroso AAM, Murata AT (eds) Matologia: estudos sobre plantas daninhas. Fábrica da Palavra, Jaboticabal, pp 106–144
Hoffman ML, Buhler DD (2002) Utilizing *Sorghum* as a functional model of crop-weed competition. I. Establishing a competitive hierarchy. Weed Sci 50(4):466–472. https://doi.org/10.1614/0043-1745(2002)050[0466:USAAFM]2.0.CO;2
Jacob W (1982) A neighborhood model of annual-plant interference. Ecology 63(5):1237–1241. https://doi.org/10.2307/1938849
Jiao Y, Lau OS, Deng XW (2007) Light-regulated transcriptional networks in higher plants. Nat Rev Genet 8(3):217–230. https://doi.org/10.1038/nrg2049
Knezevic SZ, Horak MJ, Vanderlip RL (1999) Estimates of physiological determinants for *Amaranthus retroflexus*. Weed Sci 47(3):291–296. https://doi.org/10.1017/S0043174500091797
Knezevic SZ, Evans SP, Blankenship EE et al (2002) Critical period for weed control: the concept and data analysis. Weed Sci 50:773–786. https://doi.org/10.1614/0043-1745(2002)050[0773:CPFWCT]2.0.CO;2
Korres NE, Norsworthy JK, Tehranchian P et al (2016) Cultivars to face climate change effects on crops and weeds: a review. Agron Sustain Dev 36(1):12. https://doi.org/10.1007/s13593-016-0350-5
Koyro HW, Lieth H, Gul B, Ansari R, Huchzermeyer B, Abideen Z, Hussain T, Khan MA (2014) Importance of the diversity within the halophytes to agriculture and land management in arid and semiarid countries. In: Khan M (ed) Sabkha ecosystems: cash crop halophyte and biodiversity conservation. Springer, Dodrecht, pp 175–198
Lindquist JL, Kropff MJ (1996) Applications of an ecophysiological model for irrigated rice (*Oryza sativa*)-Echinochloa competition. Weed Sci 44(1):52–56. https://doi.org/10.1017/S0043174500093541
Little NG, Ditommaso A, Westbrook AS et al (2021) Effects of fertility amendments on weed growth and weed-crop competition: a review. Weed Sci 69(2):1–48. https://doi.org/10.1017/wsc.2021.1

Liu JG, Mahoney KJ, Sikkema PH et al (2009) The importance of light quality in crop–weed competition. Weed Res 49(2):217–224. https://doi.org/10.1111/j.1365-3180.2008.00687.x
Matos CC, Silva TR, Silva IR et al (2019) Interspecific competition changes nutrient: nutrient ratios of weeds and maize. J Plant Nutr Soil Sci 182(2):286–295. https://doi.org/10.1002/jpln.201800171
Mcphee CS, Aarssen LW (2001) The separation of above-and below-ground competition in plants a review and critique of methodology. Plant Ecol 152(2):119–136. https://doi.org/10.1023/A:1011471719799
Merotto A Jr, Fischer AJ, Vidal RA (2009) Perspectives for using light quality knowledge as an advanced ecophysiological weed management tool. Planta Daninha 27(2):407–419. https://doi.org/10.1590/S0100-83582009000200025
Mortensen D, Curran B, Ryan M et al (2021) Weed germination periodicity: when do weeds wake up? Pennsylvania State University. DIALOG. https://growiwm.org/wp-content/uploads/2017/03/When-Do-Weeds-Wake-Up.pdf?x23065. Accessed 27 August 2021
Norris RF (1996) Water use efficiency as a method for predicting water use by weeds. Weed Technol 10(1):153–155
Odero DC, Wright AL (2013) Phosphorus application influences the critical period of weed control in lettuce. Weed Sci 61(3):410–414. https://doi.org/10.1614/WS-D-12-00107.1
Pitelli RA (1985) Interferência de plantas daninhas em culturas agrícolas. Informe Agropecuário 11(129):16–27
Pitelli RA (2014) Competição entre plantas daninhas e plantas cultivadas. In: Monquero PA (ed) Aspectos da biologia e manejo das plantas daninhas. Rima, São Carlos, pp 61–81
Procópio SO, Santos JB, Silva AA et al (2002) Análise do crescimento e eficiência no uso da água pelas culturas de soja e do feijão e por plantas daninhas. Acta Sci Agron 24(5):1345–1135. https://doi.org/10.4025/actasciagron.v24i0.2379
Procópio SO, Santos JB, Silva AA et al (2004) Ponto de murcha permanente de soja, feijão e plantas daninhas. Planta Daninha 22(1):35–41. https://doi.org/10.1590/S0100-83582004000100005
Radosevich SR (1987) Methods to study interactions among crops and weeds. Weed Technol 1(3):190–198. https://doi.org/10.1017/S0890037X00029523
Radosevich SR, Holt JS, Ghersa CM (1997) Weed ecology: implications for management, 2nd edn. Wiley, New York, 589p
Radosevich SR, Holt JS, Ghersa CM (2007) Ecology of weeds and invasive plants: relationship to agriculture and natural resource management, 3rd edn. Wiley, Chichester, 473p
Rejmánek M, Robinson GR, Rejmánková E (1989) Weed-crop competition: experimental designs and models for data analysis. Weed Sci 37(2):276–284. https://doi.org/10.1017/S0043174500071903
Ronchi CP, Terra AA, Silva AA et al (2003) Acúmulo de nutrientes pelo cafeeiro sob interferência de plantas daninhas. Planta Daninha 21(2):219–227. https://doi.org/10.1590/S0100-83582003000200007
Sharma D, Sanghera GS, Sahu P et al (2013) Tailoring rice plants for sustainable yield through ideotype breeding and physiological interventions African. J Agric Res 8(40):5004–5019. https://doi.org/10.5897/AJAR2013.7499
Silva AA, Ferreira FA, Ferreira LR, Santos JB (2007) Biologia de Plantas Daninhas. In: Silva AA, Silva JF (eds) Tópicos em Manejo de Plantas Daninhas. Editora UFV, Viçosa, pp 17–61
Silva RRD, Reis MRD, Mendes KF et al (2013) Períodos de interferência de plantas daninhas na cultura do girassol. Bragantia 72(3):255–261. https://doi.org/10.1590/S0100-83582004000200012
Silva DRO, Aguiar ACM, Basso CJ et al (2019) Bean yield loss in response to volunteer corn. Rev Bras Ciênc Agrár 14(2):e5636. https://doi.org/10.5039/agraria.v14i2a5636
Singh M, Bhullar MS, Chauhan BS (2014) The critical period for weed control in dry-seeded rice. Crop Prot 66:80–85. https://doi.org/10.1016/j.cropro.2014.08.009
Smith H (2000) Phytochromes and light signal perception by plants-an emerging synthesis. Nature 407(6804):585–591. https://doi.org/10.1038/35036500

Soares MM, Freitas CDM, Oliveira FSD et al (2019) Effects of competition and water deficiency on sunflower and weed growth. Rev Caatinga 32(2):318–328. https://doi.org/10.1590/1983-21252019v32n204rc

Souza MF, Lins HA, Mesquita HC et al (2020) Can irrigation systems alter the critical period for weed control in onion cropping? Crop Prot 147:e105457. https://doi.org/10.1016/j.cropro.2020.105457

Stagnari F, Pisante M (2011) The critical period for weed competition in French bean (*Phaseolus vulgaris* L.) in Mediterranean areas. Crop Prot 30(2):179–184. https://doi.org/10.1016/j.cropro.2010.11.003

Swanton CJ, Nkoa R, Blackshaw RE (2015) Experimental methods for crop–weed competition studies. Weed Sci 63(sp 11):2–11. https://doi.org/10.1614/WS-D-13-00062.1

Thiel CH, De David FA, Galon L et al (2018) Physiology of weeds in intraspecific competition. J Agric Sci 10(6):334–340. https://doi.org/10.1614/WS-D-13-00062.1

Tursun N, Datta A, Budak S et al (2016) Row spacing impacts the critical period for weed control in cotton (*Gossypium hirsutum*). Phytoparasitica 44(1):139–149. https://doi.org/10.1007/s12600-015-0494-x

Weaver SE, Ivany JA (1998) Economic thresholds for wild radish, wild oat, hemp-nettle and corn spurry in spring barley. Can J Plant Sci 78(2):357–361. https://doi.org/10.4141/P97-072

Parameters of the Phytosociological Survey to Evaluate the Abundance, Distribution, and Diversity of the Weed Community

Wendel Magno de Souza, Maria Carolina Gomes Paiva, Úrsula Ramos Zaidan, Kassio Ferreira Mendes, and Francisco Cláudio Lopes de Freitas

Abstract Phytosociology describes and seeks to understand the associations and interactions of plant species with each other and with the environment, which characterize phytogeographic units. In the phytosociological study, it is necessary to consider that the vegetation varies in spatial and time scales, allowing each species to have its own tolerance to certain selection factors and respond to environmental pressures in different ways. Thus, phytosociology can be used in plant science to evaluate floristic composition in a given location, in response to some environmental variation or to some imposed pressure factor. There are some methods that are commonly used in weed science, however, many times they are confusing and difficult to interpret, so this chapter seeks to detail the methodologies and the main parameters that can be used, in order to understand and perform comparations between the species of ecosystems, emphasizing the importance of each, the main differences and possibilities of using them.

Keywords Phytosociology · Interaction between species · Weed · Floristic composition · Abundance

W. M. de Souza (✉) · M. C. G. Paiva · K. F. Mendes · F. C. L. de Freitas
Department of Agronomy, Federal University of Viçosa, Viçosa, Minas Gerais, Brazil
e-mail: francisco.freitas@ufv.br; kfmendes@ufv.br

Ú. R. Zaidan
Agência Goiana de Assistência Técnica, Extensão Rural e Pesquisa Agropecuária (Emater-GO), Goiânia, Brazil

© The Author(s), under exclusive license to Springer Nature Switzerland AG 2022

K. F. Mendes, A. Alberto da Silva (eds.), *Applied Weed and Herbicide Science*,
https://doi.org/10.1007/978-3-031-01938-8_3

1 Introduction

Knowledge of population dynamics and how environmental conditions and interactions between plant species influence patterns of coexistence and relative abundance of species are necessary to understand the complexity of plant communities (Concenço et al. 2017). The classification and characterization of plant communities can be performed through phytosociology.

The word phytosociology originated from the combination of two terms: "Phytos" meaning plant and sociology with the meaning of groups or groupings. The definition of the word phytosociology, according to its use in studies, has been modified over time (Kuva et al. 2021). According to Braun-Blanquet (1968), phytosociology is the science that seeks to understand the interaction between the environment and plant diversity in a phytogeographic region, through the composition and distribution of plant species. Rodrigues and Gandolfi (1998) define phytosociology as the branch of plant ecology that describes and seeks to understand the associations and interactions of plant species with each other and with the environment, which characterize phytogeographic units. Blasi and Frondoni (2011) postulate that phytosociology is related to botanical geography, in which, according to the different geographical regions and environmental conditions (climate, latitude, and longitude), plant species differ in composition and physiology, being grouped into associations.

The phytosociological study considers three principles: **analytical**, **synthetic,** and **sintaxonomy**. The **analytical** refers to the size of the inventory surface, characteristics of the sampling site, and variables such as abundance, density, dominance, and the sociability of plant species; the **synthetic** portrays the frequency of the species that make up the plant community; and **sintaxonomy** establishes the phytosociological hierarchy (Oosting 1956; Roberts 1981; Blasi and Frondoni 2011; Concenço et al. 2017).

In the phytosociological study, it is necessary to take into account that vegetation varies on spatial and time scales, allowing each species to have its own tolerance to certain selection factors and respond to environmental pressures in specific ways (Barbour et al. 1998). Spatial scales are defined by the size of the sample units (Concenço et al. 2017). The size of the sampling unit interferes with the analysis of spatial models of a plant population, so several models can be identified only with the change in the observation scale (Ostermann 1998).

The phytosociological study allows obtaining a momentary evaluation of the vegetation composition, obtaining data on frequency, density, abundance, relative importance index, and similarity coefficient of the species that exist in a given environment, allowing to make several inferences about the community of the species in question. Thus, evaluating the quality and quantity in the composition of plant species occupying a given environment, as an expression of all time and spatial influences, can be paramount to understand the floristic composition (Pott 2011).

For spatial and temporal comparison to be reliable, it is necessary to obtain numerical variables, that is, phytosociological parameters. For this, some points are

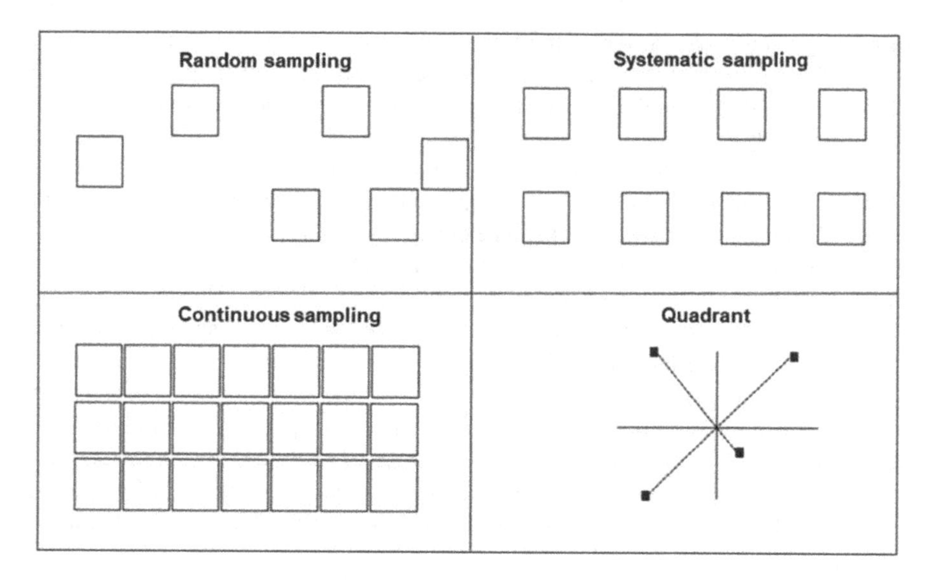

Fig. 1 Organization of sampling units by plots and quadrant. (Source: Adapted from Kuva et al. (2021))

fundamental: definition of the sampling unit, definition of the inclusion criterion, recording of numerical variables, and calculation of phytosociological parameters (Kuva et al. 2021). Thus, when starting a study of the floristic composition and structure of a plant community it is important to define the type of sampling, location, size, and quantity of sample units (Barbour et al. 1998; Concenço et al. 2017).

A sampling unit can be in plots with geometric figure of standardized dimensions, quadrant points, and standard walking (Concenço et al. 2017; Kuva et al. 2021). The arrangement of the plots may be random, systematic, or contiguous (Fig. 1). The inclusion criterion depends on the objective of the activity; and individuals covered by the sampling unit and who meet the inclusion criterion, should be registered with numerical variables (Kuva et al. 2021).

The main phytosociological parameters that characterize the horizontal structure are: density, frequency, dominance, importance value, and coverage value. There are also similarity, equitability, species diversity, and species richness. In this chapter, we will address the study methodologies of phytosociology and its parameters, when applied to agriculture, with emphasis on interaction with weed science.

2 Methods for Phytosociology Evaluation

For years several formulations and methodologies have emerged for evaluating the plant community in a given location. These methodologies allow evaluating the balance of the community in the area and the changes in responses to biotic and abiotic factors over time, associated with the area's own history (Iqbal et al. 2018). However, many of the times these methods are confusing and difficult to implement, and adaptations to the method are necessary to obtain real data according to the objective of the study.

Phytosociological studies are applied in wild environments and are also often used in weed science. As for weed science, the aim is to understand the dynamics of the weed community in response to some change whether it is a treatment, edaphoclimatic variations or imposed control method. Techniques are sought that aim at weed management while maintaining crop yield.

The main objectives of weed evaluation are the mapping, spatial dynamics, abundance, and distribution of plants in the area. According to Concenço et al. (2017), the choice between the different variants of the methodologies depends on the objective of the sampling and the characteristics of the populations (richness and distribution) to be evaluated in each agroecosystem. Given the various methodologies available in recent years, the methods of phytosociological survey by releves, use of leaked square and soil diaspore bank stand out as plausible to be used in weed science applied to agricultural systems.

2.1 Methods of the Leaked Square

The method of the leaked square of predetermined dimensions is the main method used for low-sized plants (Fig. 2). This method consists of the use of squares that are randomly released within the experimental area or unit, followed by the identification, separation, and measurement of numerical variables, such as number of plants, height, diameter, fresh and dry matter, among others (Braun-Blanquet, 1979).

The choice and sizing of the square used should take into account stability and definition, shape and fixed size, ease of viewing in the field, ease of measurement, high accuracy, and low cost (Nkoa et al. 2015). The square that provides the lowest variation between estimates within the same area has greater accuracy, and the one with the shortest sampling time is the one that generates the lowest cost (Nkoa et al. 2015).

The arrangement of the plots may be random, contiguous, or systematic (Kuva et al. 2021). The number of samples per plot and/or field will depend on the limit of variability that the researcher wishes to obtain and their observation in the area, but one should pay due to variations in the environment that can influence the weed community, in addition to the observation of the distribution of weeds, which most often occur in spots or waterball. To carry out the evaluations, the researcher can choose to fix a percentage of acceptable variation between the samples in the experimental unit, where the number of samples will only be significant when the variation between repetitions is up to 5% (Nkoa et al. 2015).

In possession of numerical variables it is possible to obtain the phytosociological parameters that are used in the knowledge of the weed community present in the area or experimental unit under evaluation, in which it allows a comparison between

Fig. 2 Leaked square (0.50 × 0.50 m) for weed sampling in wheat crop (**a**) and coffee in production (**b**). (Photos: Wendel Magno de Souza)

the different environments or treatments applied. These parameters are density, frequency, dominance, importance value, and coverage value, which can be used in weed science, where they are considered horizontal structure (Kuva et al. 2021).

2.2 *Relevé Method*

This method is based on a list of species hierarchy, where an area whose sample representativeness is equivalent to the actual condition of the area under study will be determined (Concenço et al. 2017). This minimum area is estimated by successive samples with squares of increasing sizes until the number of species tends to stabilize as demonstrated in the hypothetical example represented in Fig. 3. It is recommended that the size to be used in the samplings exceeds the minimum area size to ensure greater uniformity (Concenço et al. 2017).

The minimum area will vary according to the characteristics of the environment and the ecosystem evaluated. Mueller Dombois and Ellenberg (1974) suggested empirical values of minimum areas sampled for some environments (Table 1).

This method, as the theories surrounding its application are controversial, which can lead to unreliable sampling in constantly disturbed environments, such as agricultural areas (Concenço et al. 2013). The method also does not consider any variation in the evaluated ecosystem, since if calibration is to find the minimum area that represents the entire ecosystem, there is no need for more than one sampling, which can generate errors, because it is not possible to obtain the standard sampling error

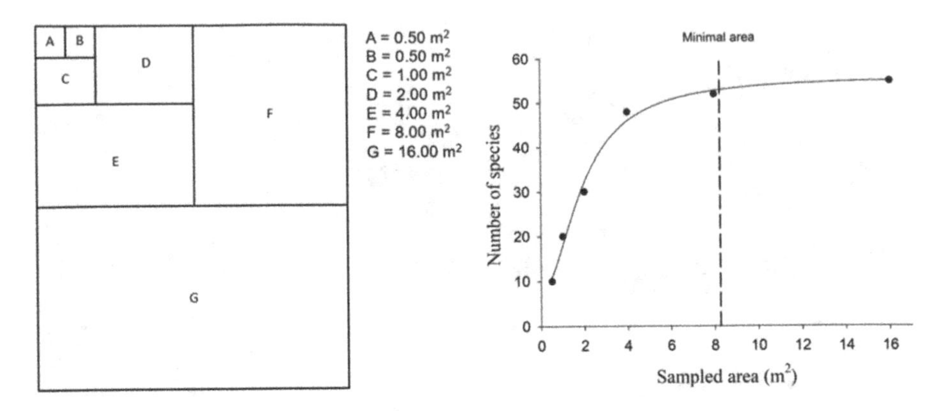

Fig. 3 Determination of the minimum area of the leaked squares required for plant sampling by the relight method. (Source: Adapted from Concenço et al. (2017))

Table 1 Empirical values of minimum sampled areas for various ecosystems

Ecosystems	Minimum area (m^2)
Forest (including tree extracts)	200–500
Forest (shrubby vegetation only)	50–200
Natural pastures	50–100
Hay plados	10–25
Pastures deployed	5–10
Weed community	25–100
Moss community	1–4
Lichen community	0.1–1

Source: Mueller Dombois and Ellenberg (1974)

for eventual statistical analyses. Another limiting factor is that there will always be a biased difference between two different sampled quadrants with the same area and therefore calibration based only on the number of species would not be sufficient to describe the community (Concenço et al. 2013).

2.3 *Database Survey of Diaspores (Propagules and Seeds) of Weeds*

Plants reproduce in a sexual and asexual way. The sexuada corresponds to the formation and dispersion of seeds, in which there is fertilization of the egg of the flower by the pollen grain and the asexual occurs through vegetative structures called propagules (stems, rhizomes, tubers, stolões, bulbs, among others). Thus, the term diaspores designate the reserves of seeds or other plant structures, that is, any part of the plant that under appropriate conditions has the ability to generate a new individual of the species.

The size and composition of the diaspore bank are different according to habitat (Carmona 1992; Monquero et al. 2014). However, any interference in the habitat, highlighting the disturbances caused by man, modify the database of diaspores in the soil, such as agricultural practices and techniques (Monquero and Christoffoleti 2005). According to Severino and Christoffoleti (2001), the diaspore bank and the vegetation present on the surface of an area are indicators of the entire soil management system and other techniques used in the area.

The database of diaspores in the soil is not homogeneous, being composed of reproductive structures of different species, deposited at different times and depths in the soil profile (Monquero et al. 2014). In addition, these structures may present different types and degrees of dormancy, and can germinate and emerge at various times for years, ensuring the longevity of the species, remaining for prolonged periods in the soil until the conditions for germination, such as dormancy breakage, are granted.

However, it is possible to survey the database of diaspores of plants in the soil, being used to establish the quantitative relationships between their populations and those of the infesting flora (Dessaint et al. 1990; Severino and Christoffoleti 2001), enabling the prediction of future infestations and in the case of weeds, allows a good planning of techniques and methods of their management.

Most weed seeds are found in the most superficial layer, where they have greater ease of access to factors to germinate such as sunlight, so some studies have shown that soil collection up to 10 cm deep is sufficient for the lifting of diaspores (Carmona 1995; Caetano et al. 2001; Silva and Dias Filho 2001). However, 90% of weed seeds in cultivated areas are in the first 20 cm of soil profile, and this depth is recommended for evaluation depending on the study objective (Buhler et al. 1997; Monquero and Christoffoleti 2005). Thus, there are some methodologies for evaluating the diaspore bank on the ground, which will be reported below.

Soil collection in most methods tends to follow a pattern, in which soil samples are collected in a certain area and depth, with the aid of some equipment of pre-established dimensions. According to Caetano et al. (2001), in the study methodologies of the diaspore bank, there is no exact definition as to the number and volume of soil to be sampled. However, samples should be collected at various points to form a composite sample, which represents the area in question. If the area has cultural remains on the surface, it should be carefully removed from these remains in order not to compromise the most superficial soil layer (Calegari et al. 2013). After collection, the samples can be destined to different laboratory methodologies for separation and/or extraction of diaspores.

The manual extraction method can be used. In this method, samples should be homogenized and diaspores can be separated with the aid of water and sieves. After separation, diaspores may be used for feasibility tests, such as the tetrazolium and germination tests (Andres et al. 2001; Monquero et al. 2014). According to Mesgaran et al. (2007), the method of manual extraction and removal may be the most accurate, however, it requires time, money and skilled labor, which makes it costly. Another relevant factor is that there is a need to identify diaspores that send the species, by trained and trained professionals, being able to use magnifying glass and other equipment. Thus, there is the possibility of using other methodologies in order to simplify extraction and separation.

A simpler method of evaluation is the emergency method, in which soil samples after homogenized are sifted and distributed in containers with known depth, provided that the conditions throughout the container area are homogeneous (Roberts 1981; Monquero and Christoffoleti 2005). Subsequently, the emergency flow is evaluated at different times, identifying, counting, and cutting the plants for drying in a greenhouse and weighing the dry matter (Caetano et al. 2001). This method should be carried out with care, especially in the sieving stage, since the diaspores have different sizes, and may be larger, such as the tubers of the species *Cyperus rotundus*, and smaller, as the seeds of *Eleusine indica*, and can be discarded at this stage. If necessary, manual picking is carried out to reduce such error, in addition to the correct choice of the sieve mesh to be used.

There is also the possibility of chemical extraction. Buhler and Maxwell (1993) proposed a chemical extraction methodology using potassium carbonate (K_2CO_3), followed by centrifugation, and was subsequently used and adapted by the scientific community. Voll et al. (2001) demonstrated an adaptation of the extraction method based on washing and flotation. The washing consisted of the elimination of clay from the soil, carried out on 0.5 mm sieves. The fraction composed of coarser soil particles, weed seeds, and straw were dried in the shade and subsequently the flotation operation was conducted. For flotation, a 75% purity dihydrate calcium chloride ($CaCl_2.2H_2O$) solution with a density of 1.40–1.42 g cm^{-3} was used for flotation. This density is higher than that of the seeds, causing them to float in the solution, occupying the top of the container. After a certain rest time the fraction of the antinatant contains seeds and straws was cast in a 0.5 mm mesh sieve, washed and dried in the shade on paper towels. Then, seeds were identified and counted with the aid of magnifying glass, counting only the viable seeds. It is noteworthy that rest after the flotation step of the samples can be replaced by centrifugation, however, there may be a reduction in the germination rate of some species, depending on the solution used (Luschei et al. 1998; Monquero et al. 2014). In this case, the pH of the medium combined with the increase in the hydrostatic pressure of centrifugation may damage the seed (Monquero et al. 2014). Finally, when calculating the number of seeds per m^2, the density of the collected soil volume should be taken into account (Monquero et al. 2014).

Both the methodologies of evaluation of the leaked square, relevé and soil diaspore bank allow estimating phytosociological parameters, comparing the species within the ecosystem or in response to some treatment. In a given weed community, not all species have the same importance, as there will be those that dominate the area, whether in growth (dry matter) or by density and frequency (Monquero et al. 2014). Thus, in the topic below are listed the way of calculating the most used parameters in weed science, because they help in the knowledge about the interaction of species with the environment.

3 Phytosociological Parameters

3.1 Frequency and Relative Frequency

The absolute frequency allows evaluating the distribution of the species of the area and indicates the number of times the species is present in the sample units (squares) in relation to the total squares obtained in the total area. This frequency can be expressed as a percentage or in decimal values (Kuva et al. 2021).

$$\text{Fr} = \left(\frac{N}{\text{TN}}\right) \times 100$$

In which: Fr = frequency of the species; N = number of sample units that contain the species; and TN = total number of sample units.

From the frequency values of each species it is possible to obtain the relative frequency.

$$\text{rFr} = \left(\frac{\text{Fr}}{\text{TFr}}\right) \times 100$$

In which: rFr = relative frequency of the species (%); Fr = frequency of the species; and TFr = sum of the absolute frequency of all species.

3.2 Abundance and Relative Abundance

Abundance provides data related to species concentration in the area (Gomes et al. 2010), if the evaluation is a description of an area, the count of the number of individuals themselves provides an assessment of abundance in this site. With the values of the numbers of individuals of the species it is possible to calculate the absolute abundance.

$$\text{Ab} = \left(\frac{\text{Ns}}{\text{TNs}}\right)$$

In which: Ab = abundance of the species; Ns; number of individuals of a given species; and TNs = sum of the number of samples containing the species.

With the values of the abundances of the species it is possible to calculate the relative abundance.

$$\text{rAb} = \left(\frac{\text{Ab}}{\text{TAb}}\right) \times 100$$

In which: rAb = Relative abundance; Ab = abundance of the species; and TAb = sum of the abundance of all species presents in the area.

3.3 Density and Relative Density

The absolute density allows to observe the amount of plants of each species per unit of area, representing the participation of the different species within the community per unit of area. When the evaluation is carried out with squares in plots, the total area of the number of sampling units used is considered as the total area of the number of sample units used. Absolute density in weed science is commonly presented by number of plants per m^2 (Monquero et al. 2014).

$$De = \left(\frac{Na}{TNa} \right)$$

In which: De = absolute species density; Na = number of individuals of species in sampled area; and TNa = sum of the abundance of all species present in the sampled area.

Relative density can be calculated from absolute density (Monquero et al. 2014) or the total number of species (Kuva et al. 2021).

$$rDe = \frac{De}{TDe} \times 100 \quad rDe = \left(\frac{Ns}{TNs} \right) \times 100$$

In which: rDe = relative density of the specie (%); De = Absolute density of the species; TDe = Sum of the absolute density of all species presents in the area; Ns = number of individuals of a given species; and TNs = sum of the total number of all species.

3.4 Dominance and Relative Dominance

Another widely used index is the absolute dominance that represents the occupancy rate of a species per unit of area. However, due to the difficult measurement of the basal area occupied by the species, which may generate inaccurate estimates (Kercher et al. 2003; Nkoa et al. 2015), this parameter can be calculated by the accumulation of dry matter of plants, which is easier to measure and greater reliability.

$$Do = \left(\frac{sOA}{TA}\right) \qquad Do = \left(\frac{sDM}{TDM}\right)$$

Basal area of occupation Dry matter

In which: Do = Absolute dominance of the species; sOA = basal area occupied by the species; TA = total basal area sampled; sDM = dry mass of the species; and TDM = Total dry mass of all species. Like the other parameters, it is possible to express absolute dominance in percentage, obtaining relative dominance in relation to the basal area occupied by all species.

$$rDo = \left(\frac{AOe}{TAOe}\right)\times 100 \qquad rDo = \left(\frac{eDM}{TDM}\right)\times 100$$

Basal area of occupation Dry matter

In which: rDo = relative dominance of the species (%); AOe = area occupied by the species; TAOe = sum of the total basal area of all identified species in the area; eDM = dry mass of the species; and TDM = sum of the total dry matter of all species in the area.

The relative frequency, relative density, relative abundance, and relative dominance allow to obtain comparative information of a given species within the balance of the plant community found in the area (Gomes et al. 2010). However, the highest values of any of these parameters do not always indicate that the species has an advantage over the other ones. An example is that the species may have a high absolute frequency, however, due to its growth habit or some limiting factor, it may have a low absolute dominance. Thus, there are newer parameters that are based on the interaction between the parameters, in order to promote greater efficiency in the comparison between species (Fig. 4).

In the example mentioned above it is possible to verify this difference in the species *C. rotundus* (Zaidan 2020). This plant reproduces asexually, through its tubers, basal bulb, and stolons, in addition to the production of seeds (sexually), which are all present in the soil, causing a density of elevated plants in the area in which it is infesting (Fig. 4). However, this species has a relatively low size, which provides lower dry matter accumulation per plant compared to others with a denser growth habit, such as the species *Urochloa decumbens*. Thus, in *C. rotundus* there was a high relative density and a low dominance, while in *U. decumbens* the inverse, low relative density and higher absolute dominance value occurred (Fig. 4). Thus, in order to minimize these errors in the comparations, other indexes can be used.

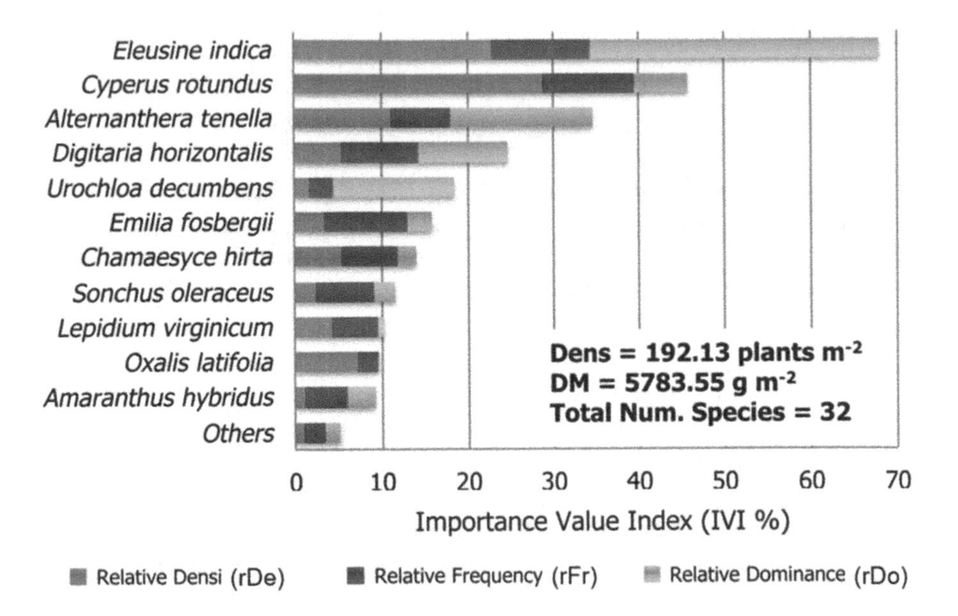

Fig. 4 Density (Dens), dry matter accumulation (DM), total number of species, relative density (rDe), relative frequency (rFr), dominance (rDo), and importance value index (IVI). (Source: Zaidan (2020))

3.5 *Importance Value Index (IVI)*

The importance value index (IVI) is the sum of the relative frequency, relative density, and relative dominance of each species ranging from 0 to 300%. This index allows us to infer which species exerts the greatest influence on the sampled area.

$$\text{IVI} = \text{rFe} + \text{rDe} + \text{rDo}$$

In which: IVI = index of importance value of the species; rFe = relative frequency (%); rDe = relative density (%); and eDo = relative dominance (%).

From the ivie values it is possible to obtain the relative importance of the species by means of the following equation.

$$\text{RI} = \left(\frac{\text{IVI}}{\text{TIVI}}\right) \times 100$$

In which: RI = relative importance of the species; IVI = Index of importance value of the species; and TIVI = sum of the importance value index of all species.

3.6 Coverage Value (CV)

The coverage value (CV) is the sum of relative density (DRe) and relative dominance of the species (DORe)

$$\mathrm{CV} = \mathrm{rDe} + \mathrm{rDO}$$

In which: CV = coverage value of the species; rDe = relative density of the species (%); and rDo = relative dominance of the species (%).

The coverage value can also be expressed in percentage, resulting in the relative coverage value (VCRe) that is calculated by the following formula.

$$\mathrm{rVC} = \left(\frac{\mathrm{VC}}{\mathrm{TVC}}\right) \times 100$$

In which: rVC = relative coverage value of the species (%); VC = species coverage value; TVC = sum of the cover value of the species.

3.7 Similarity Index (SI)

The similarity index (SI) is widely used in weed science, because it allows the comparison between the composition of species in different plots, treatments, or areas (Kuva et al. 2021). Sorense's IS can be calculated by the following equation.

$$\mathrm{SI} = \left(\frac{2a}{b+c}\right) \times 100$$

In which: a = number of species common to the two areas; b and c = total number of species in the two areas compared. Being b represented by area or plot 1 and c by area or parcel 2.

The IS ranges from 0 to 100%, and the higher the value, the greater the homogeneity of species distribution in the plots and/or area. That is, when all species are common in the sampled plot, the value of Is will be 100% (Oliveira and Freitas 2008; Gomes et al. 2010). However, in the calculation of SI is considered only the presence or absence of species, discarding density, and dominance, which can generate an Is that does not truly represent the similarity in the composition between the sampled areas (Kuva et al. 2021).

Another index that can be used is the Odum similarity coefficient (1985).

$$S = (2A \times (B - C)^{-1} \times 100$$

In which: S = Odum similarity coefficient; A = number of common species between two distinct areas or survey methodologies; B = number of species in the area or methodology 1; and B = number of species in the area or methodology 2.

This index allows comparison between two distinct areas or plots. However, in weed science it is more common to use it to compare the composition of the emergent flora (leaked square or relevé) and composition analyses of the diaspore bank in the soil.

These indexes listed above are the most commonly used in weed science, however, there are other plant-like indices developed that aid in knowledge within the infesting community, such as those listed below.

3.8 Aggregation Pattern Indexes

These indices indicate the pattern of distribution of species in a given area or region, when sampling is carried out in plots arranged in contiguous or systematic distribution (Nkoa et al. 2015; Kuva et al. 2021). In its calculation, the ratio (I) between the mean and variance of the distribution of individuals of a population is taken into account, as demonstrated in the sequence (Nkoa et al. 2015).

$$S^2 = \frac{\Sigma\left(\text{xi} - \overline{x}\right)^2}{N-1} \qquad I = \frac{S^2}{\overline{x}}$$

In which: S^2 = population variance; xi = number of individuals in the first installment; $\overline{x}$ = mean number of individuals per sample unit; N = number of sample units, and I = aggregation ratio.

By this index, the distribution pattern can be considered uniform or regular, when the ratio value is close to one ($I \cong 1$), aggregated or tending to the grouping, when the ratio value is greater than one ($I > 1$) and random when the ratio value is less than one ($I < 1$) (Monquero et al. 2014).

Another aggregation index commonly used is that proposed by Morisita (1959), being little influenced by the size of the plots, and is considered an excellent indicator of the degree of dispersion (Arruda and Daniel 2007). This index is calculated by the following formula.

$$\text{Id} = nX\frac{\Sigma\,\text{xi}^2 - N}{N\left(N-1\right)}$$

In which: Id = Morisita aggregation index; n = number of sample units; N = total number of individuals counted in all sample units; and xi^2 = is the square of the number of individuals in the first installment.

Thus, it can be considered that if Id = 1.0, the distribution pattern is considered random; if Id < 1.0, the distribution pattern is uniform, and if Id > 1.0, there is an aggregate distribution trend.

3.9 *Diversity Indices*

Species diversity is described by two main components: richness and uniformity. Richness is expressed by the number of different species detected in a given sampled area, while uniformity refers to the abundance pattern of these species (Nkoa et al. 2015). Diversity can be classified into alpha, beta, and gamma. Alpha describes the richness of species within the same area, while beta refers to richness between distinct areas and the range a region with all its habitats (several distinct areas) and communities present (Magurran 1988; Nkoa et al. 2015; Kuva et al. 2021).

To evaluate diversity, three indices are commonly used: the Margalef diversity index, the Shannon-Wiener diversity index (H'), and the Simpson dominance index (Nkoa et al. 2015), according to Table 2. The Shannon-Wiener index (H') is the most used in weed science (Kuva et al. 2021), because unlike the Margalef index, which covers only species richness, it covers the estimation of species richness and uniformity, being more complete than Simpson's, which includes the estimate of uniformity only (Nkoa et al. 2015).

The higher the index, the greater the community diversification in species composition (Kuva et al. 2021).

3.10 *Pielou Index (J)*

From the value of the Shannon-Wiener diversity index (H) it is possible to calculate the Pielou equitability index (J).

Table 2 Shannon-Wiener, Margalef, and Simpson diversity indices from the weed-infested community

Shannon-Wiener (H')	Margalef (Dmg)	Simpson (D)
$H' = -\sum_{i=1}^{n} \text{pi} \cdot \ln(\text{pi})$	$\text{Dmg} = \frac{S-1}{\ln(N)}$	$D = \Sigma \frac{\text{ni}(\text{ni}-1)}{N(N-1)}$

In which: n = number of species in the sampled area; pi = relative density of each species; ln = Nepierian logarithm; S = number of species in the sampled area (Richness); N = total number of individuals of all species; and ni = density or number of individuals of the species

$$J = \frac{H'}{\ln(S)}$$

In which: J = Pielou Index; H' = Shannon-Wiener diversity index; ln = Nepierian logarithm; and S = number of species sampled.

This index ranges from 0 to 1.00 and shows the balance in the distribution of the number of species in the infesting community. The closer its value is to one, the greater the similarity between the species.

From all phytosociological parameters it is possible to obtain a representative knowledge of the species present, and the relationship between species in the ecosystem (Adegas et al. 2010b), allowing to infer which species stand out regarding distribution, competition for the resources of the environment and which are the dominant ones in the area.

4 Weed Mapping

Mapping consists of determining the presence or absence of the species at the site. Data on the actual distribution of a weed species can be collected directly in the field or indirectly from public records, such as government documents, herbaria, and research publications and academics (Nkoa et al. 2015).

The first mapping methods were more costly and costly, in which the coverage of the evaluated area was intermittent and failed. Currently, there is the option of mapping through satellite imaging system, in which there is constant monitoring of the area with an interval of days, being made available on digital platforms (Kuva et al. 2021). This method allows covering an extensive area, solving the problem caused by old sampling, however, to obtain good results in evaluations by satellite images it is necessary to have a high infestation of weeds and clear sky (Yano et al. 2016). The visualization of any anomaly (patches of plants) in the field should be verified by surveys of the site.

To minimize problems related to the limitation of the use of aerial images by satellites, there is the possibility of using images from cameras installed in aerial vehicles. The cameras must have good resolution, and can be multispectral, hyperspectral, and thermal installed in helicopters, airplanes, and unmanned aerial vehicles—UAVs (Fig. 5). In relation to air vehicles, UAVs have been widely used, since they can be used at lower flight heights and speed, generating short-range images at different angles, besides providing less turbulence than helicopters and airplanes (Kuva et al. 2021). There is the possibility of using fixed-hand and rotor drones, however, Meneses et al. (2018) found that the images of the cameras installed in rotor drones had greater focus and sharpness. The cameras generate images of high spatial resolution, identifying the plants in the area, through various equipment of analysis of the images, based on algorithms of a database generated previously (Shirzadifar et al. 2020; Gasparovic et al. 2020).

Fig. 5 Unmanned aerial vehicle (UAV) with radio control, device coupled to a mobile phone and high-resolution digital camera. (Photos: David Longo)

However, to use the identification by algorithms and artificial neural networks, it is necessary to generate a database of images in various stages of plant growth to train and validate artificial neural networks in the distinction of weeds present in agricultural areas. Thus, later, it is possible to make the decision of the best control method to be used (Sudars et al. 2020). However, a disadvantage in large-scale use is the need to label thousands of plants in the training process, as well as a program to evaluate these data, since there is a very high number of weed species in agricultural areas (Goldman et al. 2019; Osco et al. 2021).

Using artificial neural networks it is possible to define the critical period for preventing interference (CPPI) of weeds with the crop, determining the correct moment to perform the control (Monteiro et al. 2021). A basic example of identification of weeds by algorithms is shown in Fig. 6.

The identification by artificial neural networks of melon (*Cucumis melo*) and sesame (*Sesamum indicum*) demonstrated that neural networks can predict, with high precision, the time when weed management should be started, in order to avoid productivity losses in crops with plants with photosynthetic pathway C3 (Monteiro et al. 2021). Pérez-ortiz et al. (2016) demonstrated that the segmentation algorithm system, combined with a cluster-based selection technique, demonstrated efficiency in the identification of weeds in the line and in the lead of a sunflower plantation by the use of images from unmanned aerial vehicles (UAVs). Zhang et al. (2020) observed 91.67% efficiency in a method of identification of plants of the genus Espeletia, commonly found in Colombia. These studies reinforce the advance in the identification of weeds by algorithm systems. An example of an identification system by algorithms based on color, shape, size, and differentiated textural features is represented in Fig. 7. The same made the identification of 90% of the devil's claw plants (*Proboscidea louisianica* (Mill.) Thellung) in an area cultivated with soybeans (Singh et al. 2020).

UAVs have advanced weed management based on specific maps (Gasparovic et al. 2020). With the survey it is possible to apply herbicides or other control methods in areas of localized weed infestations, adjusting the dose in herbicide applications according to the density or composition of weed species (López-Granados et al. 2016; Gasparovic et al. 2020).

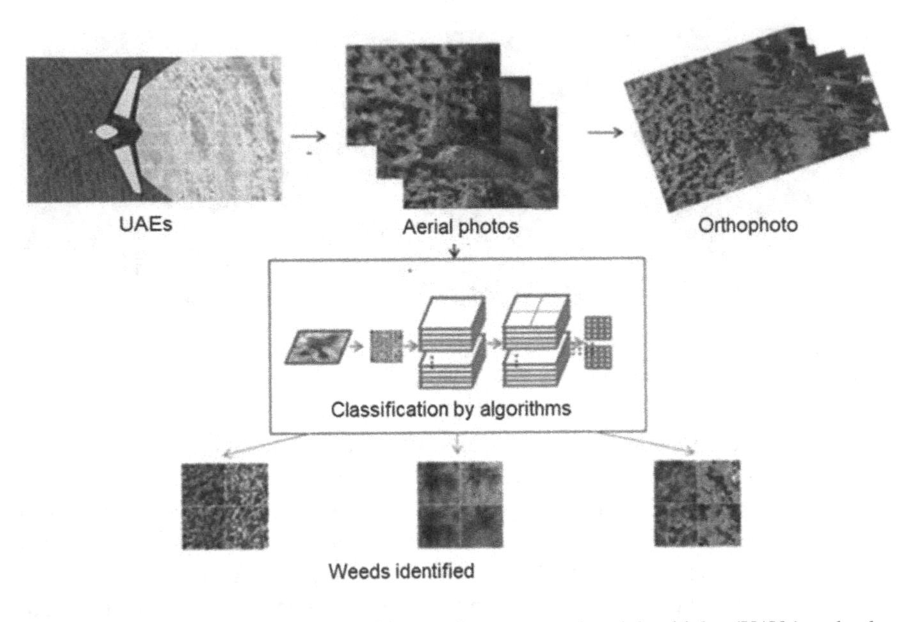

Fig. 6 Illustration of the collection of images by unmanned aerial vehicles (UAVs) and subsequent classification of weeds by algorithms. (Source: Hung et al. (2014))

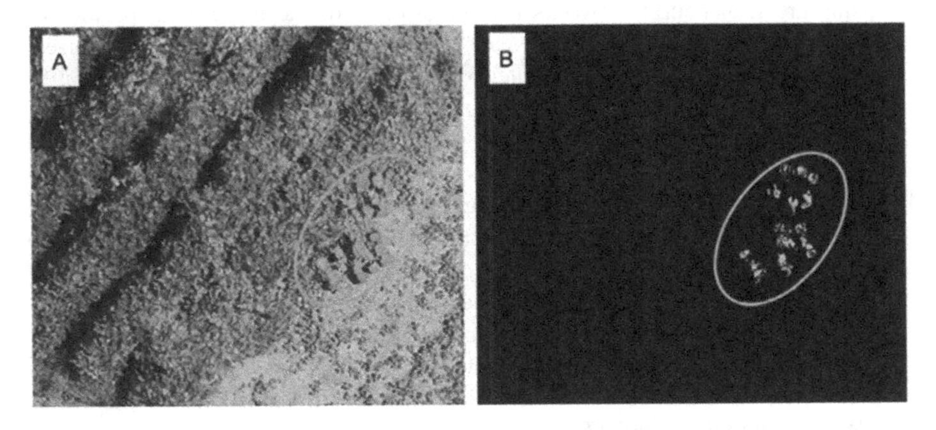

Fig. 7 Detection of devil's claw plants (*Proboscidea louisianica* (Mill.) Thellung) in an area cultivated with soybean, by high resolution image (8 mm/pixel). (**a**) Aerial image of the area with devil's claw plants highlighted in the blue circle and (**b**) detection of target plants. (Source: Singh et al. (2020))

For the knowledge of community dynamics in the area over time, the survey should be carried out at different times, for an extended period of time, with calibrated and standardized equipment in all samples. One of the problems of aerial mapping is that weeds occupy different vertical positions in the area, due to the different sizes of the same, and cannot occur by the lenses of the cameras on the weeds, because they can be covered by the leaves of other plants with different growth habits.

The use of species mapping allows monitoring of their dissemination, especially herbicide-resistant weed biotypes, in order to promote integrated management, minimizing possible dispersion to other places where they are not present, assisting in the preventive management method. Shirzadifar et al. (2020) demonstrated that spectral indices selected by algorithms can be used to detect biotypes of glyphosate-resistant weeds, and that it represents a gain in the adoption of eradication strategies in order to prevent further dissemination in new areas.

Aerial mapping through aerial images can also be used in the survey of plants in non-agricultural areas, such as water reservoirs, mainly in hydroelectric dams, dams, and others. This advance allows rapid evaluation in waters, since the identification of these plants by conventional methods are difficult to perform. Meneses et al. (2018) demonstrated that it was possible to identify different heights and densities of reed plants (*Phragmites australis*) in aquatic media, using images from UAVs.

In Brazil, the use of UAVs has gained great prominence in recent years. Some studies related to program identification, analyzing aerial images of UAVs, were conducted in sugarcane crops (Yano et al. 2016), citrus, corn (Osco et al. 2021), and soybean (Ferreira et al. 2017). Thus, this method is promising and can expand in the country in order to assist the farmer in decision making in weed management.

Finally, monitoring using UAVs makes the collection of images easier, with larger and clearer images, facilitating the processing and analysis of images by programs, generating greater reliability in the identification of the weed. It is noteworthy that UAVs have low flight life per battery, which reduces operational capacity somewhat. Alternatively, you should always have batteries and chargers in reserve.

5 Statistical Analyses Used in Phytosociology

In addition to phytosociological parameters, it is possible to perform multivariate statistical techniques to understand aspects related to weed diversity in an area (Iqbal et al. 2018). Among these techniques, it is possible to use principal component analysis and cluster analysis. It is noteworthy that these analyses seek to show how changes in habitat and anthropogenic activities affect the diversity, distribution of species, in addition to the formation of association between plants (Iqbal et al. 2015; Marshall et al. 2003; Roschewitz et al. 2005; Iqbal et al. 2018).

With multivariate analyses it is possible to evaluate the influence of soil attributes and other conditions on the composition of plants in an area. Iqbal et al. (2018) evaluated the influence of environmental factors and agricultural practices on weed composition and observed that the organic matter content, electrical conductivity, type of harvest, type of manure and practice used in weed management have a significant effect on the distribution pattern of the species of these plants. These same authors observed five associations between weed genera (1—*Emix-Vicia-Lathyrus*, 2—*Alysum-Cannabis-Lithospermum*, 3—*Oxalis-Lathyrus-Chenopodium*, 4—*Euphorbia-Cerastium-Capsella-bursa*, and 5—*Alopecurus-Mazus-Persicaria*), considering them as indicators, being the concept of indicator used to define the relationship within the same group. Esse estudo possibilitou observar as exigências de um grupo de plantas daninhas para determinado ambiente, dessa maneira é possível estabelecer conexões de características do ambiente com as possíveis espécies infestantes, podendo auxiliar no manejo das mesmas nas áreas agrícolas.

6 Phytosociology in Weed Science

The composition of the weed community in a habitat depends on the characteristics of soil, climate, growing season, and the cultural tracts used (Voll et al. 2001; Kuva et al. 2021). Thus, the composition of the weed community depends on the interaction between the phenotypic plasticity of each individual and the processes of adaptive flexibility to eventual changes in the environment (Pitelli and Pitelli 2004; Concenço et al. 2017).

In a weed community, the species do not have the same degree of importance. Each species of weed has certain biological characteristics that can influence the ability to compete for water, light, and nutrients and the degree of interference in the productivity of agricultural crops (Tyagi et al. 2018). There are dominant species, which by the greater capacity of intra- and interspecific competition by the resources of the environment, are responsible for most of the damage. The dominant species are present in the area in higher density and coverage, suppressing the development of other species. Weed species that are suppressed generally do not result in economic problems to agricultural crops (Fernández-Quintanilla et al. 1991). The degree of importance of weed species and consequently the control strategies to be adopted is given by the phytosociological parameters already detailed in this chapter.

The objective of phytosociological studies in weed science is similar to that of ecological studies. When doing a phytosociological survey of species, it should be taken into account that the nature of agronomic studies implies plots with smaller size than expected and with much stronger selection factors than those that act in the environment (Concenço et al. 2013). Furthermore, the selection factors are momentary as treatments are applied, for example, planting densities of different crops, line spacing, and herbicide application (Concenço et al. 2013).

In general, phytosociological studies of weed communities in agroecosystems allow the identification of species with higher rates of importance of plant populations in crops. However, phytosociology can contribute to the decision-making of management methods and adequate control of weed species in agricultural systems. The phytosociological study also allows analyzing the impacts of management methods and cultural treatments on the dynamics of the population of these species (Mendes et al. 2014; Zaidan 2020). According to Kuva et al. (2021) similarity can be useful in planning control strategies, considering the organization of weed communities.

The different weed management methods directly influence the composition and distribution of species in the area. Thus, it is possible to define the most efficient

Fig. 8 Diversity of weeds in the coffee line in the initial condition of the experiment (**a**). Weeding with hoe in the coffee line (**b**) and weed diversity in the coffee line, after 2 years of adoption of manual weeding by means of hoe, highlighting the species of *C. rotundus* (**c**). (Source: Zaidan (2020))

management methods, and can use them in a successive, combined, and/or alternate way. Zaidan (2020) observed that weeding by weeding of weeds in the coffee line provided a change in the composition of weeds in the area, selecting species such as *C. rotundus*, due to the reproduction habit of this species (vegetative propagation and seeds) and soil revolving due to the method used (Fig. 8).

This same author observed that the treatment in which it was performed mowing in the line with motorized brush cutter, selected low-key species, mainly the *Commelina diffusa*, which presented an importance value index (IVI) of 108%. The brush cutter tends to cut the plants at a certain height in relation to the soil, as the *Commelina diffusa* presents a habit of low growth and reproduction by vegetative propagation and seeds, this species dominated the area in relation to the other species (Zaidan 2020).

The application of herbicides in pre- or post-emergence can influence the diversity, composition, and organization of weed communities associated with a given crop. Herbicides have different spectrums of action on species control, so it tends to select the species in which it is selective, tolerant, and also resistant weed biotypes, which directly interferes with the floristic composition in the area. Examples include auxinic herbicides that are selective for monocotyledonous species, Acetyl Coenzyme A Carboxylase (ACCase) inhibitor herbicides that are selective for eudicotyledonous species and ghyphosate that does not exert control over tolerant species such as *C. diffusa* (Santos et al. 2001; Dias et al. 2013) and resistant biotypes of *Digitaria insularis*, and it is necessary to use herbicide mixtures or other control methods in the management of this species (Adegas et al. 2010a).

The knowledge of the weed community that occurs in the cultivation areas becomes the basis for the formulation of an efficient control program. The altitude and precipitation rate of the region, the predecessor crop, soil moisture, pH, and soil organic matter content can provide changes in the composition of the weed community in a given region. Mendes et al. (2014) studying the effect of oxadiazon on the dynamics of the weed community, at different times of application of water slides and incorporation of organic material, verified that the application of oxadiazon negatively affected the number and dry matter of weeds, and the species *C. rotundus* presented higher IVI, followed by *U. decumbens*, *Galinsoga parviflora*,

Bidens pilosa, and *Melampodium perfoliatum.* The greatest similarity was observed between the treatment that received the sequential application of 10 mm of water slide before and after herbicide spraying, and there was no significant interaction between the moments of application of water slides and the incorporation or not of organic material, due to the similarity of organic matter values present in the soil (Mendes et al. 2014).

According to the agricultural practices adopted and the period of cultivation of the agricultural crop, variations in the composition of weed species may also occur. In this context, Ferreira et al. (2019) studying the effects of different nitrogen doses on weed dynamics in the crop during the corn crop cycle, verified that the plant species community underwent changes in its dynamics as a function of sampling time and insertion of planting techniques, and the similarity was lower when lower nitrogen doses were applied. In studies conducted by Mi et al. (2018), it was also possible to observe that the number of species was influenced by nitrogen fertilization. The authors found that the largest number of weed species was recorded in the treatment without the presence of nitrogen, followed by treatment with high dosage.

On the other hand, it is important to point out that crop rotation is an agricultural practice that in addition to improving soil fertility, reducing erosion, and reducing the risk of crop diseases, is also an important method of weed control, since it can modify the diversity and composition of weed communities in an area. Neyret et al. (2020) investigating the effect of crop rotation with corn, rice, and rubber trees on the diversity of weed communities in northern Thailand, observed that the diversity of herbaceous species increased by 36%, while dominance decreased by 38% in agricultural areas with annual crop changes. Therefore, crop rotation can prevent the accumulation of dominant weed communities in monocultures.

The phytosociological study can be used to understand the dynamics of weeds in different crops. Several phytosociological studies were carried out in crops in Brazil, such as corn (Duarte et al. 2007), sunflower (Adegas et al. 2010b), banana (Moura Filho et al. 2015; Santos et al. 2019), sugarcane (Oliveira and Freitas 2008), pastures (Inoue et al. 2012; Inoue et al. 2013), rice (Erasmo et al. 2004), coffee (Maciel et al. 2010; Zaidan 2020), peanuts (Piazentine et al. 2020), sweet potatoes (Silva et al. 2017), and other crops. There is also the possibility of surveying aquatic weeds, aiming at measures to prevent the increase in the volume of plants in the reservoir, especially in hydroelectric reservoirs, in which desiquilibrada proliferation can cause damage to equipment harming energy generation (Rocha and Martins 2011).

In addition to these reported studies, there is the possibility of conducting several investigations related to the database of weed diaspores in the soil. Soil diaspore banks function as a source of weed communities, determining their occurrence, dynamics and succession (Bárberi and Cascio 2001). Thus, the knowledge of the dynamics of the soil diaspore bank (Fig. 9) will contribute to the prediction of the structure and dynamic patterns of the weed community, assisting farmers in integrated management (He et al. 2019).

The cultivation of various types of crops can promote the emergence of different weed communities. Therefore, agricultural areas contain different banks of weed

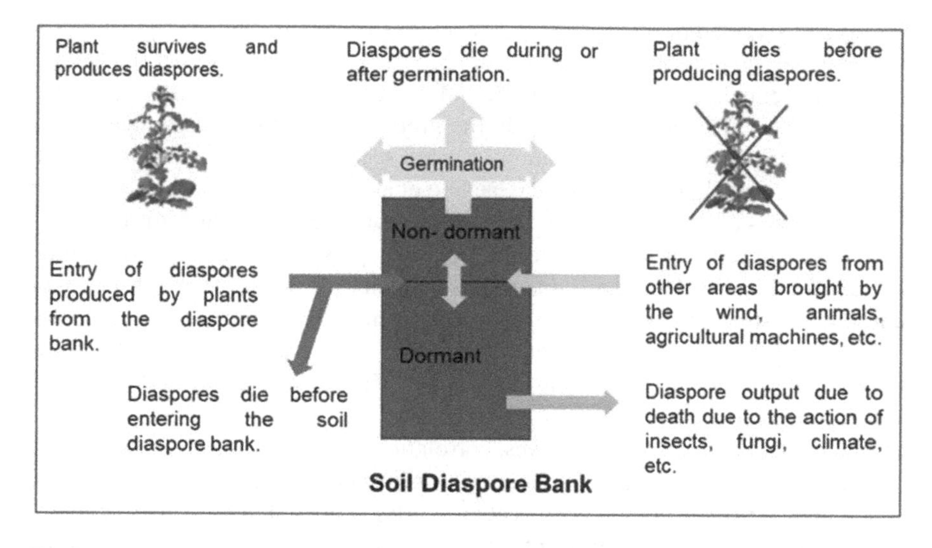

Fig. 9 Dynamics of the soil weed database. (Source: Adapted from Dias Filho (2011))

diaspores (Bellinder et al. 2004). The weed community of agricultural areas is related to the previous rotation crops, and it can be assumed that the different cultivation systems house different and complex banks of diaspores in the soil (He et al. 2019). However, in studies conducted by He et al. (2019) it was possible to observe that after the implementation of long-term monoculture patterns, the compositions of weed species in the soil diaspore bank become homogenized, regardless of the type of crop.

On the other hand, there are weed seeds that remain in the soil only for a few periods of the year, forming a temporary seed bank. These temporary seed banks are composed of species that germinate when there are favorable environmental conditions, providing the appearance of different weed communities (Santín-Montanyá et al. 2016). Geralmente, o preparo do solo atua no banco de sementes do solo promovendo ou dificultando a germinação (Trichard et al. 2013; Santín-Montanyá et al. 2016). In this context, Santín-Montanyá et al. (2016) evaluating the impact on the temporary and persistent weed seed bank of different soil tillage practices in winter cereal cultivation systems, found that soil tillage systems did not affect weed seed density and diversity parameters in the temporary seed bank, however, in the persistent seed bank, conservation cultivation techniques favored the increase in density and diversity of weed seeds compared to conventional cultivation.

The term seed bank has also been adopted to designate the reserves of viable seeds in soil, deep, and on the surface (Roberts 1981; Caetano et al. 2001). However, the composition of the seed bank is mainly determined by the cultural practices employed in the area (Caetano et al. 2001), as already reported. Thus, Caetano et al. (2001) evaluated the weed seed bank to determine the possible influences of the different management systems used in citrus crop at depths of 0–10 and 0–20 cm, verified that the seed bank is concentrated in the surface layers of 0–10 cm and the use of grid and swidden provided higher percentages of plants, while the green cover with legumes suppressed weed germination. In the practice of mowing, the vegetation is cut, being kept on the soil surface and if the vegetation presents seeds, these can germinate (Caetano et al. 2001). In the green cover with legumes, the formed plant mass can inhibit the germination of weed seeds by some allelopathic effect, physical barrier or light interception (Caetano et al. 2001).

The analysis of diaspore bank in the soil has its particularities and it is often difficult to separate and have an estimate of greater representation of the composition of plants in the area. This fact is due to the difficulty of separation and diversity of reproductive structures. Thus, there is the possibility of correlating the methodologies to verify the similarity between the diaspore bank present in the soil and the species found in the area, combining the method of the leaked square with the methods of evaluating the composition of diaspores.

Among the comparison alternatives, correlation analyses can be chosen, such as spearman rank correlation and the Odum similarity coefficient described in the phytosociological parameter topics. Kuva et al. (2008) verified that spearman's correlation and Odum similarity coefficient were inefficient to estimate the infesting flora in relation to the seed bank emergence method in sugarcane cultivation, with maximum values of 60%, depending on the area. This fact can be explained by the conditions necessary in the field for seeds to germinate, in addition to the presence of diaspores of vegetative structure, not accounted for in the seed bank evaluation method. Although emergency techniques may not accurately portray the volume of seeds in the soil, they can more accurately predict which plants can settle under field conditions (Espeland et al. 2010).

In view of the above, the phytosociological survey of weeds when applied in agricultural areas may also present some disadvantages. As these parameters were originally designed to characterize plant communities in natural environments, adaptations to the agricultural context were necessary, such as establishing basic steps in a phytosociological analysis; prioritize the formulas less affected by factors that may alter the phytosociological analysis; and to use parameters not only to evaluate the composition of current weeds in a given area, but also to analyze the soil diaspore bank (Concenço et al. 2017), as described in this chapter.

Another disadvantage of phytosociological survey is the difficulty for data collection in the field and for its processing (Concenço et al. 2017). The set of parameters adopted in phytosociology is also not standardized, with this, it will hardly be possible to compare studies conducted by different researchers, since the formulas and procedures are rarely similar (Concenço, et al. 2017). However, this chapter sought to standardize phytosociological parameters and their applications with emphasis on weed science.

7 Key Concepts

- Phytosociology.
- Plant diversity.
- Phytosociological parameters.
- Weed community.
- Methods of the leaked square.
- Relevé method.
- Diaspore bank of the soil.
- Importance Value Index (IVI).
- Similarity index (SI).
- Diversity Indices.
- Weed mapping.
- Identification by algorithms.
- Composition of weed species.
- Statistical analyses used in phytosociology.
- Phytosociology in weed science.

8 Concluding Remarks

Phytosociology is an extremely important science tool in generating knowledge about the abundance, distribution, and diversity of the weed community and possible interactions of weeds in responses to biotic and abiotic factors, which can assist farmers and researchers in decision-making in the adoption of weed management techniques and strategies, in order to minimize their interference with agricultural crops.

Phytosociological survey methodologies for weed science have been modified over the years. However, there are still many doubts and controversies regarding reliability and representativeness in existing methods. Thus, this chapter sought to highlight the main aspects related to phytosociology applied in weed science, highlighting the main parameters that can be used in the analysis of a community infesting in a given area, in which the response may be to a imposed treatment or to a place where there is interest in knowing the floristic composition.

However, it is worth mentioning that this branch of science still needs further studies, in order to seek the standardization of methodologies and different evaluation alternatives that are of greater reliability. The use of aerial images captured by high resolution cameras in the mapping and identification of weeds is shown to be a promising tool with a high degree of accuracy, being increasingly used, both in scientific research and in areas of agricultural production in the most appropriate management of weeds.

References

Adegas FS, Gazziero DLP, Voll E et al (2010a) Alternativas de controle químico de *Digitaria insularis* resistente ao herbicida glyphosate. Papper presented at the Congresso Brasileiro da Ciência das Plantas Daninhas, Ribeirão Preto-SP, 19-23 jul 2010

Adegas FS, Oliveira MF, Vieira OV et al (2010b) Levantamento fitossociológico de plantas daninhas na cultura do girassol. Planta Daninha 28(4):705–716. https://doi.org/10.1590/S0100-83582010000400002

Andres A, Avila LA, Marchezan E, Menezes VG (2001) Rotação de culturas e pousio do solo na redução do banco de sementes de arroz vermelho em solo de várzea. Rev Bras Agrociência 2:85–88

Arruda L, Daniel O (2007) Florística e diversidade em um fragmento de floresta estacional semidecidual aluvial em Dourados-MS. Floresta. https://doi.org/10.5380/rf.v37i2.8649

Bárberi P, Cascio BL (2001) Long-term tillage and crop rotation effects on weed seedbank size and composition. Weed Res 41:325–340. https://doi.org/10.1046/j.1365-3180.2001.00241.x

Barbour MG, Burk JH, Pitts WD et al (1998) Terrestrial plant ecology, 3rd edn. Menlo Park, Benjamin-Cummings

Bellinder RR, Dillard HR, Shah DA (2004) Weed seedbank community responses to crop rotation schemes. Crop Prot 23:95–101. https://doi.org/10.1016/S0261-2194(03)00174-1

Blasi C, Frondoni R (2011) Modern perspectives for plant sociology: the case of ecological land classification and the ecoregions of Italy. Plant Biosyst 145:30–37. https://doi.org/10.1080/11263504.2011.602747

Braun-Blanquet J (1968) L'école phytosociologique zuricho-montpelliéraine et la S.I.G.M.A. Vegetatio 16:2–78

Braun-Blanquet J (1979) Fitossociologia: bases para el estudio de las comunidades vegetales, Madrid

Buhler DD, Maxwell BD (1993) Seed separation and enumeration from soil using K_2CO_3 centrifugation and image analysis. Weed Sci 41:298–302. https://doi.org/10.1017/S0043174500076207

Buhler DD, Hartzler RG, Forcella F (1997) Implications of weed seed bank dynamics to weed management. Weed Sci 45:329–336. https://doi.org/10.1017/S0043174500092948

Caetano RSX, Christoffoleti PJ, Victoria Filho R (2001) Banco de sementes de plantas daninhas em pomar de laranjeira pera. Sci Agric 58:509–517. https://doi.org/10.1590/S0103-90162001000300012

Calegari L, Martins SV, Campos LC et al (2013) Avaliação do banco de sementes do solo para fins de restauração florestal em Carandaí, MG. J Árvore 37:871–880. https://doi.org/10.1590/S0100-67622013000500009

Carmona R (1992) Problemática e manejo de bancos de sementes de invasoras em solos agrícolas. Planta Daninha 10:5–16. https://doi.org/10.1590/S0100-83581992000100007

Carmona R (1995) Banco de sementes e estabelecimento de plantas daninhas em agroecossistemas. Planta Daninha 13:3–9. https://doi.org/10.1590/S0100-83581995000100001

Concenço G, Farias PM de, Quintero NFA, et al (2017) Phytosociological Surveys in Weed Science: Old Concept, New Approach. In: Youl saf Z (ed) Plant Ecology – Traditional Approaches to Recent Trends, IntechOpen, Rijeka- Croatia, pp 121–146

Concenço G, Tomazi M, Correia IVT et al (2013) Phytosociological surveys: tools for weed science? Planta Daninha 31:469–482. https://doi.org/10.1590/S010083582013000200025

Dessaint F, Chadoeuf R, Barralis G (1990) Etude de la dynamique communauté adventice: III. Influence à long terme dês techniques culturales sur la composition spécifique du stock semencier. Weed Res 30:19–30. https://doi.org/10.1111/j.1365-3180.1990.tb01716.x

Dias Filho MB (2011) Manejo de plantas daninhas. Degradação de pastagens: processos, causas e estratégias de recuperação. Belém-PA, pp 70–93

Dias ACR, Carvalho SJP, Christoffoleti PJ (2013) Fenologia da trapoeraba como indicador para tolerância ao herbicida glyphosate. Planta Daninha 31:185–191. https://doi.org/10.1590/S0100-83582013000100020

Duarte AP, Silva AC, Deuber R (2007) Plantas infestantes em lavouras de milho safrinha, sob diferentes manejos, no médio Paranapanema. Planta Daninha 25:285–291. https://doi.org/10.1590/S0100-83582007000200007

Erasmo EAL, Pinheiro LLA, da Costa NV (2004) Levantamento fitossociológico das comunidades de plantas infestantes em áreas de produção de arroz irrigado cultivado sob diferentes sistemas de manejo. Planta Daninha 22:195–201. https://doi.org/10.1590/S0100-83582004000200004

Espeland EK, Perkins LB, Leger EA (2010) Comparison of seed Bank estimation techniques using six weed species in two soil types. Rangel Ecol Manag 63:243–247. https://doi.org/10.2111/REM-D-09-00109.1

Fernández-Quintanilla C, Saavedra MS, Garcia L (1991) Ecologia de lãs malas hierbas. In: Garcia Torres L, Fernández-Quintanilla C (eds) Fundamentos sobre malas hierbas y herbicidas. Mundi-Prensa, Madrid, pp 49–69

Ferreira SA, Freitas DM, Silva GG et al (2017) Weed detection in soybean crops using ConvNets. Comput Electron Agric 143:314–324. https://doi.org/10.1016/j.compag.2017.10.027

Ferreira EA, Paiva MCG, Pereira GAM et al (2019) Fitossociologia de plantas daninhas na cultura do milho submetida à aplicação de doses de nitrogênio. Rev Agric Neotrop 6:109–116. https://doi.org/10.32404/rean.v6i2.2710

Gasparovic M, Zrinjski M, Barković D et al (2020) An automatic method for weed mapping in oat fields based on UAV imagery. Comput Electron Agric 173:105385. https://doi.org/10.1016/j.compag.2020.105385

Goldman E, Herzig R, Eisenschtat A et al (2019) Precise detection in densely packed scenes. Paper presented at the Conference on computer vision and pattern recognition, Long Beach, pp 5227–5236

Gomes GLGC, Ibrahim FN, Macedo GL et al (2010) Cadastramento fitossociológico de plantas daninhas na bananicultura. Planta Daninha 28:61–68. https://doi.org/10.1590/S0100-83582010000100008

He Y, Gao P, Qiang S (2019) An investigation of weed seed banks reveals similar potential weed community diversity among three different farmland types in Anhui Province, China. J Integr Agric 18:927–937. https://doi.org/10.1016/S2095-3119(18)62073-8

Hung C, Xu Z, Sukkarieh S (2014) Feature learning based approach for weed classification using high resolution aerial images from a digital camera mounted on a UAV. Remote Sens 6:12037–12054. https://doi.org/10.3390/rs61212037

Inoue MH, Silva BE, Pereira KM et al (2012) Levantamento fitossociológico em pastagens. Planta Daninha 30:55–63. https://doi.org/10.1590/S0100-83582012000100007

Inoue MH, Iskierski D, Mendes KF et al (2013) Levantamento fitossociológico de plantas daninhas em pastagens no município de Nova Olímpia-MT. Agrarian 6:376–384

Iqbal M, Khan SM, Khan MA et al (2015) Exploration and inventorying of weeds in wheat crop of the district Malakand, Pakistan. Pak J Weed Sci 21:435–452. http://www.wssp.org.pk/vol-21-3-2015/14.%20PJWSR-99-2015.pdf

Iqbal M, Khan SM, Khan MA et al (2018) A novel approach to phytosociological classification of weeds flora of an agroecological system through cluster, two way cluster and indicator species analyses. Ecol Indic 84:590–606. https://doi.org/10.1016/j.ecolind.2017.09.023

Kercher SM, Frieswyk CB, Zedler JB (2003) Effects of sampling teams and estimation methods on the assessment of plant cover. J Veg Sci 14:899–906. https://doi.org/10.1111/j.1654-1103.2003.tb02223.x

Kuva MA, Pitelli RA, Alves PLCA et al (2008) Banco de sementes de plantas daninhas e sua correlação com a flora estabelecida no agroecossistema cana-crua. Planta Daninha 26:735–744. https://doi.org/10.1590/S0100-83582008000400004

Kuva MA, Salgado TP, Alves PLC (2021) Índices fitossociológicos aplicados na ciência e na gestão das estratégias de controle de plantas daninhas. In: Martins AA, Murata AT (eds) Matologia: Estudo sobre plantas daninhas. Jaboticabal, Sao Paulo, pp 60–105

López-Granados F, Torres-Sánchez J, Serrano-Pérez A et al (2016) Early season weed mapping in sunflower using UAV technology: variability of herbicide treatment maps against weed thresholds. Precis Agric 17:183–199. https://doi.org/10.1007/s11119-015-9415-8
Luschei EC, Buhler DD, Dekker JH (1998) Effect of separating giant foxtail (*Setaria faberi*) seeds from soil using potassium carbonate and centrifugation on viability and germination. Weed Sci 46:545–548. https://doi.org/10.1017/S0043174500091074
Maciel CDDG, Poletine JP, AMD ON et al (2010) Levantamento fitossociológico de plantas daninhas em cafezal orgânico. Bragantia 69:631–636. https://doi.org/10.1590/S0006-87052010000300015
Magurran AE (1988) Ecological diversity and its measurements. Princeton University Press, London, 179p
Marshall EJP, Brown VK, Boatman ND, Lutman PJW et al (2003) The role of weeds in supporting biological diversity within crop fields. Weed Res 43:77–89. https://doi.org/10.1046/j.1365-3180.2003.00326.x
Mendes KF, Reis RM, Reis MR et al (2014) Dinâmica de plantas daninhas após aplicação de oxadiazon com simulação de lâminas d'água e incorporação de material orgânico. Agrária 9:65–71. https://doi.org/10.5039/agraria.v9i1a3679
Meneses NC, Baier S, Reidelstürz P et al (2018) Modelling heights of sparse aquatic reed (*Phragmites australis*) using structure from motion point clouds derived from rotary-and fixed-wing unmanned aerial vehicle (UAV) data. Limnologica 72:10–21. https://doi.org/10.1016/j.limno.2018.07.001
Mesgaran MB, Mashhadi HR, Zand E et al (2007) Comparison of three methodologies for efficient seed extraction in studies of soil weed seedbanks. Weed Res 47:472–478. https://doi.org/10.1111/j.1365-3180.2007.00592.x
Mi W, Gao Q, Sun Y et al (2018) Changes in weed community with different types of nitrogen fertilizers during the fallow season. J Crop Prot 109:123–127. https://doi.org/10.1016/j.cropro.2018.01.014
Monquero P, Christoffoleti PJ (2005) Banco de sementes de plantas daninhas e herbicidas como fator de seleção. Bragantia 64:203–209. https://doi.org/10.1590/S0006-87052005000200006
Monquero PA, Hirata ACA, Pitelli RA (2014) Métodos de levantamento da colonização de plantas daninhas. In: Monquero PA (ed) Aspecto da biologia e manejo das plantas daninhas. RiMa, São Carlos, pp 103–128
Monteiro AL, Freitas SM, Lins et al (2021) A new alternative to determine weed control in agricultural systems based on artificial neural networks (ANNs). Field Crops Res 263:108075. https://doi.org/10.1016/j.fcr.2021.108075
Morisita M (1959) Measuring of the dispersion and analysis of distribution patterns. Memoir Fac Sci Kyushu Univ Ser Biol 2:215–235
Moura Filho ER, Macedo LPM, Silva ARS (2015) Levantamento fitossociológico de plantas daninhas em cultivo de banana irrigada. Holos 2:92–97. https://doi.org/10.15628/holos.2015.1006
Mueller Dombois D, Ellenberg H (1974) Aims and methods of vegetation ecology. Wiley, New York
Neyret M, Rouw A, Colbach N et al (2020) Year-to-year crop shifts promote weed diversity in tropical permanent rainfed cultivation. Agric Ecosyst Environ 301:107023. https://doi.org/10.1016/j.agee.2020.107023
Nkoa R, Owen MDK, Swanton CJ (2015) Weed abundance, distribution, diversity, and community analyses. Weed Sci 63:64–90. https://doi.org/10.1614/WS-D-13-00075.1
Odum EP (1985) Ecologia. Interamericana, Rio de Janeiro, p 434
Oliveira AR, Freitas SP (2008) Levantamento fitossociológico de plantas daninhas em áreas de produção de cana-de-açúcar. Planta Daninha 26:33–46
Oosting HJ (1956) The study of plant communities: an introduction to plant ecology. W.H. Freeman, San Francisco, p 440

Osco LP, Arruda MDS, Gonçalves DN et al (2021) A CNN approach to simultaneously count plants and detect plantation-rows from UAV imagery. ISPRS J Photogramm Remote Sens 174:1–17. https://doi.org/10.1016/j.isprsjprs.2021.01.024

Ostermann OP (1998) The need for management of nature conservation sites designated under Natura 2000. J Appl Ecol 35:968–973. https://doi.org/10.1111/j.1365-2664.1998.tb00016.x

Pérez-ortiz M, Peña JM, Gutiérrez PA et al (2016) Selecting patterns and features for between-and within-crop-row weed mapping using UAV-imagery. Expert Syst Appl 47:85–94. https://doi.org/10.1016/j.eswa.2015.10.043

Piazentine AE, Carrega WC, Costa MR et al (2020) Levantamento fitossociológico na cultura do amendoim. South Am Sci 1:2031. https://doi.org/10.17648/sas.v1i1.31

Pitelli RA, Pitelli RLCM (2004) Biologia e ecofisiologia das plantas daninhas. In: Vargas L, Roman ES (eds) Manual de Manejo e Controle de Plantas Daninhas. Embrapa Uva e Vinho, Bento Gonçalves, pp 29–56

Pott R (2011) Phytosociology: a modern geobotanical method. Plant Biosyst 145:9–18. https://doi.org/10.1080/11263504.2011.602740

Roberts HA (1981) Seed banks in the soil. In: Roberts HA (ed) Advances in applied biology. Academic Press, Cambridge, pp 1–55

Rocha DC, Martins D (2011) Assessment of aquatic plants from Alagados dam, Ponta Grossa-PR. Planta Daninha 29:237–246. https://doi.org/10.1590/S0100-83582011000200001

Rodrigues R, Gandolfi S (1998) Restauração de florestas tropicais: subsídios para uma definição metodológica e indicadores de avaliação de monitoramento. In: Dias LE, JWV M (eds) Recuperação de áreas degradadas. Aprenda Fácil, Viçosa, pp 203–215

Roschewitz I, Gabriel D, Tscharntke T et al (2005) The effects of landscape complexity on arable weed species diversity in organic and conventional farming. J Appl Ecol 42:873–882. https://doi.org/10.1111/j.1365-2664.2005.01072.x

Santín-montanyá MI, Martín-lammerding D, Zambrana E et al (2016) Management of weed emergence and weed seed bank in response to different tillage, cropping systems and selected soil properties. Soil Tillage Res 161:38–46. https://doi.org/10.1016/j.still.2016.03.007

Santos IC, Silva AA, Ferreira FA et al (2001) Eficiência de glyphosate no controle de *Commelina benghalensis* e *Commelina diffusa*. Planta Daninha 19:135–143. https://doi.org/10.1590/S0100-83582001000100016

Santos G, Maia V, Aspiazú I et al (2019) Weed interference on 'Prata-Anã' banana production. Planta Daninha 37:019222533. https://doi.org/10.1590/S0100-83582019370100150

Severino FJ, Christoffoleti PJ (2001) Banco de sementes de plantas daninhas em solo cultivado com adubos verdes. Bragantia 60:201–204. https://doi.org/10.1590/S0006-87052001000300007

Shirzadifar A, Bajwa S, Nowatzki J et al (2020) Field identification of weed species and glyphosate-resistant weeds using high resolution imagery in early growing season. Biosyst Eng 200:200–214. https://doi.org/10.1016/j.biosystemseng.2020.10.001

Silva DSM, Dias Filho MB (2001) Banco de sementes de plantas daninhas em solo cultivado com pastagens de *Brachiaria brizantha* e *Brachiaria humidicola* de diferentes idades. Planta Daninha 19:179–185. https://doi.org/10.1590/S0100-83582001000200004

Silva J, Cunha JLXL, Teixeira JS et al (2017) Levantamento fitossociológico de plantas daninhas em cultivo de batata-doce. Revista Ciência Agrícola 15:45–52

Singh V, Rana A, Bishop M et al (2020) Unmanned aircraft systems for precision weed detection and management: prospects and challenges. In: Sparks DL (ed) Advances in agronomy. Academic Press, Cambridge, pp 93–134

Sudars K, Jasko J, Namatevs I et al (2020) Dataset of annotated food crops and weed images for robotic computer vision control. Data Brief 31:105833. https://doi.org/10.1016/j.dib.2020.105833

Trichard A, Alignier A, ChauveL B (2013) Identification of weed community traits response to conservation agriculture. Agric Ecosyst Environ 179:179–186. https://doi.org/10.1016/j.agee.2013.08.012

Tyagi VC, Wasnick VK, Choudhary M et al (2018) Weed management in Berseem (*Trifolium alexandrium* L.): a review. Int J Curr Microbiol Appl Sci 7:1929–1938. https://doi.org/10.20546/ijcmas.2018.705.226

Voll E, Torres E, Brighenti AM et al (2001) Dinâmica do banco de sementes de plantas daninhas sob diferentes sistemas de manejo do solo. Planta Daninha 19:171–178. https://doi.org/10.1590/S0100-83582001000200003

Yano IH, Alves JR, Santiago WE et al (2016) Identification of weeds in sugarcane fields through images taken by UAV and random forest classifier. IFAC Pap Online 49:415–420. https://doi.org/10.1016/j.ifacol.2016.10.076

Zaidan UR (2020) Sistemas conservacionistas de manejo integrado de plantas daninhas na cultura do café. Doctoral thesis, Federal University of Viçosa, Viçosa

Zhang C, Atkinson PM, George C et al (2020) Identifying and mapping individual plants in a highly diverse high-elevation ecosystem using UAV imagery and deep learning. ISPRS J Photogram Remote Sens 169:280–291. https://doi.org/10.1016/j.isprsjprs.2020.09.025

Methods of Control and Integrated Management of Weeds in Agriculture

Vicente Bezerra Pontes Junior, Antonio Alberto da Silva, Leonardo D'Antonino, Kassio Ferreira Mendes, and Bruna Aparecida de Paula Medeiros

Abstract Weeds are one of the main problems currently faced in global agriculture. Among some important aspects to be known, weed management and control methods stand out here, as well as their integration, which consists of Integrated Weed Management (IWM). Understanding the importance of these methods and how they should be implemented, as well as their interactions, is an essential tool in cultivation systems, from the most conventional to the most agroecological. Each measure aims, directly or indirectly, to prevent the establishment and development of weeds, as well as to allow the proper establishment of the crop of interest. Moreover, the excessive use of herbicides over the years has promoted the increase of resistant weed cases, a problem that generates economic and environmental damage. With this, the IWM emerges as an alternative to avoid and control these resistant species, being necessary its implementation, with the objective of promoting the sustainability of agroecosystems and ensuring the possibility of use of agricultural and environmental resources by future generations. In this chapter will be discussed in detail the preventive, cultural, physical, mechanical, biological, and chemical methods, their importance and disadvantages, in addition to the IWM and its enormous contribution to weed control.

Keywords Weed management and control · Reduction of weed interference · Integration between methods

V. B. Pontes Junior (✉) · L. D'Antonino · B. A. de Paula Medeiros
Federal University of Viçosa, Viçosa, Minas Gerais, Brazil
e-mail: leonardo@ufv.br; bruna.medeiros@ufv.br

A. Alberto da Silva · K. F. Mendes
Department of Agronomy, Federal University of Viçosa, Viçosa, Minas Gerais, Brazil
e-mail: aasilva@ufv.br; kfmendes@ufv.br

© The Author(s), under exclusive license to Springer Nature Switzerland AG 2022
K. F. Mendes, A. Alberto da Silva (eds.), *Applied Weed and Herbicide Science*,
https://doi.org/10.1007/978-3-031-01938-8_4

1 Introduction

The choice of weed control methods should take into account the species present in the area of interest (culture of economic interest and weeds), the local conditions of labor and equipment, without forgetting the environmental and economic aspects. Control methods range from controlling the plants with your hands to using sophisticated equipment that uses controlled fire and electrocution (Oliveira and Brighenti 2018). The reduction of weed interference, considering a crop, should be made to a level at which the interference losses are equal to the increase in the cost of control, that is, that do not interfere in the economic production of the crop. Weed control methods are showed in Fig. 1.

Among the methods, it is important to highlight the methods of weed management and control. According to Carvalho (2013), management is understood as a non-punctual and strategic procedure, which aims to reduce the potential for weed interference in the short, medium, and long term, that is, a process over time and planned. While control is a non-strategic one-off intervention, which aims at the rapid elimination of the weed community, eliminating them or hindering their growth. In this chapter, preventive and cultural methods are considered management methods, while the control methods are physical, mechanical, chemical, and biological.

Management methods can be considered as indirect control measures, as they are practices that do not directly reach weeds. They are carried out by producers even for purposes other than weed control, such as fertilization, liming, purchases of

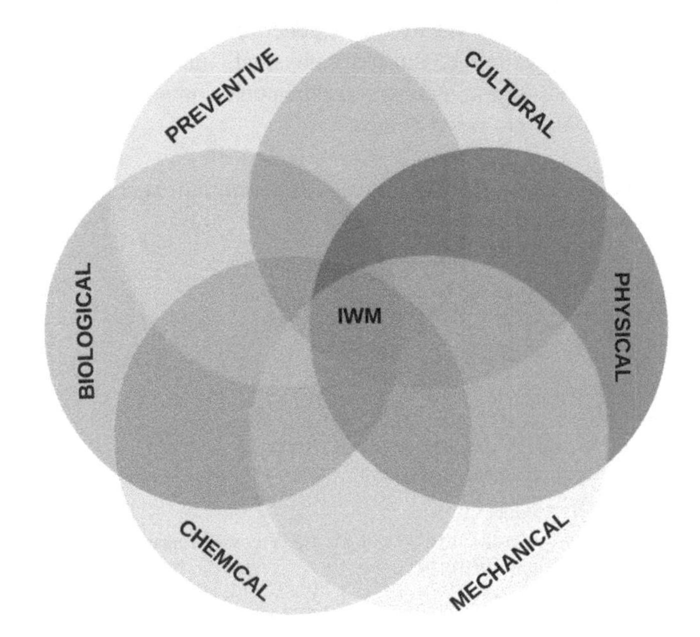

Fig. 1 Weed control methods. *IWM*, integrated weed management

certified seeds, localized irrigation system, crop rotation, among others. In a way, management methods bring other benefits to the crop besides the elimination of competition with weeds, such as soil preparation suitable for their growth, availability of water, nutrients, and higher productivity. On the other hand, control methods are considered direct measures, because they will act after incidence, when the weed has the potential to cause interference.

In the sustainability of agricultural systems, it is important to integrate control methods by observing the characteristics of the soil, climate, and socioeconomic aspects of the producer. The realization of the integration compatible, environmentally and economically, demands deep knowledge of the available strategies, promoting balance with soil and water management measures, in addition to the control of pests and diseases. When adopting any control measure, the means in which weeds meet should be treated as an ecosystem capable of responding to any change imposed, thus not limited to herbicide application or the use of any other method alone. The use of isolated methods will hardly control all weeds present in an area. On the other hand, chemical control in isolation can still promote the selection of resistant weed biotypes due to the selection pressure of the method, which makes it more difficult and costly to control these resistant species. Therefore, integrating management and control methods, promoting integrated management, is the best way to prevent such problems from occurring. In addition, we will seek to encourage the improvement of the quality of life, both of the farmer directly involved, and of the entire population that will benefit from the production chain, as well as the preservation of the environment.

In view of the above, the different methods of weed management and control will be addressed in this chapter, as well as their benefits and disadvantages. Integrated Weed Management (IWM) and its importance in the current environmental and economic context will also be detailed.

2 Preventive Management

Preventive management consists in the use of practices aimed at preventing the introduction, establishment, and/or dispersion of certain weed species in areas not yet infested by them. These areas can be a country, a state, a municipality, or a glebe of land on the property.

At federal and state levels, there are laws regulating the entry of seeds into the country or state and their internal commercialization. In these legislations, there are the tolerable seed limits of each weed species and also the list of seeds prohibited by crop or group of crops.

At the local level, it is the responsibility of each farmer or cooperatives to prevent the entry and dispersion of one or more weed species, which could become serious problems for the region. In summary, the human element is the key to preventive control. The efficient occupation of the agroecosystem space by culture decreases the availability of factors suitable for weed growth and development, and can be considered an integration between prevention and the cultural method.

Some practices can and should be performed in order to avoid the dispersion of weed seeds to other areas. Such practices are importantly implementing to prevent future additions to the cost of weed control. Some of them are mentioned below:

2.1 Cleaning of Equipment and Machinery

In a production system, various equipment and machinery are used in crops, having direct contact with the components present there, such as soil, crop, and weeds. Some weed species have very small and light seeds, being easily loaded to other places when joining the implements (e.g., *Amaranthus* spp.). Tractors, brush cutters, railings, plow, harvesters, hoes, among others should be carefully cleaned after their use, especially if it is of shared use (cooperatives, associations, or rental swelters). The lack of care in cleaning equipment and machinery can cause the dispersion of species that are difficult to control. As an example, there is the introduction, in pastures of the State of Mato Grosso, Brazil, of seeds of Palmer amaranth (*Amaranthus palmeri*), an exotic species to Brazil, possibly coming from harvesters imported from Argentina, which were not cleaned properly (Gazziero and Silva 2017). The biotypes of this species showed in Brazil multiple resistance to glyphosate, inhibitor of the enzyme 5-enolpyruvylshikimate-3-phosphate synthase (EPSPs) and herbicides inhibitors of the enzyme Acetolactate Synthase (ALS) (Gonçalves Netto et al. 2016), being highly aggressive, difficult to control and with high reproductive potential.

2.2 Use of Certified Seeds and Seedlings

After defining the crop that will be implemented in the area, it is necessary to use seeds and/or seedlings from a certified site for its production, which meet the current legislation on the minimum amount allowed for each weed species along with the seeds or with the soil. Seed and seedling production is ultimately a planting system that is also at the mercy of the presence of any pest in its cultivation area. Therefore, weed seeds can mix with produced seeds or be present in the soil that will be used as substrate for seedlings.

2.3 Cleaning of Irrigation Channels

Irrigation systems are an important component in a production that aims at high productivity, providing water (and even nutrients and pesticides) to crops, especially in drier periods. However, they can also be a means of dispersing weed seeds, mainly species that have dry and light reproductive structures, or with lower density than water (Carvalho 2013), characteristics common to many species. In a study conducted in Nebraska, USA, evaluating the dispersion of weed seeds by irrigation channel, Wilson Junior (1980) found that, during the period evaluated, more than 48,000 ha^{-1} seeds were dispersed in the sampled fields from irrigation channels. The main species found was barnyardgrass (*Echinochloa crus-galli*). Dispersion occurs when there is infestation near the pipes or hoses, in which diaspores are transported in the water that is conducted to the crops.

Cleaning these irrigation channels is a very important practice to prevent weed infestation in different locations in the area. Performing the analysis of water quality, selecting appropriate filters and knowing the entire irrigation system, from planning to its maintenance, are some practices that can prevent this problem from occurring (Almeida 2009).

2.4 Quarantine and Movement of Animals and Persons

Some species of weeds, such as southern sandbur (*Cenchrus echinatus*), bristly starbur (*Acanthospermum hispidum*), and starbur (*Acanthospermum australe*) have specific structures in their seeds that can adhere to the animal's skin (zoochory) or to the clothing of people (anthropochory), and can be transported to different locations. The morphological structures of weed seeds are very well described in another chapter.

Another form of dispersal by animals is through their feces. In a study conducted by Lacey et al. (1992), it was possible to verify that 18% of the seeds of spurge (*Euphorbia esula)* ingested by goats and ewes were recovered in fecal material. Although these recovered seeds have lower germination potential, the authors recommend a quarantine of 5 days before the animals return to the field where there is no incidence of the species.

Even after passing through the digestive tract of the animals, weed seeds may present high germination and viability rates. In an experiment carried out by Viero et al. (2018), rice seeds (*Oryza sativa*) and barnyardgrass (*E. crus-galli*) were dosed in six cattle, and their feces were collected during the days. Although there was a reduction in physiological quality and the germination rate was lower than in the soil, on the third day of collection, around 90% of the initial quantity dosed for both species had already been recovered. The authors stated that the animals must undergo a minimum quarantine of 6–7 days after ingesting rice and rice grass, respectively.

In another experiment, Schaedler et al. (2021) evaluated the recovery and germination of 12,112 glyphosate-resistant Italian ryegrass seeds (*Lolium multiflorum*) after passing through the digestive system of cattle. Only 9.1% of the total dosed was recovered. However, 4 days after ingestion, the germination potential was 18%, which, for a resistant weed biotype, may represent an infestation that is difficult to control. These results demonstrated that cattle can be a weed dispersal agent, including resistant biotypes, from their feces. Therefore, organic fertilizers from animal manure should be carefully inspected and evaluated before use in planting.

2.5 *Inspection of Manure and Seedling Clods*

A measure related to the previous item, the animal manure used for crop fertilization should be inspected, due to the possibility of weed seeds and propagules in its composition. It is recommended to use sterilized manure, but if it is own production, allow the temperature of the compound to reach between 60 and 70 °C in the first 15–20 days, a very important step for its sterilization (Nunes et al. 2007). The objective of this is to reduce the viability of viable seeds and/or propagules in the final composition of manure to be used in the crop. Similarly, seedling roasting should be carefully inspected, as well as follow the recommendations mentioned in item 2.2 (Use of certified seeds and seedlings) mainly on the acquisition of certified seedlings.

2.6 *Windbreaker*

Many weed species present dispersion via anemochory. For example, horseweed seeds (*Conyza* spp.) can reach 140 m above ground level, causing a wind current of 20 m s^{-1} to transport these seeds within a radius of 500 km (Shields et al. 2006). Another example is the spread of glyphosate-resistant sourgrass (*Digitaria insularis*) in Brazil. With an initial focus on the State of Paraná, it quickly spread to other states, such as Mato Grosso do Sul, Mato Grosso, Bahia, and Pará, which, among other factors, was due to transportation via anemochory (Lopez-Ovejero et al. 2017). The objective of the use of windbreaks is to prevent the entry of these seeds into the crop, as well as pathogens and insects, creating a physical barrier that protects the culture of these enemies, as exemplified in Fig. 2.

3 Cultural Management

Cultural management consists of a set of practices that ensures rapid emergence, in addition to good growth and development of crops. Thus, cultural management consists in the use of common practices, such as excellent soil preparation or desiccation of vegetation in the case of no-tillage, seeds of high genetic, physiological and sanitary quality, good water and soil management, such as crop rotation, adequate spacing to cultivar used, use of green cover, among others. These practices contribute to reduce the seed bank of weeds in the soil. Thus, the ecophysiological characteristics of crops and weeds are used, in order to benefit the establishment and development of crops.

Fig. 2 Use of windbreak in vine. (Picture: Magna Soelma B. Moura. Source: Teixeira et al. (2013))

Measures aimed at carrying out cultural management are recommended to promote high crop productivity, which will therefore make it more competitive against weeds. Thus, it is up to the producers to know the interactions and needs that the plant requires in its production cycle, as well as to implement, according to its conditions, other practices that will help in the success of management. Fertilization and liming can be considered cultural management practices, which will depend on each cultivated crop and soil analysis of the area. Other measures used are mentioned below:

3.1 Crop Rotation and Succession

Each agricultural crop is usually infested by weeds that have similar crop requirements or have the same growth habits. Examples include: barnyardgrass (*Echinochloa* spp.) in rice crops; true yellow calico plant (*Alternanthera tenella*) in corn crops; mustard (*Brassica juncea*) in wheat crops; and slender amaranth (*Amaranthus deflexus*) in sugarcane. Some species are difficult to chemically control in certain crops, such as nightshade (*Solanum americanum*) in tomato crop (Lima and Mendes 2021) and shoo-fly plant (*Nicandra physaloides*) in potato culture (Kehl et al. 2021). In this context, crop rotation and succession are important practices in weed control, because they start from the principle of not repeating the crop planted in the previous cycle in the next cycle, promoting the planting of another species with different habits. Rotation consists of the alternating use of different plant species in the same area, over time, in a planned way, avoiding using the same species in the same season next year (Gonçalves et al. 2007). For example, planting soybeans in the summer of an agricultural year and wheat in winter; and planting corn in the summer of the other year and oats in winter. Succession is the repetitive cultivation of species in the same area over time. For example, formation of pasture for fodder in the summer of an agricultural year and oats in winter, repeating next year the same crops.

Rotation and succession crop break the weed life cycle, preventing their dominance in the area. When the same cultural techniques are applied then, year after year, in the same soil, the interference of these weeds increases greatly. When the main objective is weed control, the choice of crop in rotation and succession should fall on plants with growth habit and very contrasting cultural characteristics.

In a study conducted with corn-brachiaria consortium, higher rates of palisade grass (*Urochloa brizantha*) promoted the reduction in the weed population (Castagnara et al. 2011). In a corn and pumpkin intercropping, there was a reduction in the biomass of redroot amaranth (*Amaranthus retroflexus*) and bindweed (*Convolvulus arvensis*) compared to single corn (Fujiyoshi et al. 2007). These results demonstrate that the crop consortium promotes weed suppression without compromising crop productivity, provided that the species are chosen and managed correctly for each region.

3.2 Spacing Variation

The variation of the spacing between lines and/or plants in the line can contribute to the reduction of weed interference on the crop, depending on the architecture of cultivated plants and weed species. The reduction of spacing generally provides competitive advantage to most crops over shading-sensitive weeds. In this case, with the reduction of the spacing between rows, as long as it does not exceed the maximum limit, there is an increase in the interception of light by the canopy of cultivated plants. This effect is dependent on factors such as the type of species to be cultivated, morphophysiological characteristics of genotypes, weed species present in the area, time, and climatic conditions at the time of its emergence, in addition to environmental conditions. As an example, cotton is planted in a densely (48 cm) shape allows for faster canopy closure, as it promotes greater accumulation of dry matter than conventional planting (96 cm) (Rosolem et al. 2012). This makes cotton cultivation more competitive in relation to weeds and promotes cultural control of the weeds.

3.3 Planting Density

Related to the previous topic, the decrease in plant spacing increases planting density, being defined as the number of plants per unit of area. In addition to providing greater productivity gain, it promotes greater weed suppression by providing greater light interception by the crop canopy, reducing the amount that reaches the soil, mainly inhibiting positive photoblastic weeds, such as hairy fleabane (*Conyza* spp.) and arrowleaf sida (*Sida rhombifolia*). It will also depend on the cultivated species, morphophysiological characteristics of genotypes, weed species present in the area and time and climatic conditions at the time of its emergence, in addition to environmental conditions. Kolb et al. (2012) found higher suppression of weeds with higher planting density in wheat crops, in which the density of 600 plants m^{-2} reduced by 30% the infestation of weeds compared with the density of 400 plants m^{-2}. In another study, soybean planting density of 250,000 plants ha^{-1} caused a reduction of almost 50% of the dry matter of Palmer amaranth (*A. palmeri*) compared to the density of 125,000 plants ha^{-1}, in addition to reducing the production of weed seeds (Korres et al. 2020) (Fig. 3).

3.4 Green Cover

Green cover crops are generally very competitive with weeds. Lupin, vetch, annual ryegrass, turnip, oats, and rye are used in southern Brazil. There is the possibility of consorting the species used in the green cover, such as crotalaria consorted with millet. It is also recommended as autumnal management (off-season), preventing the soil from being discovered and thus infestation by weeds, generating control cost at the beginning of the next harvest. The main effect is the reduction of the seed bank combined with the improvement of the soil's physical-chemical and biological conditions; however, these plants may possess also inhibiting power over others. In a study conducted in Sete Lagoas, Minas Gerais, Brazil, among some species with potential for use as green cover, velvet bean showed higher potential for weed suppression, possibly due to competitive pressure and possible allelopathic effects (Favero et al. 2001).

In addition, they can reduce the infestations of some weeds after desiccation or mowing and be incorporated into the soil, and should be well chosen for each case. The presence of dead cover (physical control that will be addressed in a specific topic) creates conditions for the installation of a dense and diversified microbiota in the soil, especially in the surface layer with a high amount of microorganisms responsible for the elimination of dormant seeds through deterioration and loss of viability, in addition to decreasing the thermal amplitude and incidence of diseases (Linhares et al. 2018). In a study conducted in the State of São Paulo, Brazil, the sunn hemp, used as a green cover, promoted the suppression of different weed species, except for purple nutsedge (*Cyperus rotundus*) (Gomes et al. 2014). In the same study, sorghum promoted control of tiririca when incorporated into the soil, evidenced that the species used as green cover and the management of its phytomass and the weeds present will influence the efficiency of weed control. The use of green rye cover increased the period prior to interference in cotton crop, compared to the absence of cover (Korres and Norsworthy 2015). In the year of the implementation of the experiment, the control should be performed 4 weeks after planting

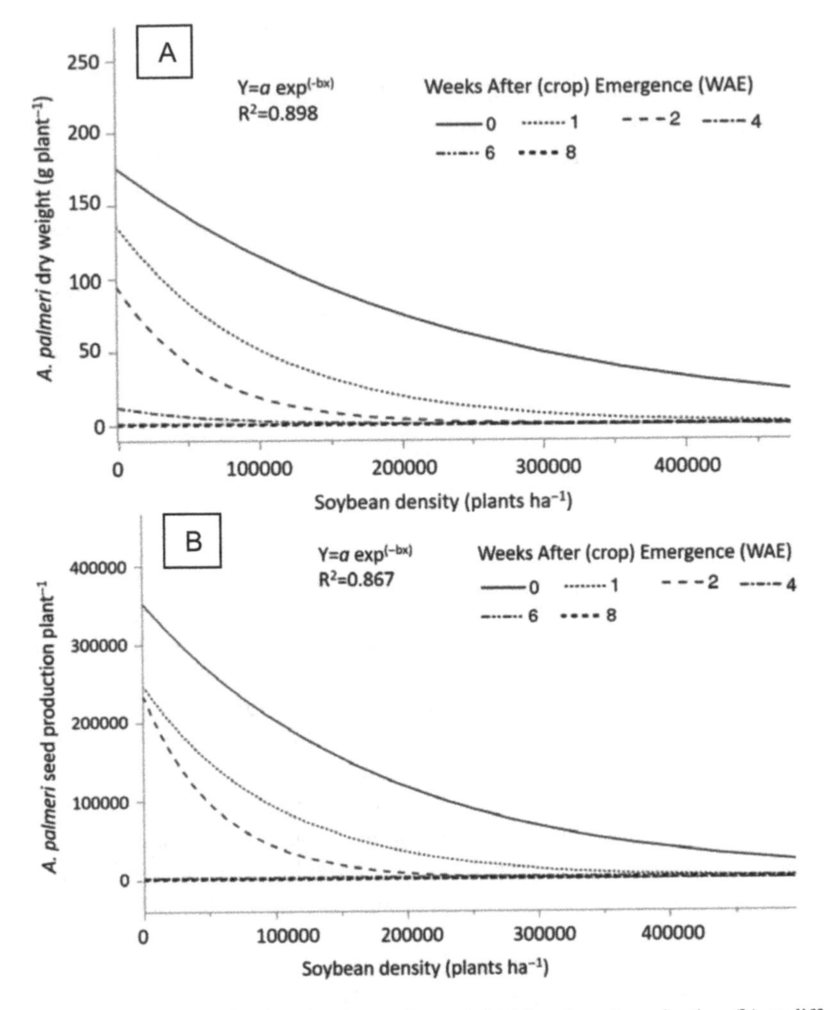

Fig. 3 Effect of soybean planting density on dry weight (**a**) and seed production (**b**) at different times of *Amaranthus palmeri* infestation. (Source: Adapted from Korres et al. (2020))

and, after the establishment of rye, the control was not necessary until 7 weeks after planting, reducing its cost.

Figure 4 demonstrates different managements in the coffee line, using green cover with Congo grass (*Urochloa ruziziensis)* and manual weeding, in which the soil was kept in the clean. The green cover decreased the density and dry matter of weeds after 2 years of the implementation of the experiment compared to manual weeding, because of the formation of straw, which will promote physical control and suppress weed development (Zaidan 2020).

Fig. 4 Different cover managements in the coffee line. (**a**) green cover with *Urochloa ruziziensis* and (**b**) manual weeding, in which the soil was kept in the "clean". (Source: Zaidan (2020))

3.5 Crop Consortium

The crop consortium is the simultaneous cultivation of two or more crops in the same production site, sown at the same time or with most of their cycles sharing the same resources (Hernani et al. 2012). In addition to providing increases in yield, it assists in the maintenance of the physicochemical and biological characteristics of the soil and in the suppression of weeds, by optimizing the space available for cultivation, allowing cultivated plants to cover the soil, avoiding the arrival of solar radiation in the lower layers, besides providing live and dead cover, important in moisture retention and nutrient cycling (Hernani et al. 2012). As a widely used example, there is the intercropping between corn and beans, in which corn, C4-cycle grass, presents rapid development, requiring greater water, light and nutrient swell; and beans, C3-cycle legumes, less demanding in light, promote biological nitrogen fixation, assisting in corn fertilization. The yield gains for these intercropped crops are higher compared to their monoculture (Pereira Filho et al. 1997; Saldanha et al. 2017). An example of a consortium is *EMBRAPA's Santa-Fé* system, which consists of intercropping between grain crops (corn, sorghum, millet, and soybean) with forage species (pincipalmente of the genus *Urochloa*), in which there is grain production in the harvest and forage production in the off-season. One of the advantages of this system is the suppression of weeds after the formation of forage straw, promoting an efficient and beneficial physical control for the annual crop that will be planted in the next cycle (Kluthcouski et al. 2000).

3.6 Crop-Livestock-Forest Integration

It is a specific production strategy for the agrossilvipastoril system, using crop rotation and intercropping, with the simultaneous presence of the livestock and agricultural component (crop–livestock), the livestock component with the forestry (forest–livestock), the agricultural component with the forestry component (crop–forest) or all these components present (agricultural and livestock components in rotation and intercropping with the forest component), constituting Crop-Livestock-Forest Integration (Balbino et al. 2011; Kichel et al. 2014; Vinholis et al. 2020). This system promotes the same management benefits seen in the topics on crop rotation and intercropped cultures, besides allowing less infestation of toxic weeds to animals, such as flower-of-souls (*Senecio brasiliensis)* and fireweed (*Senecio madagascariensis)*, causing losses between 1.13 and 1.58 million head of cattle annually in Brazil (Brighenti et al. 2017). The crop–livestock promoted the reduction of the seed bank of creeping marmeladegrass (*Urochloa plantaginea*) and dayflower (*Commelina benghalensis*), as well as the viability of these seeds, in a study carried out in the State of Paraná, Brazil (Voll et al. 1995, 1997). A similar result was observed by Ikeda et al. (2007), verifying reduction in the weed seed bank, in grazing succession pasture—tillage—pasture.

3.7 Season and Region of Planting

The planting season will interfere in the development of the plants, as each species will be influenced by the rainfall regime, photoperiodism, and temperature of the region. Consequently, it will interfere with your ability to compete with weeds. In addition, the regions have different climatic conditions, besides having different rainfall regimes and times of drought, causing certain cultures adapted in one region to tend not to adapt in others. Thus, knowing the needs and characteristics of crops and weeds, as well as the climatic conditions of the region where planting will be done is essential to define which crop to use, at which time of year sowing will be done, which stage of development will require more water and light (if irrigation system will be required), probable weed species that will be found at certain times of the year and how to promote control, among others. In the State of Paraná, the planting of corn (second harvest) is done in February and March, and the sooner planting is done, the corn will present better grain development and yield, due to higher accumulations of degree-days, while delays in planting promote the increase of the crop cycle, not being interesting for this system (Shioga and Gerage 2010). Therefore, a higher growth rate can promote weed suppression with canopy closure.

3.8 Use of Suitable Cultivars

With the advent of breeding programs, there was the emergence of cultivars of cultivated species recommended for the different environmental conditions found. These cultivars began to be created according to the purpose, either to adapt to different regions, to be resistant to diseases, pests, insects and/or herbicides. Therefore, the choice of suitable cultivar became a criterion to be taken into account at the time of planting planning. The selection of cultivars that will present better development and competitive capacity against weeds is fundamental to succeed in production. As an example, the different soybean cultivars present in Brazil are used. The cultivar BRS110IPRO is recommended for the states of São Paulo, Paraná, and Santa Catarina, while BRS7270IPRO is recommended for the states of Minas Gerais, Goiás, and Mato Grosso do Sul (EMBRAPA 2021), among many other existing cultivars, with different planting times, water needs, and light hours in its cycle.

3.9 Drip Irrigation System

Irrigation aims to provide water for crops in periods of drought, protected crops or production in small areas. Among the various types of irrigation, drip irrigation aims to provide localized water for the crop, avoiding waste, reducing costs and making available as little as possible for weeds. Therefore, a good drip irrigation system, well planned, with the channels always clean, supplying the need for the crop, will allow the cultivated plants to present good development, making them more competitive against weeds, as it will limit the availability of water for them. Drip irrigation is very important in crops with slow initial growth, for example, vegetables such as zucchini, eggplant, onion, and cabbage (Barcellos Júnior 2021; Moraes 2021; Mota and Mendes 2021; Pontes Junior et al. 2021). Moreover, water restriction reduces the biomass of the aerial part of weeds, as reported by Lima et al. (2016) for the species Indian waltheria (*Waltheria indica*) and rattlebox (*Crotalaria retusa*).

3.10 Intercalate Culture

The intercrop pouter is used temporarily (usually annual species) in perennial crops or during the fallow between one crop and another. It has advantages similar to those observed in crop rotation and intercropping, being able to add economic value, maintain or improve the physical-chemical properties of the soil, protect the soil by providing cover in the off-season, suppressing weed infestation, reducing the amount of solar radiation that arrives in the most superficial layers and decreasing competition for water, light, and nutrients. In a study conducted in Vitória da

Conquista, BA, the coffee consortium with silky oak (*Grevillea robusta*) promoted lower density and relative frequency of weed species compared to coffee monoculture (Silva et al. 2006). In another study conducted with coffee, the consortium of this with macadamia nut (*Macadamia integrifolia*) promoted a reduction of 82% in the incidence of weeds compared to coffee monoculture (Silva et al. 2013).

4 Physical Control

Physical control consists of the use of physical barriers that inhibit weed germination or development. Examples of physical control are: mulching (coverage), controlled fire, flooding, and electricity.

4.1 Mulching

Technique consisting of soil cover, creating a barrier that almost completely prevents the passage of solar radiation. Mulching reduces the incidence of weeds and allows the crop to settle without competition for light, water, and nutrients, favoring growth and subsequent production. However, as a disadvantage, it is not recommended in areas infested with species that passes through this barrier, such as the morning glory (*Ipomoea* spp.) and flannel weed (*Sida* spp.), which, because they have large seeds, can pierce the plastic if it does not have the proper thickness (Cavalieri 2012). It is an efficient, low-cost method and when handled properly does not harm the environment. As an example, that can be used in mulching, we have dead cover, plastic cover, and biodegradable cover (Fig. 5).

4.1.1 Dead Cover

It is a technique that consists of the use of plant residues in soil cover. In no-tillage, soil cover with plant residues from the previous crop is of great use. This planting system is used in extensive areas of soybean, corn, and wheat. The straw layer formed by desiccation inhibits the germination or development of low-light-sensitive weeds, in addition to promoting changes in soil temperature that hinder the breaking of dormancy of weed seeds (Martins et al. 2016). Dead cover can still present useful allelopathic effects in the control of certain weed species, in addition to other important effects on crops implanted in the area. The use of ruziziensis grass, tanzania grass, and marandu grass straws provided higher bean grain yield and dry matter reduction of weeds (Sousa et al. 2020). Moreover, it protects the soil against erosion, allows the maintenance of good moisture content, as seen by Freitas et al. (2019) in cowpea in a no-tillage system, and the decomposition of straw can

Fig. 5 Weed control in chives with black, white polyethylene films and recycled paper in UFV, Viçosa, Minas Gerais, Brazil. (Photo: Sara Rafaela Salazar Matias (2019). Source: Mendes (2021))

promote improvement in the physical, chemical, and biological characteristics of the soil.

4.1.2 Plastic Cover

It is a technique that consists in the use of polyethylene plastic film as soil cover, promoting the effect known as solarization. This should be done 60–75 days before planting, in the hottest months of the year, using polyethylene film on the soil surface. The plastic cover causes the increase in temperature and thermal amplitude. In a study carried out in the pepper culture, Coelho et al. (2013) verified a temperature variation of 28–40 °C of soil in conventional planting. In moist soil, weed seeds germinate and then die due to excessively high temperature up to 5 cm deep, which can limit their efficiency. Another advantage observed is the efficiency of water use. In melon cultivation, the use of polyethylene plastic film reduced water consumption by 23% in conventional planting and 21% in no-tillage (Teófilo et al. 2012). As a disadvantage, it is an unfeasible process in large areas and, although efficient in weed control, can promote environmental contamination, since the material used can take many years to be degraded.

4.1.3 Biodegradable Cover

With the disadvantage of polyethylene plastic in relation to its degradability, materials have been created that promote soil cover and are not aggressive to the environment. In the control of weeds in tomato, it was verified that the use of kraft paper and biodegradable plastic were efficient, emerging as possible substitutes for polyethylene plastic film (Anzalone et al. 2010). The use of cotton scraps, through pressing and heating process, produced a biodegradable film with weed control efficiency similar to black polyethylene plastic film, in watermelon and melon plantation in the USA (Johnson et al. 2014), emerging as a potential substitute. The use of biodegradable coverage is already a reality, in which the environmental component is favored without affecting agronomic performance. Research with new materials is already being carried out in order to verify the best relationship between weed control efficiency and greater degradability of the material used (Malinconico et al. 2008).

4.2 Controlled Fire

The use of fire is a very old technique in agriculture. From the beginning it was mainly used to clean the area, preparing a new planting. Even today it is used inappropriately, such as in pastures in the Cerrado, mainly for promoting uncontrollable fires, causing the loss of native fauna and flora, in addition to eliminating microorganisms from the soil. Controlled fire was also widely used in sugarcane plantations to facilitate pre-harvest operations. However, there are new technologies that allow the use of this resource in a controlled manner and reducing the possibility of causing environmental damage.

The use of controlled fire is a technique still little widespread in Brazil, although there are already equipment ranging from manual applicators to equipment coupled to tractors, with complex operating system, many of them used in the USA and Europe (Silva et al. 2018a). Fire causes irreversible physiological effects, such as protein denaturation, water loss, and enzyme inactivation (Silva et al. 2018b). Favarato et al. (2016), using flamethrowers, found that 10 s of exposure was the most efficient time in weed control. In aquatic environments, weed control using high temperatures—called thermal control—is an alternative management integration aimed at complementing mechanical control, especially in water reservoirs. Marchi et al. (2005) obtained efficient control of aquatic species, common water hyacinth (*Eichhornia crassipes*), small flowered (*Urochloa subquadripara*), water lettuce (*Pistia stratiotes*), and eared salvinia (*Salvinia auriculata*). The use of the flamethrower depends on several factors such as temperature, exposure time, and energy consumption. The advantages of the use of controlled fire are the non-selective control of weeds, besides the fact that it does not leave chemical residues in the environment and control a wide spectrum of weeds, including those that have resistance to herbicides. As disadvantages, the high cost of fossil fuel for its operation and has no residual effect, requiring more applications when there is reinfestation.

4.3 Flood

In flat and level soils, flooding is an effective weed control method, as in rice trays. Perennial species that are difficult to control, such as purple nut sedge (*Cyperus rotundus*), bermudagrass (*Cynodon dactylon*), kikuyugrass (*Pennisetum clandestinum*), and many annual species, are eradicated under prolonged flooding. The flood has no effect on weeds that develop in soggy soils, such as barnyard grass (*Echinochloa* spp.) and California arrowhead (*Sagittaria montevidensis*). This practice causes the death of sensitive plants, due to the suspension of oxygen supply to their roots. The limiting factors of this method, in most cases, are the cost of soil leveling and the large amount of water needed for its implantation, however, it is an important weed control method used in irrigated rice.

4.4 Electricity

Like the use of controlled fire, the use of electricity is still a technology little widespread in Brazil. This fact is due to the high cost of the equipment, the high energy demand, the operational performance, and the size of the specific equipment (Becker et al. 2020). After application, the electrical discharge is conducted throughout the plant due to the high water content present, causing increased temperature, disruption of cell walls, and protein denaturation (Becker et al. 2020). Research is being carried out to make this technology accessible to agriculture, because it does not leave chemical residues in the soil, it is a non-selective method and immediate action in weeds, without residual effect (Brighenti et al. 2018). As a disadvantage, it can promote the death of the soil microbiota, besides not being recommended when the plants are dry and at the hottest times of the day, as it can cause outbreaks of fire (Costa et al. 2018). One of these equipments is Electroherb™ (Zasso Latam), which is being used in weed control in urban areas, has potential for the management of desiccation of cover crops and in the directed control of weeds between crop lines (Costa et al. 2018) (Fig. 6).

5 Mechanical Control

Mechanical control consists of the use of tools, equipment and/or implements in weed control. The practices of mechanical control promote total arranquio or cutting of the aerial part of the plant, preventing it from developing. As a disadvantage, it is precisely in the fact that it is possible to regrowth of the remaining structures of the plant in the soil, besides the possibility of retituding the vegetative parts of plants that develop by this mechanism, further increasing their infestation. The

Fig. 6 Use of Electroherb in oat crop desiccation and weed control in agroecological areas. (Source: Adapted from Costa et al. (2018))

mechanical methods of weed control are manual start-up, manual weeding, mowing, and mechanized cultivation.

5.1 *Hand Pluck*

It is the oldest and simplest method of weed control. Even today it is used for control in home gardens, gardens, and weeding among plants of online crops, when the main control method is the use of hoe. Also used on a commercial scale, when other control methods are unfeasible, such as the removal of purple nutsedge tubers (*Cyperus rotundus*) in commercial gardens (Fig. 7).

5.2 *Manual Weeding*

Technique made with hoe, being very effective and still widely used in Brazilian agriculture, especially in mountainous regions, where there is subsistence agriculture, and for many families, this is the only source of work. However, in more intensive agriculture, in larger areas, the high cost of labor and the difficulty of finding workers at the necessary time and in the desired amount make this method only complementary to other methods, and should be carried out when weeds are still young and the soil is not too humid. In the management of the tiririca, the revolving caused by the hoe can aggravate the problem by bringing the soil surface the dormant tubers present in the layers below the surface.

Fig. 7 Manual start-up of purple nutsedge tubers (*Cyperus rotundus*) in a commercial vegetable garden in Rio Paranaíba, Minas Gerais, Brazil. (Source: Marcelo Rodrigues dos Reis)

5.3 Mowing (Manual or Mechanical)

In crops with greater space between lines, such as orchards and coffee trees, manual or mechanical mowing is a very important method to control weeds, especially in declivoso terrain, where erosion control is fundamental. The space between the rows of the crops is kept mowed and, by means of other control methods, the row of plants, at the level, is kept clean. Also in wasteland, roadside and pasture, mowing is a method of weed control of the most important.

5.4 Mechanized Cultivation

Mechanized cultivation, made by cultivators pulled by animals or tractors, is of wide acceptance in Brazilian agriculture, being one of the main methods of weed control in properties with smaller planted areas. The main limitations of this method are: a) difficulty in controlling weeds in the crop line; and b) low efficiency: when performed in rainy conditions (wet soil), it is inefficient to control weeds that reproduce by vegetative parts. However, all annual species, when young (2–4 pairs of leaves), are easily controlled in heat conditions and dry soil. The cultivation breaks the intimate relationship that exists between root and soil, suspends water absorption and exposes the root to unfavorable environmental conditions. According to the relative size of cultivated plants and weeds, the displacement of the soil over the line, through special cultivating hoes, can cause the enterrio of seedlings and, therefore, promote the control of weeds in the line.

The use of knife roller in Rio Grande do Sul, Brazil, decreased soil tillage costs and can be used in rainy periods (Silva et al. 2012). The authors also stated that this implementation provided greater emergence of weeds, which would facilitate further control with herbicides. In an experiment carried out in the State of Paraná, the articulated brush cutter satisfactorily controlled beggarticks (*Bidens* spp.) and wild poinsettia (*Euphorbia heterophylla*) in organic soybean cultivation.

It is noteworthy that the improvement of the use of machines, using artificial intelligence to identify, select, and remove weeds present in the field is already a reality. In an experiment in cabbage planting, a camera was coupled to the tractor, promoting computer vision that detected and differentiated weeds from the crop, causing a reduction of 77% and 87% of weed reduction at 16 and 23 days after planting, respectively (Tillett et al. 2008). The control was performed with a rotating disc, activated at the time of weed detection.

Technological innovations inserted in agricultural implements (best addressed in chapter apart) grow and continue to improve, allowing the prospect of less dependence on chemical control or sufficient application of products, avoiding environmental damage and improving the life of the producer.

6 Biological Control

Biological control consists of the use of natural enemies (fungi, bacteria, viruses, insects, birds, fish, among others) capable of reducing the weed population, reducing their ability to compete. This is maintained through the population balance between the natural enemy and the host plant. Allelopathic inhibition of weeds should also be considered as biological control. This type of control will be efficient when the parasite is highly specific, that is, once eliminated the host, it should not parasitize other species. In general, the efficiency of biological control is doubtful when it is used in isolation, because it controls only one species and another is favored, which is a normal trend in field conditions.

Research with biological control of weeds involves successive steps:

- Selection of weed species to be controlled
- Selection of more efficient natural enemies
- Study and evaluation of the ecology of various natural enemies
- Determination of host specificity
- Monitoring the introduction and establishment of the biological control agent in the field
- Evaluation of effectiveness at different times of the year, in order to relate infection levels with the reduction of host population density.

There are three biological control strategies that can be implemented in a weed control system: classical or inoculative strategy, flood strategy, and aumentative strategy.

6.1 Classical or Inoculative Strategy

Applicable in conditions where the weed species is exotic to the site of incidence, geographically separated from its natural enemies. Therefore, it is necessary that this natural enemy has coevolved together with the weed species and is highly specific to the species that one wishes to control, that is, it cannot be a guest of other weeds. Therefore, it is idealized not to eradicate the species, thus allowing the natural enemy to always have conditions of survival. After performing the first stages of research described above, there is the introduction of the biological control agent in the area of incidence of the weed and its constant monitoring. The objective of the classical or inoculative strategy is to reduce the weed population to levels where they do not cause economic damage to the producer, promoting a balance between plant populations and biological control agents. This agent must self-perpetuate and have natural dispersal capacity.

Examples of classical strategy in the world can be cited: in Australia, the control of cacti or prickly pear (*Opuntia* spp.) with the larvae of the insect *Cactoblastis cactorum*; and in Hawaii, the thorn largeleaf lantana (*Lantana camara*) was controlled by the insects *Agromisa lantanae* and *Crocidosema lantanae*. A survey of fungal pathogens associated with the species beggarticks (*Bidens pilosa* and *Bidens subalternans*) found the association of ten fungi in both species (*Cercospora bidentis*, *C. maculicola*, *Entyloma bidentis*, *E. compositarum*, *E. guaraniticum*, *Neoerysiphe cumminsiana*, *Plasmopara halstedii*, *Podosphaera xanthii*, *Uromyces bidentis*, and *Uromyces bidenticola*) and three more only in *B. subalternans* (*Colletotrichum bidentis*, *Pseudocercosporella bidentis*, and *Sphaceloma bidentis*), emerging as potential natural enemies of these species in different regions of the world, through the inoculative strategy (Guatimosim et al. 2015).

Strains of the fungus *Corynespora cassiicola* caused leaf spots and severe injuries only in Brazilian peppertree (*Schinus terebinthifolius*), in an experiment involving 24 plant species. This weed is infested in several tropical and subtropical regions, such as the USA (Hawaii and Florida) and Australia. Therefore, *Corynespora cassiicola* is a potential natural enemy of *S. terebinthifolius* (Macedo et al. 2013).

6.2 Inundative Strategy

It is the periodic and punctual use of biological control agents, whenever there is infestation of the weed to be controlled. It occurs in a similar way to the application of herbicides, with formulated products, registered, having active ingredient (in this case the pathogen) and application methods. For this reason it is called bioherbicide (in other literatures, bioerbicidal or bio-herbicides), despite the possibility of the use of bioherbicides in the aumentative strategy. This bioherbicide is applied to weeds, causing the emergence of the disease caused by the pathogen and consequent plant death. Since this pathogen does not survive in the plant remains of the dead plant and has no alternative host, there will be no dissemination and residual effect in other productive cycles. Therefore, a new application will be necessary whenever specific weed infestation arises (Carvalho 2013; Silva et al. 2018a).

In the USA, the fungus *Colletotrichum gloeosporioides* can be used to control the virginia jointvetch (*Aeschynomene virginica*) in soybean and corn; the natural herbicide is registered as Collego. And in citrus orchards, to control *Morrenia odorata*, the fungus *Phytophthora palmivora*, named Devine, is used. In a study conducted in Bahia, the eucalyptus leaf extract promoted the reduction of radicella emission from the embryo of purplenut sedge (*C. rotundus*) seeds, emerging as a bioherbicide potential to control this weed (Silva et al. 2020).

In Brazil, isolates of *Fusarium graminearum* have been studied as a biological control agent of elodeas (*Egeria densa* and *Egeria najas*), aquatic plants that cause problems in hydroelectric reservoirs. The photoperiod influences the control efficiency of weed species by the fungus, and temperatures above 30 °C have provided better control of *Egeria* (Borges Neto et al. 2005).

6.3 Augmentative Strategy

Strategy used when natural enemies already exist at the infestation site. This técnicase characterizes the periodic establishment of these enemies, in order to maintain an acceptable level of control in the productive cycle in parts of the cultivated area. When bioherbicides are used, the application is made in lower intensity and frequency than in the inundative strategy. Another possibility is the use of vertebrates as biological control agents, such as fish, sheep, and birds (Carvalho 2013; Silva et al. 2018a). Some growers have used sheep to control weeds in coffee crops.

However, some species do not have good palatability and are refused during grazing. The use of tilapia, carp, and other herbivorous fish are possible in the control of aquatic plants. Miyazaki and Pitelli (2003) verified control of up to 100% of aquatic species, *Egeria densa*, *Egeria najas,* and *Ceratophyllum demersum* by small-scaled pacu (*Piaractus mesopotamicus*).

Biological control proves to be highly efficient in controlling key weeds within a crop. However, the advantage ends up becoming a disadvantage due to this high specificity, since there is a range of weed species in an area. At the same time, research related to biological control is scarce, costly and will not always present satisfactory results, in addition to Brazilian legislation, until recently, hindering the registration of products of this class. However, with the increasing selection of resistant weed biotypes, in addition to the concern for the environment and sustainability, it becomes an interesting method for not leaving chemical residues in the soil. It is efficient, then, when associated with other control methods and will be recommended for weed species of proven difficult control by mechanical, physical, and/or chemical methods.

7 Chemical Control

7.1 *Brief History*

Chemical control of weeds is done with the use of herbicides. Herbicides are biological or chemical substances capable of killing or suppressing weed growth (Roman et al. 2007). It is the most used weed control method. Research aimed at chemical control of weeds was initiated between 1897 and 1900, when Bonnet (France), Shultz (Germany), and Bolley (USA) showed action of copper leaves on some broad leaves. In 1908, ferrous sulfate was evaluated by Bolley, in the USA, for wide leaf control in wheat crop. Only in 1942, Zimmerman and Hitchock, in the USA, synthesized 2,4-D. This herbicide is the basis of many other products synthesized in the laboratory (2,4-DB; 2,4,5-T, among others) and marked the beginning of chemical control of weeds on a commercial scale. From 1950, new chemical groups of herbicides emerged: amides (1952), carbamates (1951), symmetric trialines (1956), among others (Oliveira Junior 2011).

Due to the great development of the chemical weed control area, in 1956, in the USA, the Weed Science Society of America (WSSA) was created. Currently, the objectives of research in the world in the area of Weed Science are to develop environmentally safe, low-cost technologies that ensure the sustainability of crops. Thus, a special emphasis has been placed on biotechnology in the development of herbicide tolerant crops and also on the impact of these technologies on the environment.

7.2 Advantages in the Use of Herbicides

This great acceptance of the use of herbicides by producers can be attributed to the fact that chemical control of weeds provides the following advantages:

- Less dependence on labor, which is increasingly expensive, difficult to be found at the right time, in the quantity and quality required.
- Even in rainy times, chemical control of weeds is more efficient.
- It is efficient in controlling weeds in the planting line and does not affect the root system of crops.
- Allows minimal cultivation or no-tillage of crops.
- It can control vegetative propagating weeds.
- Allows planting to launch and, or, change in spacing, when necessary.

Given the importance of chemical control in the world agricultural scenario, it is necessary to consider that every herbicide is a chemical molecule that has to be handled with care, with a danger of intoxication of the applicator, mainly. Environmental pollution can also occur—water (rivers, lakes, and underground bodies of water), soil, and food when handled incorrectly. There is a need for specialized labor for herbicide application. Knowledge of plant physiology, herbicide groups, application technology, soil physical and chemical characteristics, and climatic conditions is fundamental to the success of chemical weed control. The risks of use exist, but if the recommendation and application are accompanied by a professional trained for this purpose, they can be perfectly controlled and avoided.

Chemical weed control should be done in conjunction with other control practices. The use of chemical control alone can bring problems, such as weed resistance, as well as to the selection of resistant weed biotypes, a growing problem in recent years (HEAP 2021). Therefore, the herbicide is a very important tool in integrated weed management, as long as it is used at the appropriate time and correctly.

8 Integrated Weed Management (IWM)

The integrated way of cultivation that considers all the factors that can provide the plant with greater and better production allows the efficient use of the resources of the environment. Within this context, the IWM is also included. This integrated production system has been increasingly recommended in all agricultural sectors, having, in Brazil, its base reinforced in the field of entomology, when pioneers promoted the study of the problems of the odoper in the Northeast of the country, proposing a series of measures that fit the concept of integration (Conceição 2000). The premises that understated the proposal of integrated management can be well summarized in: quality assurance of the harvested product, including the exemption of pesticide residues in food within traceability; environmental sustainability, including non-degradation of soil and contamination of air and water; economic and social sustainability in production, maintaining or increasing productivity; and ensuring a better quality of life for the farmer with regard to economic return and greater safety in activities involving the use of pesticides.

The need to implement the IWM is because of the excessive use of herbicides as the only control method. This fact generates problems:

- Such as river pollution, soil contamination, and extinction of plant species;
- Agronomics, with the selection of resistant biotypes of weeds, due to the selection pressure caused by this method;
- Because of the high cost of controlling these resistant weeds.

Therefore, the integration of control methods becomes necessary, promoting the alteration of practices normally used, aiming at the prevention of selection of resistant biotypes (López-Ovejero and Christoffoleti 2003), in addition to environmental problems. In this sense, preventive and cultural managements play an important role, as they recommend the non-dispersion of weeds and the good establishment of crops, allowing greater competitive capacity. Control methods should be used when necessary, associated with management methods and always rotated when possible.

In a study with lentils in Canada, the association of sowing rate four times higher than recommended, with the use of rotary hoe and herbicide use did not affect grain yield and promoted high weed control rates (Redlick et al. 2017). This study was carried out with herbicides of different mechanisms of action, as a response to the emergence of weeds resistant to ALS-inhibiting herbicides, widely used in lentil cultivation. Therefore, IWM emerges as an alternative in the control of resistant weed biotypes.

As seen in the previous topics, selecting the cultivar and proper spacing provides crop benefits in competition with weeds, and, when associated with herbicide use, decreases competition and weed density. In a study carried out with two rice cultivars, in two spacings, the cultivar IET-21214, with spacing between lines of 20 cm and isolated application of bispiribac-sodium provided similar control when the application of pendimenthalin and bispiribac-sodium sequentials was performed, providing savings in the cost of control, besides avoiding possible environmental contamination (Mahajan et al. 2014).

An important aspect to note is that the IWM does not necessarily need the herbicide component in control planning. In a study carried out in Manitoba, Canada, Geddes and Gulden (2018) evaluated different spacings between lines, planting densities and mechanical control in the mainline, aiming to control voluntary canola resistant to glyphosate. The authors verified that using line spacing of 19 cm, associated with planting density of 682,500 plants ha^{-1}, there was the suppression of the development of the voluntary canola. In larger row spacing (76 cm), associated with planting density of 455 plants ha^{-1} and control between rows with two cultivator passages, they obtained the same control efficiency. In both cases soybean yield was similar. In an organic no-tillage system in the USA, in succession corn-soybean, the use of green (vetch and triticale) cover crops between rows, in the spacings of 19 and 38 cm, reduced 58%, 23%, and 62% of weed biomass in Delaware, Maryland, and Pennsylvania, respectively (Wallace et al. 2017). Both vetch and triticale were cut with cultivator, producing the straw that promoted physical control.

In addition to avoiding the possibility of selecting resistant biotypes, reducing the use of herbicides to the adoption of IWM can generate a reduction in the cost of control. In a study conducted in Pennsylvania, USA, two production systems were evaluated, being a conventional system with herbicide application in total area and a system with herbicide application in the planting and cultivation lines between rows with rye as a cover species, in corn crops (Snyder et al. 2016). The authors verified that the system with herbicide application in the planting and cultivation lines between rows with rye promoted a reduction of \$33.65 dollars ha^{-1} in the cost of control, probably due to the decrease in the amount of herbicide applied.

In IWM, it is more understandable when weeds are treated not as a direct target that should be "exterminated," but as an integral part of an ecosystem in which nutrient cycling in the soil is directly involved, among other functions. They still form complex interactions with microorganisms, and through these associations guarantee the agronomic characteristics that give the soil environment greater capacity to support a sustainable cultivation. With the exception of a few species that need to be eradicated from the area (mainly herbicide-resistant biotypes), much of the infesting plant community commands nutrient dynamics in the soil, besides being a key component in the process of formation and burning of organic matter, mainly because of the role that the rhizosphere has in stimulating microbial activity. Therefore, the IWM should be a mandatory component at the time of planning a plantation, in order to preserve the environment, reduce dependence on herbicides and avoid the selection of resistant weed biotypes.

9 Concluding Remarks

Knowledge of management and control methods is fundamental in today's agriculture. However, it is also important to know the various impacts that such methods can bring to the environment. Therefore, technical care is needed to achieve maximum efficiency with minimal negative impact on soil, water, and non-target organisms, seeking the lowest dependence of chemical control and its integration with other methods. Therefore, it is necessary to associate the various methods available (preventive, cultural, mechanical, physical, biological, and chemical), taking into account the species of weed, the type of soil, the topography of the area, the equipment available on the property, the environmental conditions and the cultural level of the owner. In this regard, it should be noted that in the Integrated Management of Weeds (IWM), the herbicide is considered only one more tool in obtaining control that is efficient and economical, preserving the quality of the harvested product, the environment and human health.

Weed control, as it is being implemented in most of the national territory, has been a predatory activity with regard to the sustainability of the system. With new technologies, if the IWM is not adopted urgently, this fact tends to worsen, because in tropical and subtropical regions soil degradation is more intense, due to climatic conditions favorable to the various types of erosion, as well as increasingly the selection of resistant weed biotypes. This fact, together with the idea of eliminating almost all weeds to the point of leaving the uncovered soil, will have negative consequences for Brazilian agriculture, being virtually impossible to recover in the conventional management systems adopted.

Another fundamental aspect is that the capacity of the infesting species, in relation to culture, to compete for water, light, and nutrients, which are the factors responsible for reducing productivity, is known. Furthermore, one cannot disregard the ability of certain weed species to hinder or prevent harvesting, reduce the quality of the harvested product, and host pests and vectors of diseases and natural enemies. Finally, it is necessary to know the types of relationships between cultivated plants and weeds that allow their passive coexistence. In this sense, it is also a determining factor in the implementation of the IWM to know the density and distribution of weeds in the area, as well as the time of emergence of them in relation to the crop.

References

Almeida OA (2009) Entupimento de emissores em irrigação localizada. EMBRAPA Mandioca e Fruticultura Tropical, Cruz das Almas, p 61

Anzalone A, Cirujeda A, Aibar J et al (2010) Effect of biodegradable mulch materials on weed control in processing tomatoes. Weed Technol 24(3):369–377

Balbino LC, Cordeiro LAM, Porfirio-da-Silva V et al (2011) Evolução tecnológica e arranjos produtivos de sistemas de integração lavoura-pecuária-floresta no Brasil. Pesqui Agropecu Bras 46(10):1–12

Barcellos Júnior LH (2021) Repolho. In: Mendes KF (ed) Atualidades no Manejo de Plantas Daninhas em Hortaliças Herbáceas. Brazil Publishing, Curitiba, pp 131–153

Becker RS, Alonço AS, Francetto TR et al (2020) Inovações tecnológicas em máquinas agrícolas para controle de plantas daninhas. Tecno-Lógica 15(2):98–108

Borges Neto CR, Gorgati CQ, Pitelli RA (2005) Influência do fotoperíodo e da temperatura na intensidade de doença causada por *Fusarium graminearum* em *Egeria densa* e *E. najas*. Planta Daninha 23:449–456

Brighenti A, Lamego F, Miranda JEC et al (2017) Plantas Tóxicas em Pastagens: (*Senecio brasiliensis* e *S. madagascariensis*) - Família: Asteraceae. EMBRAPA Gado de Leite, Juiz de Fora, p 11

Brighenti AM, Oliveira MF, Coutinho Fiho SA (2018) Controle de plantas daninhas por roçada articulada e eletrocussão. In: Oliveira MF, Brighenti AM (eds) Controle de plantas daninhas: métodos físico, mecânico, cultural, biológico e alelopatia. EMBRAPA Milho e Sorgo, Sete Lagoas, pp 34–51

Carvalho LB (2013) Plantas Daninhas. Editado pelo autor, Lages, p 92

Castagnara DD, Zoz T, Berté LN et al (2011) Taxa de semeadura de *Brachiaria brizantha* consorciada com milho na incidência de plantas daninhas. Revista Brasileira de Ciências Agrárias 6(3):440–446

Cavalieri SD (2012) Árvore do conhecimento: Pimenta. https://www.agencia.cnptia.embrapa.br/gestor/pimenta/arvore/CONT000gn0jdxdz02wx5ok0liq1mq3v5rvls.html. Accessed 26 July 2021

Coelho MEH, Freitas FD, Cunha JLXL et al (2013) Coberturas do solo sobre a amplitude térmica e a produtividade de pimentão. Planta Daninha 31:369–378

Conceição MZ (2000) Manejo integrado em defesa vegetal. In: Zambolim L (ed) Manejo Integrado: Doenças, pragas e plantas daninhas. Viçosa, UFV, pp 1–80

Costa NV, Rodrigues-Costa ACP, Coelho ÉMP et al (2018) Métodos de controle de plantas daninhas em sistemas orgânicos: breve revisão. Rev Bras Herb 17(1):25–44

EMBRAPA - Empresa Brasileira de Pesquisa Agropecuária. Portifólio Embrapa de cultivares de soja: Sistema intacta. https://www.embrapa.br/documents/1355202/1529289/Portf%C3%B31io+Sistema+Intacta/60ec412e-b9c9-4d07-8fab-a5b3a102b58a?version=1.0#:~:text=Para%20usar%20adequadamente%20a%20tecnologia,de%20mercado%20convencional%20e%20RR1.&text=a%20tecnologia%20intacta%20RR2%20PRo%E2%84%A2%20tem%20o%20objetivo%20de,e%20ambientais%20para%20a%20agricultura. Accessed em: 27 Maio 2021

Favarato LF, Souza JL, Guarconi RC et al (2016) Flamethrower application time in weed control. Planta Daninha 34(2):327–332

Favero C, Jucksch I, Alvarenga RC et al (2001) Modificações na população de plantas espontâneas na presença de adubos verdes. Pesqui Agropecu Bras 36:1355–1362

Freitas RM, Dombroski JL, Freitas FCL et al (2019) Water use of cowpea under deficit irrigation and cultivation systems in semi-arid region. Rev Bras Eng Agríc Ambient 23:271–276

Fujiyoshi PT, Gliessman SR, Langenheim JH (2007) Factors in the suppression of weeds by squash interplanted in corn. Weed Biol Manag 7(2):105–114

Gazziero DLP, Silva AF (2017) Caracterização e manejo de *Amaranthus palmeri*. EMBRAPA Soja, Londrina

Geddes CM, Gulden RH (2018) Candidate tools for integrated weed management in soybean at the northern frontier of production. Weed Sci 66(5):662–672

Gomes DS, Bevilaqua NC, Silva FB et al (2014) Supressão de plantas espontâneas pelo uso de cobertura vegetal de crotalária e sorgo. Rev Bras Agroecol 9(2):206–213

Gonçalves Netto A, Nicolai M, Carvalho SJP et al (2016) Multiple resistance of *Amaranthus palmeri* to ALS and EPSPS inhibiting herbicides in the state of Mato Grosso, Brazil. Planta Daninha 34(3):581–587

Gonçalves SL, Gaudencio CA, Franchini JC et al (2007) Rotação de culturas. EMBRAPA Soja, Londrina, p 10

Guatimosim E, Pinto HJ, Pereira OL et al (2015) Pathogenic mycobiota of the weeds *Bidens pilosa* and *Bidens subalternans*. Trop Plant Pathol 40(5):298–317

Heap I The international herbicide-resistant weed database. http://www.weedscience.org/Home.aspx. Accessed 29 Maio 2021

Hernani LC, Souza LCF, Ceccon G Árvore do conhecimento: Sistema Plantio Direto. Agência Embrapa de Informação Tecnológica. http://www.agencia.cnptia.embrapa.br/gestor/sistema_plantio_direto/arvore/CONT000fx4zsnby02wyiv80u5vcsvyqcqraq.html. Accessed 26 Maio 2012

Ikeda FS, Mitja D, Vilela L et al (2007) Banco de sementes no solo em sistemas de cultivo lavoura-pastagem. Pesqui Agropecu Bras 42:1545–1551

Johnson WC, Ray JN, Davis JW (2014) Rolled cotton mulch as an alternative mulching material for transplanted cucurbit crops. Weed Technol 28(1):272–280

Kehl LGH, Mendes KF, Reis MR (2021) Batata. In: Mendes KF (Org.). Atualidades no Manejo de Plantas Daninhas em Hortaliças Fruto. Brazil Publishing, Curitiba, pp 43–91

Kichel AN, Costa JAA, Almeida RG et al (2014) Sistemas de Integração Lavoura-Pecuária-Floresta (ILPF): experiências no Brasil. Boletim de Indústria Animal 71(1):94–105

Kluthcouski J, Cobucci T, Aidar H et al (2000) Sistema Santa Fé - Tecnologia Embrapa: integração lavoura-pecuária pelo consórcio de culturas anuais com forrageiras, em áreas de lavoura, nos sistemas direto e convencional. EMBRAPA Arroz e Feijão, Santo Antônio de Goiás, p 28

Kolb LN, Gallandt ER, Mallory EB (2012) Impact of spring wheat planting density, row spacing, and mechanical weed control on yield, grain protein, and economic return in Maine. Weed Sci 60(2):244–253

Korres NE, Norsworthy JK (2015) Influence of a rye cover crop on the critical period for weed control in cotton. Weed Sci 63(1):346–352

Korres NE, Norsworthy JK, Mauromoustakos A et al (2020) Soybean density and Palmer amaranth (*Amaranthus palmeri*) establishment time: effects on weed biology, crop yield, and economic returns. Weed Sci 68(5):467–475

Lacey JR, Wallander R, Olson-Rutz K (1992) Recovery, germinability, and viability of leafy spurge (*Euphorbia esula*) seeds ingested by sheep and goats. Weed Technol 6(3):599–602

Lima AC, Mendes KF (2021) Tomate. In: Mendes KF (ed) Atualidades no Manejo de Plantas Daninhas em Hortaliças Fruto. Brazil Publishing, Curitiba, pp 255–287

Lima MFP, Dombroski JLD, Freitas FCL et al (2016) Weed growth and dry matter partition under water restriction. Planta Daninha 34:701–708

Linhares CMDS, Freitas FCL, Ambrósio MMQ et al (2018) Efeito de coberturas do solo sobre a podridão cinzenta do caule em *Vigna unguiculata*. Summa Phytopathol 44(2):148–155

López-Ovejero RF, Christoffoleti PJ (2003) Recomendações para prevenção e manejo da resistência a herbicidas. In: Christoffoleti PJ (ed) Aspectos de resistência de plantas daninhas a herbicidas. Londrina, Associação Brasileira de Ação a resistência de Plantas aos herbicidas (HRAC-BR), pp 45–79

Macedo DM, Pereira OL, Wheeler GS et al (2013) *Corynespora cassiicola* f. sp. *schinii*, a potential biocontrol agent for the weed *Schinus terebinthifolius* in the United States. Plant Dis 97(4):496–500

Mahajan G, Poonia V, Chauhan BS (2014) Integrated weed management using planting pattern, cultivar, and herbicide in dry-seeded rice in Northwest India. Weed Sci 62(2):350–359

Malinconico M, Immirzi B, Santagata G et al (2008) An overview on innovative biodegradable materials for agricultural applications. In: Moeller HW (ed) Progress in polymer degradation and stability research. Nova Science, New York, pp 69–114

Marchi SR, Velini ED, Negrisoli E et al (2005) Utilização de chama para controle de plantas daninhas emersas em ambiente aquático. Planta Daninha 23(2):311–319

Martins D, Gonçalves CG, Silva Junior ACD (2016) Coberturas mortas de inverno e controle químico sobre plantas daninhas na cultura do milho. Rev Cienc Agron 47(4):649–657

Mendes KF (2021) Cebolinha. In: Mendes KF (ed) Atualidades no Manejo Integrado de Plantas Daninhas em Hortaliças Herbáceas. Curitiba, Brazil Publishing, p. 75–88

Miyazaki DMY, Pitelli RA (2003) Evaluation of the biocontrol potential of pacu (*Piaractus mesopotamicus*) for *Egeria densa*, *E. najas* and *Ceratophyllum demersum*. Planta Daninha 21:53–59

Moraes HMF (2021) Abóbora. In: Mendes KF (ed) Atualidades no Manejo Integrado de Plantas Daninhas em Hortaliças Fruto. Brazil Publishing, Curitiba, pp 9–35

Mota LM, Mendes KF (2021) Berinjela. In: Mendes KF (ed) Atualidades no Manejo Integrado de Plantas Daninhas em Hortaliças Fruto. Brazil Publishing, Curitiba, pp 37–57

Nunes MUC, Santos JR, Santos TC (2007) Tecnologia para biodegradação da casca de coco seco e de outros resíduos do coqueiro. EMBRAPA Tabuleiros Costeiros, Aracajú, p 6

Oliveira MF, Brighenti AM (2018) Controle de plantas daninhas: métodos físico, mecânico, cultural, biológico e alelopatia. EMBRAPA Milho e Sorgo, Sete Lagoas, p 196

Oliveira Junior RS (2011) Introdução ao Controle Químico. In: Oliveira Junior RS, Constantin J, Inoue MH (eds) Biologia e Manejo de Plantas Daninhas. Omnipax, Curitiba, pp 125–139

Ovejero RFL, Takano HK, Nicolai M, Ferreira A, Melo MS, Cavenaghi AL, Christoffoleti PJ, Oliveira RS (2017) Frequency and dispersal of glyphosate-resistant sourgrass (*Digitaria insularis*) populations across Brazilian agricultural production areas. Weed Sci 65(2): 285–294

Pereira Filho IA, Ramalho MAP, Cruz JC (1997) Consórcio milho-feijão. EMBRAPA milho e sorgo, Sete Lagoas, p 27

Pontes Junior VB, Mendes KF, Silva AA et al (2021) Cebola. In: Mendes KF (ed) Atualidades no Manejo Integrado de Plantas Daninhas em Hortaliças Tuberosas. Brazil Publishing, Curitiba, pp 151–183

Redlick C, Syrovy LD, Duddu HS et al (2017) Developing an integrated weed management system for herbicide-resistant weeds using lentil (*Lens culinaris*) as a model crop. Weed Sci 65(6):778–786

Roman ES, Beckie H, Vargas L et al (2007) Como funcionam os herbicidas: da biologia à aplicação. Berthier, Passo Fundo, p 159

Rosolem CA, Echer FR, Lisboa IP (2012) Acúmulo de nitrogênio, fósforo e potássio pelo algodoeiro sob irrigação cultivado em sistemas convencional e adensado. Rev Bras Ciênc Solo 36:457–466

Saldanha ECM, Silva Júnior ML, Alvez JDN et al (2017) Consórcio milho e feijão-de-porco adubado com NPK no nordeste do Pará. Global Sci Technol 10(1):20–28

Schaedler CE, Scalcon RM, Viero JLC et al (2021) Endozoochorous seed dispersal of glyphosate-resistant *Lolium multiflorum* by cattle. J Agric Sci 3(4):243–248

Shields EJ, Dauer JT, Vangessel MJ et al (2006) Horseweed (*Conyza canadensis*) seed collected in the planetary boundary layer. Weed Sci 54(6):1063–1067

Shioga PS, Gerage AC (2010) Influência da época de plantio no desempenho do milho safrinha no estado do Paraná, Brasil. Rev Bras Milho Sorgo 9(3):236–253

Silva LCV, Braulio CS, Correia AJ, Oliveira AS, Sousa CBDC, Vieira JDLS, Machado JP, Novaes APN (2020) Efeito alelopático do extrato foliar de eucalipto na germinação de sementes de tiririca (*Cyperus rotundus* L.) Brazilian J Animal Environ Res 4(1):1315–1320

Silva SDO, Matsumoto SN, Bebé FV et al (2006) Diversidade e frequência de plantas daninhas em associações entre cafeeiros e grevíleas. Coffee Sci 1(2):126–134

Silva JJC, Theisen G, Andres A et al (2012) Avaliação do uso de rolo-faca no preparo do solo pós-colheita do arroz irrigado em áreas da Planície Costeira do RS. EMBRAPA Clima Temperado, Pelotas, p 28

Silva VDC, Perdoná MJ, Soratto RP et al (2013) Ocorrência de plantas daninhas em cultivo consorciado de café e nogueira-macadâmia. Pesqui Agropecu Trop 43:441–449

Silva AF, Concenço G, Aspiazú I et al (2018a) Métodos de controle de plantas daninhas. In: Oliveira MF, Brighenti AM (eds) Controle de plantas daninhas: métodos físico, mecânico, cultural, biológico e alelopatia. Sete Lagoas, EMBRAPA Milho e Sorgo, pp 11–33

Silva MR, Marques TS, Kurachi SAH et al (2018b) Método de controle físico de plantas daninhas com alta temperatura – Flamejamento. In: Oliveira MF, Brighenti AM (eds) Controle de plantas daninhas: métodos físico, mecânico, cultural, biológico e alelopatia. Sete Lagoas, EMBRAPA Milho e Sorgo, pp 176–198

Snyder EM, Curran WS, Karsten HD et al (2016) Assessment of an integrated weed management system in no-till soybean and corn. Weed Sci 64(4):712–726

Sousa GD, Pereira LS, Oliveira GS et al (2020) Produtividade do feijão-caupi cultivado após plantas de cobertura com e sem aplicação de herbicidas em pós-emergência. Colloq Agrariae 16(5):57–66

Teixeira AHC, Moura MSB, Angelotti F (2013) Árvore do conhecimento: uva de mesa. https://www.agencia.cnptia.embrapa.br/gestor/uva_de_mesa/Abertura.html. Accessed 12 May 2022

Teófilo TDS, Freitas FCL, Medeiros JD et al (2012) Eficiência no uso da água e interferência de plantas daninhas no meloeiro cultivado nos sistemas de plantio direto e convencional. Planta Daninha 30(3):547–556

Tillett ND, Hague T, Grundy AC (2008) Mechanical within-row weed control for transplanted crops using computer vision. Biosyst Eng 99(2):171–178

Viero JLC, Schaedler CE, Azevedo EBD et al (2018) Endozoochorous dispersal of seeds of weedy rice (*Oryza sativa* L.) and barnyardgrass (*Echinochloa crus-galli* L.) by cattle. Cienc Rural 48(8):e20170650

Vinholis MD, Souza Filho HDS, Carre M et al (2020) Adoção de sistemas de Integração Lavoura-Pecuária-Floresta (ILPF) em São Paulo. EMBRAPA Pecuária Sudeste, São Carlos, p 57

Voll E, Gazziero DK, Karam D (1995) Dinâmica de populações de *Brachiaria plantaginea* (Link) Hitch. sob manejos de solo e de herbicidas. I. Sobrevivência. Pesqui Agropecu Bras 30(12):1387–1396

Voll E, Karam D, Gazziero DL (1997) Dinâmica de populações de trapoeraba (*Commelina benghalensis* L.) sob manejos de solo e de herbicidas. Pesqui Agropecu Bras 32(6):571–578

Wallace JM, Williams A, Liebert JA et al (2017) Cover crop-based, organic rotational no-till corn and soybean production systems in the mid-Atlantic United States. Agriculture 7(4):34

Wilson Junior RG (1980) Dissemination of weed seeds by surface irrigation water in Western Nebraska. Weed Sci 28(1):87–92

Zaidan UR (2020) Sistemas conservacionistas de manejo integrado de plantas daninhas na cultura do café. These, Federal University of Viçosa, Viçosa

Retention, Absorption, Translocation, and Metabolism of Herbicides in Plants

Kassio Ferreira Mendes, Kamila Cabral Mielke, Leonardo D'Antonino, and Antonio Alberto da Silva

Abstract Herbicides depend on the absorption, translocation, metabolism, and the plant's sensitivity to the herbicide and/or its metabolites for their biological activity. To act efficiently, the herbicide needs to penetrate the plant cuticle, reach the cytoplasm of the cell, and then exert its effect (contact herbicides) or be translocated to the target site (systemic herbicides). Plants can absorb herbicides through their leaves, stems, flowers, fruits, and also through their roots. For post-emergent herbicides, leaves are the main penetration route, while for pre-emergent, roots and young seedling structures are the most important routes. However, the morphological characteristics and herbicide selectivity of some plants can directly influence herbicide uptake. The main reason why crops and some weeds are selective to herbicides is because of their ability to metabolize the herbicide into a non-toxic form. Environmental factors such as temperature, light, relative humidity of air, and soil also influence the availability of herbicides at the sites of uptake and translocation to the site of action. In this sense, the chapter discusses the mechanisms of retention, absorption, translocation, and metabolism of herbicides in plants and their importance for efficient weed control and crop selectivity.

Keywords Herbicide behavior · Plant morphology · Foliar uptake · Selectivity · Systemic herbicide · Contact herbicide

K. F. Mendes (✉) · A. Alberto da Silva
Department of Agronomy, Federal University of Viçosa, Viçosa, Minas Gerais, Brazil
e-mail: kfmendes@ufv.br; aasilva@ufv.br

K. C. Mielke · L. D'Antonino
Federal University of Viçosa, Viçosa, Minas Gerais, Brazil
e-mail: kamilamielke@ufv.br; leonardo@ufv.br

© The Author(s), under exclusive license to Springer Nature Switzerland AG 2022

K. F. Mendes, A. Alberto da Silva (eds.), *Applied Weed and Herbicide Science*,
https://doi.org/10.1007/978-3-031-01938-8_5

1 Introduction

The biological activity of a herbicide in the plant is a function of the retention, absorption, translocation, metabolism, and sensitivity of the plant to this herbicide and/or its metabolites. Therefore, the mere fact that a herbicide reaches the leaves of the plant and, or, is applied to the soil where the plant develops is not enough for it to exercise its action. There is a need for it to penetrate the plant, translocate, and reach the organelle where it will act. Atrazine, for example, when applied to the soil, penetrates through the roots, translocates to the leaves, and then reaches and penetrates the chloroplasts, where it acts, destroying them. On the other hand, 2,4-D needs to be absorbed, translocated, and further metabolized to exert its herbicide action.

Herbicides can penetrate plants through their aerial (leaves, stems, flowers, and fruits) and underground (roots, rhizomes, stolons, tubers, etc.), of young structures such as radicles and stem, and also by seeds. The main route of penetration of herbicides in the plant is a function of a series of intrinsic and extrinsic (environmental) factors.

When herbicides are applied directly to the aerial part of the plant (post-emergence), the leaves are the main penetration route. In turn, the roots, the young structures of seedlings (radicle and stem), and seeds are the most important penetration pathways for herbicides applied and/or incorporated into the soil. The stem (bark) of trees or shrubs can also be a route of penetration of herbicides, especially when one wishes to control only a few plants, within a mixed population, or when, in a reforestation, one wants the tree strains not to resprout after felling.

The absorption of herbicides by roots or leaves is influenced by the availability of products in the absorption sites and by environmental factors (temperature, light, relative humidity, and soil moisture), which also influence the translocation of these to the place of action.

Thus, herbicide behavior in the plant is fundamental in weed control efficiency, selectivity of the compound agricultural crops, and weed resistance mechanism not related to the herbicide site of action.

2 Retention and Absorption of Herbicides by the Plant

2.1 Leaf Retention and Absorption

The retention of a herbicide is the maximum permanence of droplets on the surface of a plant. It is necessary to maximize herbicide distribution by determining the amount of active ingredient available to reach the place of action. It is dependent on the complex interfacial interaction of spray drops and plant surface. Factors considered important for retention of the sprayed herbicide include (Yao et al. 2014):

1. Physical-chemical properties of spray solutions
2. Diameter spectra and impact speed of spray drops
3. Plant surface characteristics, shape, and orientation of the target leaves and density of the canopy of the plant

To minimize these factors, application technology is extremely important to ensure good plant cover, in addition to the use of surfactants to improve herbicide swell on the plant surface.

Leaf absorption of a herbicide requires that the product be retained in the leaf and remain there for a sufficient period of time, until it is absorbed. However, for this process to occur, it is not only the variability of species and genotypes and the age of plants that influence this process, but also the environmental conditions such as moisture, light, temperature, and soil type of the area in question. The physical-chemical properties of the molecule also influence the amount and route of absorption and translocation of the herbicide (Monquero and Hirata 2014).

Plant morphology influences the amount of herbicide intercepted and retained. Among the aspects related to plant morphology, the stage of development (plant age), shape and leaf limb area stand out, the angle or orientation of the leaves in relation to the spray jet and specialized structures such as trichomes (hair). The cross-section of a sheet is represented in Fig. 1. The leaves, like all aerial structures of plants, are covered by a dead (non-cellular), lipophilic layer, called cuticle. Although to a lesser extent, it also exists in the roots, which is why many factors also influence both the penetration of herbicides by both leaves and roots.

After deposition of the droplet containing the herbicide molecules on the plant or tissue, the cuticle presents itself as the first barrier on the way to the inside of the cell. This structure evolved in vegetables to provide protection to processes such as water loss, insect attack, high radiation, among others (Reis et al. 2021). The cuticle covers all cells of the epidermis of the plant, including the stomata guard cells and the cells surrounding the substomatic chamber. In leaves and stems, the cuticle is cerose, so that compounds that have affinity with water (hydrophilic) have difficulty

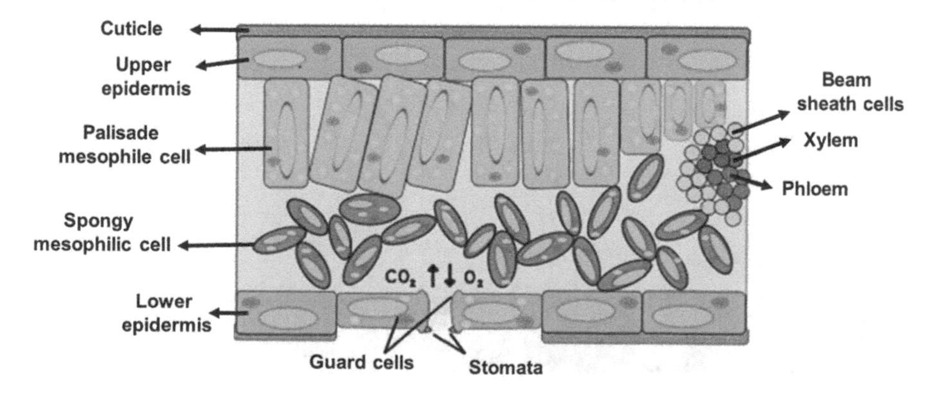

Fig. 1 Cross-section of a leaf (schematic), showing the guard cells, beam sheath cells, xylem and phloem, palisade, and spongy mesophile cell. (Source: Adapted from Almeida and Almeida 2018)

passing through this layer. On the other hand, compounds that have no affinity with water and are soluble in oil (hydrophobic) can dissolve the cuticle and no longer return to the outside of the leaf, crossing with relative ease this barrier (Roman 2005). Generally, the cuticle is 0.1–13 μm thick and contains three components: an insoluble cutin matrix, cuticular waxes, and epicuticular waxes (Fig. 2). While the general chemical composition of the cuticle is lipophilic, it also has hydrophilic components (pectin) (Cobb and Reade 2010).

Between the cuticular layer and the cytoplasmic membrane, there is the cell wall, which is formed of pectin-impregnated cellulose fibrils (Fig. 2). The surface pattern of the cuticular layer is quite variable. It can be the form of granules, dish (or disc), superimposed layers and can be semifluid or fluid. The chemical composition of the epicuticular coating is very variable among plant species, but some components are common. In general, this layer is a complex mixture of long-chain alkanes (21–37 carbons), alcohols, keones, aldehydes, esters, fatty acids, etc. (Ferreira et al. 2005). As a result of the variability of its components, the degree of polarity of cuticles varies greatly, that is, more than more water-permeable amounts of water it becomes. The cerose layer surrounding the cuticle is richer in less polar compounds than cutin, which has variable polarity groups, functioning as a cation exchange resin. In the presence of water, the cutin is believed to increase in volume (by ingrading), by separating the wax particles, thus increasing their permeability.

It is known that there is a rather complex interaction between the chemical nature of the applied product and the leaf surface. There are two main types of surfaces: one easily wettable (rich in alcoholics) and another of more difficult wetting (rich in alkanes). The characteristics of the applied solution, the age of the plant, the polarity of the compound, the surface tension of the syrup, among others, are important in this interaction.

After overtaking the cuticle, the herbicide finds the second barrier to absorption, the apoplast (a part that contains water, which bathes the cells), where the reverse of the cuticle situation occurs, that is, hydrophilic compounds easily overcome this barrier, while hydrophobics enter it and move slowly. Additionally, water moves through the apoplast on its way to the leaves, where it is mostly lost by perspiration,

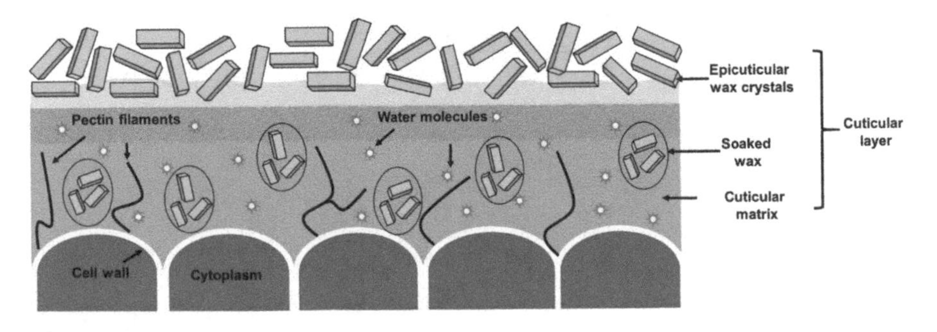

Fig. 2 Schematic representation of the main components of the cuticular layer. (Source: Adapted from Stagnari 2007)

taking along the hydrophilic compounds to the edges of the leaves. The third barrier to the penetration of the herbicide molecule into the cell is the cell wall which is composed mainly of cellulose. Like pectin, cellulose is hydrophilic and may have cuticle extensions. The fourth and final barrier is the cell membrane that surrounds the cell, composed mainly of lipid substances (phospholipids) and some proteins. Hydrophobic herbicides easily penetrate/cross the cell membrane, while hydrophilic herbicides do so slowly (Roman 2005).

The passage of a herbicide molecule through the cuticular layer is a physical process that can be influenced by a number of factors, such as hydrogen ionic potential (pH), environmental factors (light, temperature, relative humidity), particle size and herbicide concentration, cuticle thickness, cerosity and leaf hair, use of surface activating agents (surfactants), and others. As for organic herbicides, derived from weak acids, a lower pH of the syrup increases the absorption of the herbicide, because it reduces its polarity. However, non-dissolving herbicides (amides, esters, etc.), the pH of the solution has little or no effect on penetration.

The absorption of a herbicide is not necessarily related to the thickness or weight of the cuticle, but to the lipid constitution and the degree of impediment of the passage of solutes. There is evidence that herbicide penetration decreases with increasing leaf age. Although the physical and chemical constitution and thickness can be practically the same, the new leaf cuticle is more permeable to water than that of old leaves (Grover and Cessna 1991).

Environmental factors, together, such as air temperature, relative humidity, light, and moisture content in soil and plant, influence herbicide activity in the aspects of absorption, translocation and degree of detoxification. High temperature and luminosity conditions, or low relative humidity and soil moisture, generally promote the formation of more impermeable and thick cuticles. In addition to affecting the amount of wax, light and temperature influence its composition (Shepherd and Griffiths 2006).

High temperature can improve absorption by causing greater fluidity of lipids from the cuticular layer and cell membrane and, consequently, faster absorption of the herbicide. For example, *Abutilon theophrastiplants* treated with acifluorfen at 20–15 °C (day/night) exhibited 70% more epicuticular wax deposition on the leaf surface than plants at a temperature of 32–22 °C (Hatterman-Valenti et al. 2011). The same study showed that higher temperatures alter the composition of the cuticle, reducing ester content by 25%, increasing the hydrocarbon content by 11%. The reduction of wax deposition was correlated with higher efficacy of protoporphyrinogen oxidase (PROTOX) inhibitor herbicides when the temperature increased, confirming the hypothesis of better herbicide efficiency as the cuticle composition changes.

Low relative humidity (RH) negatively influences leaf absorption of the herbicide. RH usually affects herbicide absorption through its impact on cuticle hydration and droplet drying rate after herbicide application (Price 1982). In high RH, however, the effects of high temperature on the drying of drops are reduced due to increased leaf retention time, therefore increased absorption of the herbicide. Plants grown in high RH generally develop softer cuticles than plants grown in low RH, which tend to have thicker cuticles and therefore lower herbicide penetration. In general, RH has a greater effect on herbicide absorption than the temperature (Varanasi et al. 2016). The activity of ammonium glufosinate in species of *Hordeum vulgare* cv. "Samson" and *Setaria viridis* was significantly decreased by low RH. For example, when grown at 22/17 °C, the *S. viridis* survived the potentially lethal dose of 100 g/ha to 40% of RH and accumulated 70% of the dry biomass of the plants in relation to the control treatment, but was killed at 95% of RH (Anderson et al. 1993).

The degree of impermeability of the cuticle can be attributed to the increase in its thickness, to the change in the composition of the waxes or to the increase in the formation of epicuticular waxes. The nature of the response to the different environmental conditions varies with the plant species. Biotypes of the genus *Ipomoea* present great variability in the deposition of epicuticular wax in the leaves. The glyphosate-tolerant biotypes of the species *Ipomoea grandifolia* and *I. indivisa* presented greater plastic capacity to increase the amount of epicuticular wax in their leaves under water stress, which may indicate an adaptive mechanism (Trezzi et al. 2020). Experiments in a controlled environment showed that *Abutilon theophrasti* cultivated under water stress or low temperature had higher leaf epicuticular wax deposition compared to plants grown in soil with moisture in field capacity or high temperature regime. The intensity of light did not affect wax deposition; however, the increase in light intensity decreased the ester content and increased the secondary alcohol content of leaf epicuticular waxes (Hatterman-Valenti et al. 2011).

Stomata may be involved, in two ways, with the penetration of herbicides into the leaves. First, the cuticle on the guard cells looks thinner and more permeable to substances than the cuticle over other epidermal cells. Second, the pulverized solution could, in the stake, move through the pore of an open stomata into the substomatic chamber, and from there to the cytoplasm of the cells of the leaf parenchyma. However, infiltration by stomata is not possible unless the surface tension of the pulverized solution is greatly reduced by the use of surfactants in the spray formulation or tank. Most surfactants currently in use act by increasing cuticular penetration and cannot reduce surface tension adequately to allow stomatic penetration. However, the development of organosilicone-based surfactants are able to reduce surface tension to the point that stomata infiltration occurs. They can also induce a mass flow of the solution sprayed through the stomatal pore and also increase cuticular penetration.

After interception, for each herbicide, there should be a critical period without occurrence of rains until sufficient absorption of the same. The loss of the herbicide or its activity depends on the occurrence of rain (intensity and duration) in this range, the method and application technology, climatic conditions, and plant species involved (Pires et al. 2000; Jakelaitis et al. 2001). The influence of rain on herbicide efficiency is also related to the formulation. Differences between glyphosate formulations may result in variations in efficacy, especially in the face of unfavorable environmental conditions, such as the occurrence of unexpected rains after application. Dalazen et al. (2020) evaluated glyphosate salts (isopropylamine, potassium, and ammonium salt) in different periods of simulated rainfall after herbicide application (30, 120, and 240 min, and without rain), and presence or absence of saflufenacil and a non-ionic adjuvant. The occurrence of simulated rainfall after 240 min reduced glyphosate efficiency by 15%, 30%, and 60% in potassium, isopropylamine, and ammonium salt formulations, respectively. The addition of adjuvant improved the efficiency of potassium salt glyphosate by 40%.

Rain can cause considerable losses of herbicides from plant leaves. Anionic salts (negative charges), e.g., sodium salts, do not penetrate rapidly, are not absorbed by the cuticle surface and are soluble in water, and can be washed if rain occurs up to more than 24 h after application. Cationic salts (positively charged), such as paraquat, are water soluble, but are quickly absorbed and therefore less subject to washing by rain. Lipophilic herbicides (usually formulated as EC or flowable) are poorly soluble in water, but are rapidly absorbed into cuticle lipids and poorly washed by rain.

2.2 *Absorption by the Stem*

Absorption of herbicides may occur by the stem of young plants (during emergence) and adult plants. In young plants, it is an important entry site for many herbicides applied to the soil that are active in seeds, during germination and seedling emergence. The seedling stem during emergence has a very little developed cuticle, devoid of the wax layer, making it more permeable to herbicides, this being a herbicide entry route in many grass species. Furthermore, the barrier that striae of Caspary represents in the root is not present in these tissues.

The penetration of herbicides through the bark of woody plants is another option that can be used in practice. However, periderm is a protective tissue that replaces the epidermis after the death of its cells. Periderm cells contain tannin and are highly suberized. Other constituents commonly found in these cells are fatty acids, lignin, celluloses, and terpenes. Based on its structure and composition, the periderm should present low water permeability and also to herbicides applied in the shoots, especially polar ones. Lenticels are structures that cross the periderms and are therefore important routes for the penetration of herbicides through the stem. The growth of the stem, in diameter, causes small ruptures in the bark, which facilitate the penetration of herbicides.

For better performance of herbicides applied to the bark of trees, they should be prepared in lipophilic formulations, using oil as a vehicle, in addition to being applied in high concentrations (5–10%). These products are sprayed or brushed on the stem of the plant. More efficient practical alternative would be to inject the herbicide with its own equipment with an injection gun to the exchange rate region (xylem, and/or phloem). In this case, the herbicide will be mechanically introduced through the bark. This process has been used in some reforestation companies, using imazapyr 20–30 days before the eucalyptus trees fall, in order to avoid the regrowth of the strains. However, eucalyptus plants treated with imazapyr promote root exudation of the herbicide. Due to the great persistence of this herbicide in some soils, exuded residues have caused poisoning in seedlings after transplantation during forest reform. In order to work around the problem, some companies have made the application of glyphosate in the regrowth, provided that the height of the branches is between 40 and 90 cm, which allows greater absorption and translocation of the herbicide due to greater leaf area, ensuring efficient control (Ferreira et al. 2010). Less used, but also efficient is the application of glyphosate on the stump immediately after cutting the trees. However, the efficiency of the herbicide is greatly decreased if the application exceeds 24 after felling, because the formation of callose in the wound prevents the penetration of glyphosate into the tissues of the stumps.

2.3 Absorption by Roots

Many herbicides applied to the soil are absorbed by the roots. The entry of herbicides through the roots is not as limited as by the leaves, since no significant layer of wax or cuticle is present in the parts of the roots where most herbicide absorption occurs (Reis et al. 2021). The most important route of entry is the passage of the herbicide along with the water through the root hairs at the ends of the roots. Root hair is responsible for the significant increase in the available area, the absorption of water, salts, and herbicides.

The availability of herbicides to roots is a function of the physical-chemical properties of herbicides and soil and the spatial distribution of these compounds and roots in the soil. Herbicides have to come into contact with the root, which can occur by the growth of the herbicide or by the diffusion of the herbicide in the gaseous state and, in solution with water, to the area of absorption of the roots. Many herbicides with different molecular structures, sizes, and solubility are readily absorbed by the roots.

The root system of the upper plants presents an extremely large absorption surface, with high permeability to water and solutes (salts). Although young roots are also covered by a waxy layer, and the older ones are strongly suberized, water, and solute penetration usually occurs. Some herbicides will be absorbed by the roots and will tend to remain in the membranes and lipid bodies of the epidermis, while herbicides with some water solubility can move by three main pathways toward the plant's vascular system (Fig. 3). The apoplastic pathway provides a route through free spaces and cell walls of the epidermis and cortex. The endoderm is characterized by Caspary striae, a suberized layer that forces all herbicides to move through the symplast to enter the vascular system. The symplastic route involves cell-to-cell transport across plasmodesmos. Plasmodesmos are channels of the cytoplasm lined by the plasma membrane that traverse cell walls. These channels allow herbicides to move from cell to cell without going through the cell wall. The transmembrane route involves a movement through cells and cell walls combining the symbioplastic and apoplastic movement (Nissen et al. 2021a).

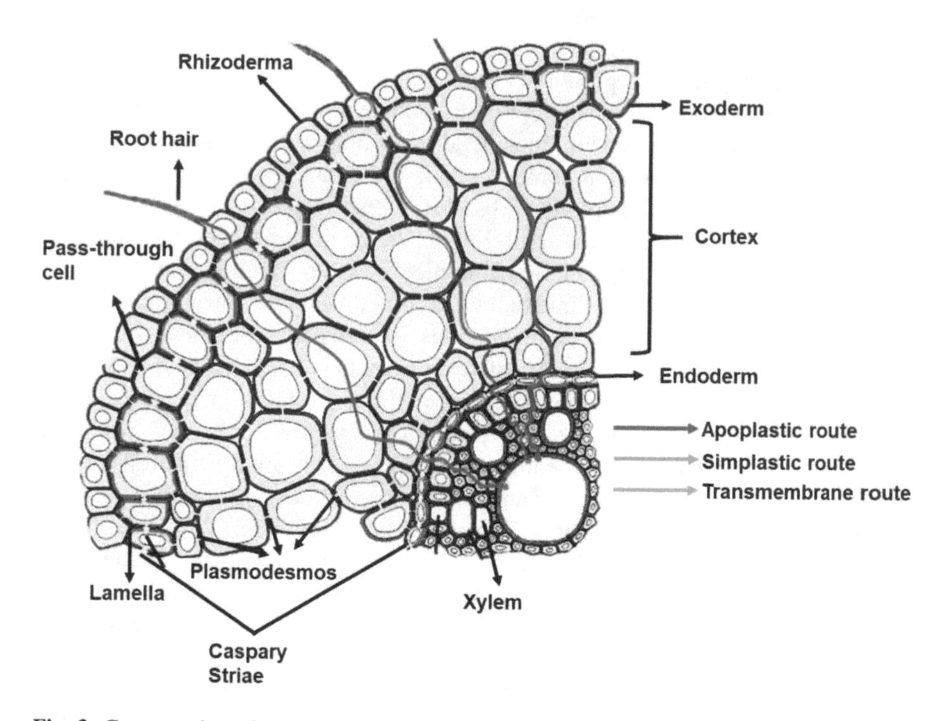

Fig. 3 Cross-section of a root, showing its main structures. The apoplastic pathway provides a porous path, but can be interrupted by the striations of Caspary in the endo and exoderm. But there may be some passage of water and solutes through the striae of Caspary. The simplastic pathway takes place through plasmodesmos and the cytosol of cells. (Source: Adapted from Steudle 2000)

The absorption of herbicides by the roots is characterized by an initial phase of high absorption rate during the first 30 min up to 2 h, followed by a slower absorption phase. For example, the absorption rate of 2,4-D increases rapidly soon after application and then decreases in this rate until it becomes null, then negative due to loss by root exudation.

As for the concentration of the herbicide, within certain limits, there is a linear relationship between the concentration of the available product and its penetration through the root. Linearity is lost when the herbicide exerts toxic effect on the plant. Although some studies demonstrate a close relationship between transpiration and absorption, there is evidence to the contrary.

The absorption of herbicides by the root may also be limited by herbicide bindings or adsorption in the cellular components. Triazines and ureas, for example, can be adsorbed, in part, by the roots. The correlation between transpiration and absorption is valid for polar herbicides; however, there are non-polar herbicides that are also readily absorbed by the roots.

For polar herbicides translocated via xylem, the transpiration current correlates with its transport to the aerial part of the plant, establishing a concentration gradient between the outside of the root (soil solution) and the inside of the plant (current of assimilated). High temperature and irradiance, low relative air humidity, high soil temperature, and high soil water potential are conditions that favor transpiration and, consequently, the absorption of polar herbicides. Also the physical-chemical properties of herbicides, such as lipophilicity and pK_a, in addition to the pH of the soil solution, influence absorption. In general, according to Donaldson et al. (1973) the rate of herbicide absorption correlates with the octanol/water partition coefficient (K_{ow}), with the more lipophilic herbicides (higher K_{ow}) being absorbed faster.

The influence of soil pH on the solubility of a herbicide is another important factor that influences root absorption. Some herbicides will be neutral, that is, without load, regardless of soil pH. Pendimethalin and S-metolachlor are examples of these types of herbicides. The dose recommendations of these herbicides are made strictly based on soil texture and percentage of organic matter. For other herbicides, the water solubility changes depending on the pH of the soil solution. The herbicides of weak acid characteristics contain a functional group that transforms the initial molecule from neutral to negative depending on the pH of the soil. The neutral form of the herbicide is less soluble in water and more lipophilic. This means that the sorption by organic matter will be higher. The negatively charged form of the herbicide is more soluble in water and is less likely to be sorbed by soil organic matter. This increase in water solubility means that there is more bioavailable herbicide in the soil solution, therefore more herbicide to be absorbed by weed roots (Nissen et al. 2021a).

Differentiation in herbicide absorption can be observed in resistant weed biotypes. *Euphorbia heterophylla* populations resistant and susceptible to imazethapyr were evaluated for the role of differential leaf penetration as a determinant of herbicide resistance.

The surface analysis of the leaf cuticle by scanning electron microscopy revealed higher wax density in the leaf cuticles of resistant biotypes than in the sensitive biotype. Thus, it is suggested that biotypes are resistant to imazethapyr because increasing the wax content of its cuticle reduced herbicide penetration into the plant (Plaza et al. 2006). Riar et al. (2011) observed that resistant *Lactuca serriola* biotypes absorbed less the applied 2,4-D and retained more herbicide in the treated part of the leaf compared to the susceptible biotype. *Amaranthus palmeri* biotype resistant to glyphosate absorbed 18% of ^{14}C-glyphosate 12 h after treatment while the sensitive biotype absorbed 25% (Palma-Bautista et al. 2019a). This fact shows the reduced absorption as a mechanism of resistance of weeds not related to the site of action of herbicides.

3 Herbicide Translocation

There are several reasons why it is important to study herbicide translocation. Young plants that are not able to regenerate through their underground organs can be killed by contact herbicides, when complete coverage of the aerial part by the sprayed herbicide solution occurs. However, those plants that are able to regenerate through bulbs, rhizomes, stolons, tubers, etc. require that a certain amount of the product be able to translocate and reach these recovery organs, in order to produce efficient control. For most soil-applied herbicides, translocation is also of great importance. Many herbicides are absorbed by the roots or underground parts of the stem and are translocated to other areas, such as growth point and chloroplasts, to exert their effective herbicide action. The translocation of herbicides, from the absorption site to the action site, can be carried out mainly by two routes: phloem (simplastic) and xylem (apoplastic). The physicochemical characteristics of the herbicide influence its translocation.

3.1 Simplastic Translocation

The phloem, through living cells, transports the photoassimilates to all parts of the plant, the translocation of herbicides being called simplastic. The simplasto is called as the total mass of living cells of a plant, forming a continuous set through the intercommunications of the cytoplasm, called plasmodesmos. Ions and molecules can move from cell to cell through intercom structures, called plasmodesmos, until they reach the companion cells, from where they are transposed to phloem, without crossing the barriers to permeability, which are cytoplasmic membranes (Taiz et al. 2017).

Herbicides move passively in phloem due mainly to solubility characteristics, K_{ow}, acid ionization constant (pK_a), and vapor pressure (VP). Lipophilicity (Log K_{ow}) represents the affinity of a molecule for a lipophilic environment or for non-polar solvents. In phloem, there are predominance of substances of higher lipophilic characteristics.

Herbicides applied post-emergence are typically weak acids, and this characteristic influences translocation via phloem. Most of these herbicides have an ionizable group (such as a carboxylic acid group COOH) as an integral part of the molecule, although there are herbicides that dissociate as weak bases (atrazine, for example). These herbicides have the ability to acquire or donate hydrogen ions, depending on the pH of the solution in which they are (Oliveira Júnior and Bacarin 2011). Within the plant, there are pH gradients, which are established to conduct numerous physiological processes. Plants have specialized carrier proteins that pump hydrogen ions from the cytoplasm through the cell membrane into intercellular spaces. This increases the pH (alkaline) of the cytoplasm, while reducing the pH (acid) of the solution outside the cell. With this, weak acid herbicides outside the cell acquire hydrogen ions and become lipophilic, increasing their ability to move through the plasma membrane and in the cytoplasm undergo dissociation reactions, losing protons and becoming more hydrophilic, preventing them from leaving the cell (ion trap due to the proton pump). This creates a concentration gradient that will allow the entry of the rest of the molecules (Christoffoleti and Nicolai 2016).

The information collected over the years about the herbicide characteristics and phloem mobility indicated that if a compound is intermediate in its membrane permeability, it will have some mobility in the phloem. Membrane permeability is

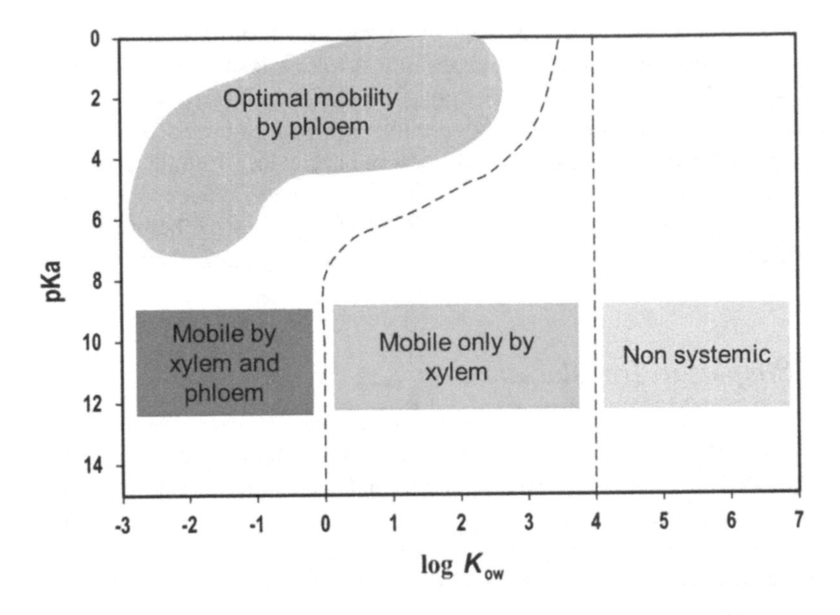

Fig. 4 Relationship between dissociation and lipophilicity required to group herbicides related to the systemic translocation pathway. (Source: Adapted from Bromilow et al. 1990)

estimated by log K_{ow} values. Herbicides with log values K_{ow} between −1 and 1 should have mobility in phloem after leaf applications (Nissen et al. 2021b). However, there are other acids, such as aryloxyphenoxy propionate, which have log values K_{ow} between 3 and 4.5, which are more lipophilic and therefore have limited mobility in phloem. In general, compounds that have high polarity (log $K_{ow} < 0$) and strong ionization (pK_a < 2), such as glyphosate (log K_{ow} = −2.77 to −3.22), are phloem mobile, although important amounts move peloxylem (apoplast). Figure 4 shows a relationship between ionization properties (pK_a) and lipophilicity (log K_{ow}) (Oliveira Júnior and Bacarin 2011).

3.2 Apoplastic Translocation

The other transport system is xylem (apoplast), formed by the set of dead cells, including cell walls and intercellular spaces, which form a continuous system where water and solutes move freely. The movement of solutes and photoassimilates inside the upper plants can be defined basically in two directions, descending and ascending.

In the descending movement, the assimilates and solutes move at an average distance corresponding to 2.5 times the diameter of the cell, before reaching the smaller phloem vessels. Part of this distance occurs through the apoplastic system. Once these assimilates moves into two vessels, in the opposite direction to the concentration gradient, it is assumed that this movement occurs at the expense of metabolic energy (active transport). Companion cells and parenchymate cells, which accompany phloem cells, are involved in the loading flow of these vessels. Cells with very dense protoplasm and scores on the inside of the cell wall allow greater contact surface between the simplastic system and the apoplastic system. These cells are known as transfer cells and appear to work in loading phloem vessels and transferring phloem to xylem.

In the upward movement, ions and molecules can spread through the intercellular spaces and cell walls of the cortex. The movement along this route to the interior of the root is blocked by the longitudinal walls of the "Caspary striae," in the endoderm. However, somehow not yet defined, it was assumed that the substances (ions or molecules) broke through this barrier and penetrated the simplastic system of cells. It is known today, however, that Caspary striae are not present in the root apexes of young endodermal cells and in the basal region of the developing lateral roots (Ciamporová 1992), which may represent an important route of passage from apoplast herbicides to the simplast. These substances can then move from cell to cell through the simplastic system, or leak into the parenchyma xylem and be transported in the acropeto direction by the transpiration current. To move in the xylem and reach the shoot, the herbicide must be soluble in water. Therefore, herbicides applied to the soil must be translocation via xylem, being necessarily soluble in water. In the case of herbicide application that moves via xylem over the foliage, the water and nutrient stream transport the product to the edges of the leaves (Roman 2005). Table 1 shows some herbicides and their translocation pattern in plants.

In general, environmental conditions favorable to transpiration (low relative humidity, high temperatures, and adequate water supply in the soil) are also favorable to herbicide translocation. The influence of low temperatures on the efficacy of glyphosate was studied in the biotypes of *Conyza sumatrensis* resistant and sensitive to glyphosate (Palma-Bautista et al. 2019b). The low temperatures (15/5 °C day/night) increased the absorption and translocation of glyphosate, suppressing the resistance mechanisms, improving its effectiveness in the control of resistant weeds. Soil moisture has a direct influence on the absorption and translocation of the herbicide by the plant. Studies carried out by Miller and Norsworthy (2018) evaluated the influence of soil moisture on the absorption and translocation of florpyrauxifen-benzyl in *Echinochloa crus-galli*, *Sesbania herbacea*, and *Cyperus esculentus*. In all weed species, higher absorption and translocation to shoots occurred under conditions of high soil moisture (60% of field capacity) compared to low soil moisture (7.5% of field capacity).

4 Herbicide Metabolization in Plants

Plants show large differences in their ability to metabolize herbicides, and these differences can often be correlated with the tolerance or susceptibility of plant species (Reis et al. 2021). Differential metabolism is probably the most common of the

Table 1 Translocation pattern of some herbicides in plants

Mechanism of action	Representative herbicide	Time of application	Translocation
Photosystem II inhibitors (PSII)	Atrazine	Pre-emergence	Xylem
	Diuron	Pre-emergence	Xylem
	Bentazon	Pre-emergence	Non-translocated (contact)
Carotenoid synthesis inhibitors	Clomazone	Pre-emergence	Xylem
	Isoxaflutole		Xylem and phloem
Inhibitors of the formation of long chain fatty acids	Acetochlor	Pre-emergence	Xylem and phloem
Cell division inhibitors	S-metolachlor	Pre-emergence	Low translocation (xylem)
Protoporphyrinogen oxidase inhibitors (PROTOX)	Fomesafen	Post-emergence	Non-translocated (contact)
	Sulfentrazone	Pre-emergence	Xylem
Acetolactase synthase inhibitors (ALS)	Imazapic	Pre and Post-emergence	Xylem and phloem
	Cloransulam	Post-emergence	Xylem and phloem
Photosystem I inhibitors (PSI)	Diquat	Post-emergence	Non-translocated (contact)
Glutamine synthetase inhibitors (GS)	Ammonium glufosinate	Post-emergence	Non-translocated (contact)
Auxin mimics	2,4-D	Post-emergence	Xylem and phloem
Inhibitors of 5-enolpyruvylshikimate-3-phosphate synthase (EPSPs)	Glyphosate	Post-emergence	Xylem and phloem
Acetyl-coenzyme A carboxylase inhibitors (ACCase)	Diclofop-methyl	Post-emergence	Phloem
	Clethodim	Post-emergence	Xylem and phloem

Source: Adapted from Marchi et al. (2008) and Oliveira Júnior and Bacarin (2011)

mechanisms that contribute to the tolerance and/or selectivity of herbicides in agricultural crops. In plants, this mechanism is characterized by the conversion of lethal molecules into less toxic compounds, to be stored where they do not affect the survival of plant cells. A plant species containing such mechanisms can alter or degrade the herbicide molecule through biochemical reactions producing non-toxic products (Oliveira Júnior 2001).

Plants use a three-phase process to convert herbicides into molecules with reduced phytotoxicity. Phase I is characterized by the process of chemical modification of the herbicide molecule through oxidation, reduction, and hydrolysis. In Phase II, the herbicide molecule or metabolite originated from Phase I is conjugated to endogenous substrates, such as sugars, amino acids, or glutathione. In Phase III, the substrates from the conjugation of Phase II are sequestered in the vacuoles, cell wall, or transformed into insoluble residues (Sterling et al. 2021). However, some authors divide the metabolic process into four phases, one of which is characterized by secondary transformation and transport of the herbicide to the vacuole (Eerd et al. 2003; Yuan et al. 2007).

4.1 Phase I: Chemical Modification of the Molecule

Phase I reactions can reduce or modify herbicide phytotoxicity, predispose molecules to Phase II reactions, and increase polarity. In the activation phase, the non-polar lipophilic molecules of herbicides become hydrophilic polar compounds by hydrolysis or oxidation (Kreuz et al. 1996). Because activation phase reactions are not synthetic, the herbicide does not dramatically increase in size, although functional groups are added to a herbicide molecule (Cobb and Reade 2010).

The most important enzymes involved in this phase are cytochrome P450 monoxygenases (Yuan et al. 2007). These proteins are found in many kingdoms, including animals, fungi, and bacteria, although those found in plants outnumber and diversity those found in other organisms (Powles and Yu 2010).

Diclofop-methyl metabolism was one of the most well-characterized cytochrome P450 activities in terms of crop tolerance to herbicides. Diclofop-methyl was the first selective inhibitor of specific paragramine acetyl-Coenzyme A carboxylases (ACCase). The herbicide is a pro-herbicide in which the active diclofop acid attached to the methyl group helps the herbicide to be absorbed by plants, and is then rapidly de-esterified to release the active diclofop acid (Dimaano and Iwakami 2020). Diclofop-methyl-tolerant wheat produces three-ring diclofhydroxyllate isomers that are later conjugated with glucose (Tanaka et al. 1990). And grass species are able to tolerate the application of 2,4-D due to herbicide metabolism by cytochrome P450 enzymes (Dodge 1990).

Different reactions are observed in herbicide metabolism by plant cytochrome P450 (Fig. 5). P450 can cause different reactions even with a single herbicide. For example, CYP71AH11 and CYP81B2 in tobacco (*Nicotiana tabacum*) cause alkyl-hydroxylation and N-demethylation in chlorotoluron (Yamada et al. 2000). In the case of CYP76B1 in *Helianthus tuberosus*, the di-N-demethylation reaction followed by mono-N-demethylation is confirmed by inactivating the herbicide (Robineau et al. 1998). Demethylation is often observed in sulfonylurea. Iwakami et al. (2014) showed that overexpression of two cytochrome P450 genes (*CYP81A12* and *CYP81A21*) in *Arabidopsis thaliana* increased plant tolerance to two Acetolactato Synthase inhibitors (ALS) (bensulfuron-methyl and penoxsulam). These cytochrome P450 genes showed resistance by O-demethylation. Aryl hydroxylation is also a typical reaction in herbicide metabolism. In addition to a standard arylhydroxylation reaction, the atypical aryl-hydroxylation reaction called NIH-shift is observed in wheat microsomes when treated with diclofop-methyl,

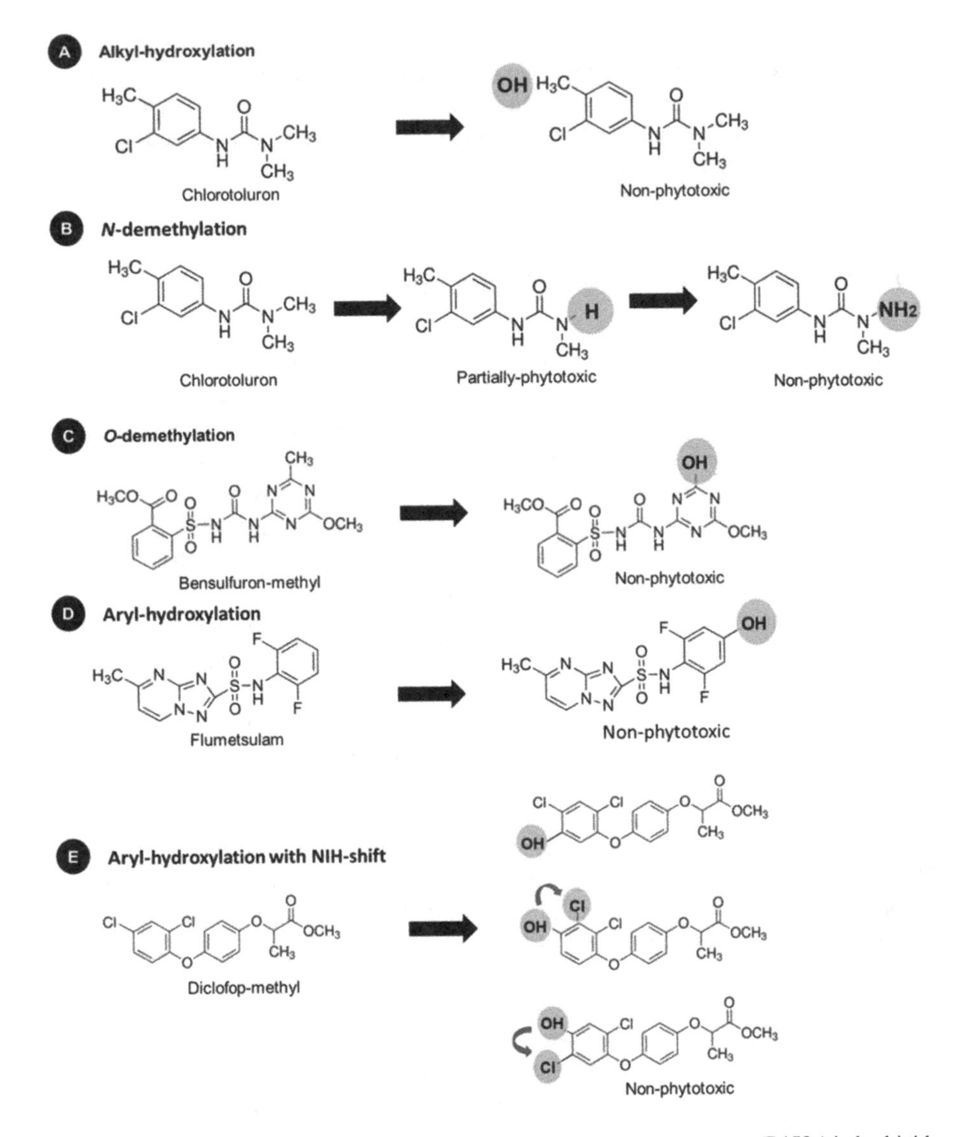

Fig. 5 (**a–e**) Main reactions performed by cytochrome P450 monooxygenase (P450s) in herbicide metabolism. (Source: Adapted from Dimaano and Iwakami 2020)

occurring the migration of chlorine from the herbicide molecule (Fig. 5e) (Zimmerlin and Durst 1990).

Phase I reduction reactions are also catalyzed by the P450 enzyme, often occurring in anaerobic conditions. In the metabolization of a herbicide, reduction is a very rare type of reaction in plants. The most common reduction reaction is when a nitro group present in phenyl ring is reduced to an amine group, which happens with trifluralin. This reaction is catalyzed by aryl-nitro reductases and requires a reduction cofactor, such as NADPH (Sterling et al. 2021). The third type of Phase I reaction is hydrolysis, common in plants, and involves breaking a molecule by adding H or OH from water. Hydrolysis is common in herbicides that have anter, amide, or nitrile groups. It is catalyzed by enzymes of the hydrolase classes, such as esterases and carboxylesterases (Riechers et al. 2010).

In contrast to the role of P450 in plant selectivity to herbicides, this is a common mechanism of weed resistance to herbicides where metabolic deactivation occurs by which the herbicide's active ingredient is transformed into non-phytotoxic metabolites (Yu and Powles 2014). *Lolium rigidum* is a weed that can offer resistance to the most important herbicides through herbicide metabolism along with mechanisms related to the herbicide's site of action. Resistance to ACCase inhibitors is related to overexpression of cyp72A subfamily cyp72A cytochrome P450, playing a crucial role in conferring metabolic resistance to herbicides in *L. rigidum* (Gaines et al. 2014). In *Alopecurus myosuroides* resistant to ALS inhibitors, three candidate genes of cytochrome P450 (CYP71A, CYP71B, and CYP81D) were found in highly overexpressed conditions (Gardin et al. 2015). In *Beckmannia syzigachne*-resistant fenoxaprop-p-ethyl, two cytochrome P450 genes (*CYP87A3* and *CYP71D7*) also showed to be involved in herbicide metabolism (Pan et al. 2016).

4.2 Phase II: Conjugation

Phase II is characterized by conjugation reactions, where the herbicide molecule or metabolite derived from Phase I is conjugated with sugars, amino acids, or tripeptide glutathione, increasing solubility in water while reducing phytotoxicity (Carvalho et al. 2009). Although the activation of the toxic molecule in Phase I plays a crucial role in other enzyme detoxifications in the conjugation phase, if the compound already has a functional group suitable for the conjugation phase, then metabolism can continue without the need for Phase I (Coleman et al. 1997).

The conjugation reaction is mediated by enzymes such as Glycosyltransferases (GT), MalonylTransferase (MT), or Glutathione S-Transferase (GST) (Carvalho 2013). It has been suggested that GST enzymes play a crucial role in herbicide metabolism in weeds (Yuan et al. 2007). GST is presented in its reduced form in plants and acts as a radical scavenger against oxidative stress (Coleman et al. 1997). The selectivity of herbicides and the natural tolerance to trialine in corn and soil are due to the rapid metabolism of these herbicides, conjugating with GST (Shimabukuro et al. 1971). In weeds, some cases of resistance based on the metabolization of GST have been reported. Resistance to atrazine by *Abutilon theophrasti* biotype is mediated by glutathione (Anderson and Gronwald 1991). Metabolic resistance to GST-mediated atrazine has also been reported in *Amaranthus palmeri* biotypes (Nakka et al. 2017).

Herbicides can also be conjugated with glucose. Herbicides containing phenolic groups N-arylamine or carboxylic acid as well as those metabolized into phenols, aniline, or acids during Phase I can be made in sugar conjugates. The enzyme responsible for catalyzing this reaction is GT (Sterling et al. 2021). In the crops, some phytoprotectors (safeners) have demonstrated to increase the activity of GT to provide selective control of weeds (Yuan et al. 2007). Although the role of GT in the detoxification of herbicides in cultivated plants is known, little is known about the role of GT enzymes in herbicide detoxification in weeds (Ghanizadeh and Harrington 2017).

Conjugation can also occur with amino acids, and the most common conjugates are aspartate or glutamate. These conjugates may still be biologically active but are relatively immobile. There is evidence that amino acid conjugates are excreted in the cell wall. Conjugation with amino acids is particularly common with herbicides derived from phenoxyacetic acid, such as 2,4-D. A peptide bond is formed between the herbicide carboxylic acid residue and the amino acid amine group (Sterling et al. 2021).

4.3 Phase III: Kidnapping

Phase III is known as compartmentalization phase. In this process, the water-soluble conjugates produced in Phase II may undergo a secondary conjugation (with malonic acid), be sequestered in the vacuole, or incorporated into cell wall regions. In secondary conjugation or conjugation with malonyl CoA, a malonic acid residue is attached to a Phase II sugar conjugation product by an ester binding catalyzed by malonyl-CoA-transferases. This process may increase the vacuolar sequestration of the conjugate (Sterling et al. 2021).

The mechanism of incorporation of herbicide metabolites into the cell wall is poorly understood but has been documented with many herbicides. They are found in non-extractible fractions (bonded residue) remaining from solvent extraction processes used in metabolism studies (Rigon et al. 2020). The formation of these bonded residues has been observed in metabolism studies of herbicides containing aromatic or heterocyclic rings. These herbicides include phenolic acids, some triazines (atrazine and prometryn), among others (Cobb and Reade 2010).

The sequestration of herbicide metabolites in the vacuole is mediated by transport proteins called ABC transporters (ATP-binding cassette) present in the tonoplast or plasmalemma and are dependent on metabolic energy (ATP) for the excretion or immobilization of the metabolite (Carvalho 2013). The most prominent example in weeds is the rapid vacuolar sequestration of glyphosate and paraquat (Jugulam and Shyam 2019). *Conyza canadensis* resistant to glyphosate has been shown to be able to metabolize most of the glyphosate in the cytoplasm for vacuoles at a faster rate than in a susceptible biotype (Ge et al. 2010). Subsequently, it was reported that the reduced rate of glyphosate translocation in *Lolium* spp. was associated with vacuolar glyphosate sequestration, with a positive correlation between the level of resistance to glyphosate and the rate of vacuolar sequestration of this herbicide (Ge et al. 2012).

5 Methods of Evaluating Herbicide Behavior in Plants

5.1 Absorption

The leaf morphology of plants affects the amount of herbicide intercepted and retained, while leaf anatomy determines the process by which herbicides will be absorbed. High trichome density, higher cuticle thickness on the adaxial surface and low stomatic density on the adaxial surface were the main potential barriers detected in the retention and absorption of herbicides in *Conyza bonariensis* and *Crotalaria incana*. The main obstacle in leaves was the high content of epicuticular wax (Procópio et al. 2003).

Herbicide retention analysis in leaves is a way to identify problems related to herbicide absorption. In this analysis, a solution containing the herbicide and a fluorescein solution is applied to plants in the stage of three to four leaves, and the spray is calibrated according to the requirements of the tested herbicide. After the solution is dried in the foliage (~10 min), the plants are immersed in NaOH solution for washing the tissue. These tissues will be kiln dried, and fluorescein absorbance will be measured using a spectrofluorometer in lengths of 490/510 nm to measure fluorescein concentration. After measurement, the material is weighed, and the results are expressed in μL of the solution in g^{-1} of dry matter (Gauvrit 2003).

The contact angle of the herbicide in the leaf is also a way to analyze its absorption. In this study, the youngest leaf of the fully expanded plant is removed at the base of the petiole and placed in a horizontal position on a wooden blade. The leaf is treated with drops applied in the center of the adaxial surface containing the solution of the tested herbicide. The pattern of the drops is then observed under a horizontal microscope, and photographic images are captured. Digital analyses are developed digitally using specific software (Grangeot et al. 2006). A tribenuron-methyl-resistant *Sinapis alba* biotype had reduced retention in leaves by 24% compared to the sensitive biotype. The mean surface contact angle on the adaxial surface of the resistant biotype was 3.2 times lower compared to the sensitive biotype, which is the factor underlying the reduced retention of the herbicide in resistant biotype (Rosario et al. 2011).

In morphoanatomical analyses of leaves, scanning electron microscopy (SEM) or light optical microscopy are used. In the analysis of anatomy, the second fully expanded leaf of the plant is collected and 1 cm^2 of the leaf blade is sampled in the central region. Then the material is fixed in FAA 50 (formaldehyde + acetic acid + alcohol 50%) (Johansen 1940). The permanent slides are prepared according to the methods of the ethanol dehydration series (70%, 85%, 95%, and 100% v/v) and paraffin inclusion. Transverse histological sections are performed on rotating microtomes, and the material is stained according to the specificity of the analysis. In these histological analyses, leaf characteristics can be observed, such as epidermis, parenchyma, cuticle, and conductive vessels (Barroso et al. 2015). The anatomical structures are illustrated in Fig. 6.

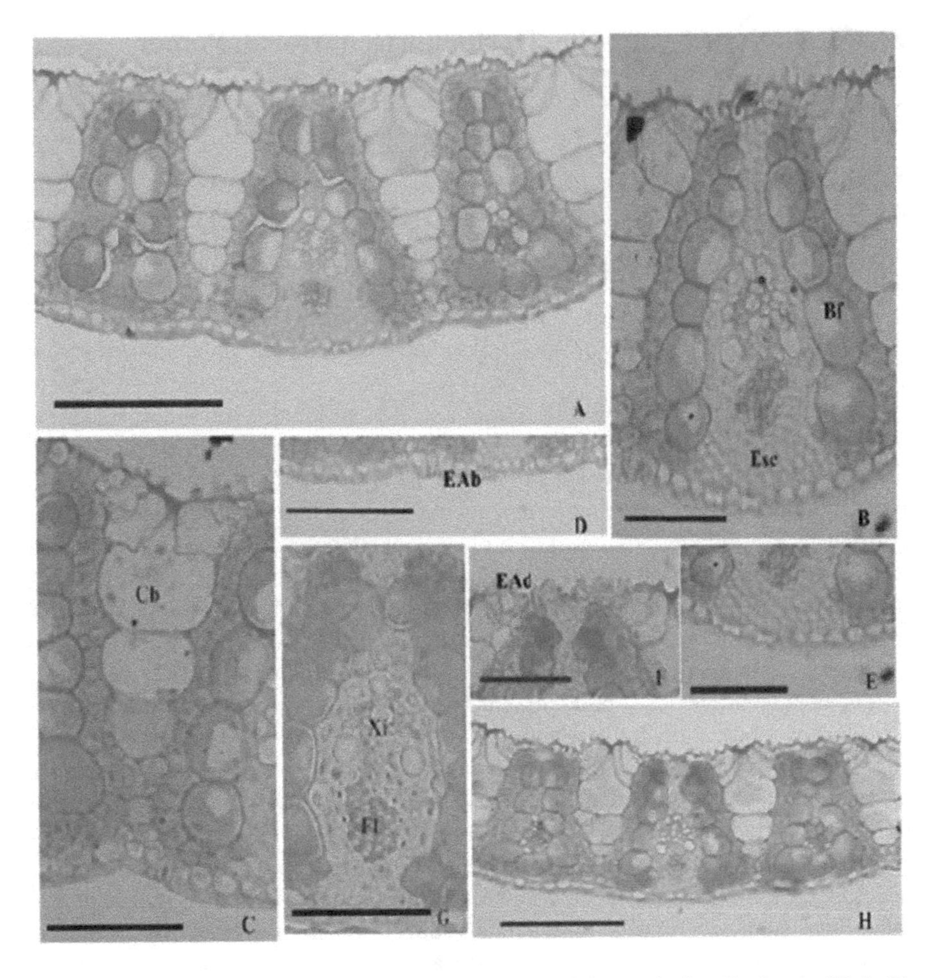

Fig. 6 (**a–h**) Cross-sections of *Zoysia japonica* leaf blade. Bf: vascular bundle sheath. CB: bulliform cells. EAb: epidermis of abaxial surface. EAd: epidermis of the adaxial surface. Esc: sclenomy. Fl: Phloem. Pair: parenchyma. Xi: Xylem. A: Central rib. H: Part between the central rib and the edge of the blade. Bar = 50 mm (M). (Source: Marques et al. 2016)

In the morphological evaluation of leaves, it is possible to observe structures such as tricha, stomata and epicuticular waxes, using SEM. After fully expanded leaf collection, 0.25 cm^2 samples of the leaf are fixed in conservative solutions (such as glutaraldehyde solutions 25%) potassium phosphate buffer (0.1 mol) or "Karnovsky" solution, which is a preparation of paraformaldehyde and glutaraldehyde. Subsequently the samples are dehydrated in an increasing series of ethanol (25%, 50%, 75% and 100% v/v). Then, the material is dried to the critical point in order to remove the ethyl alcohol from the leaf samples, replacing it with liquid CO_2, and then transforming it into the gaseous state. After drying, the samples are placed in aluminum cylinders and metallized with gold (Grangeot et al. 2006). The images are then captured in SEM as exemplified in Fig. 7.

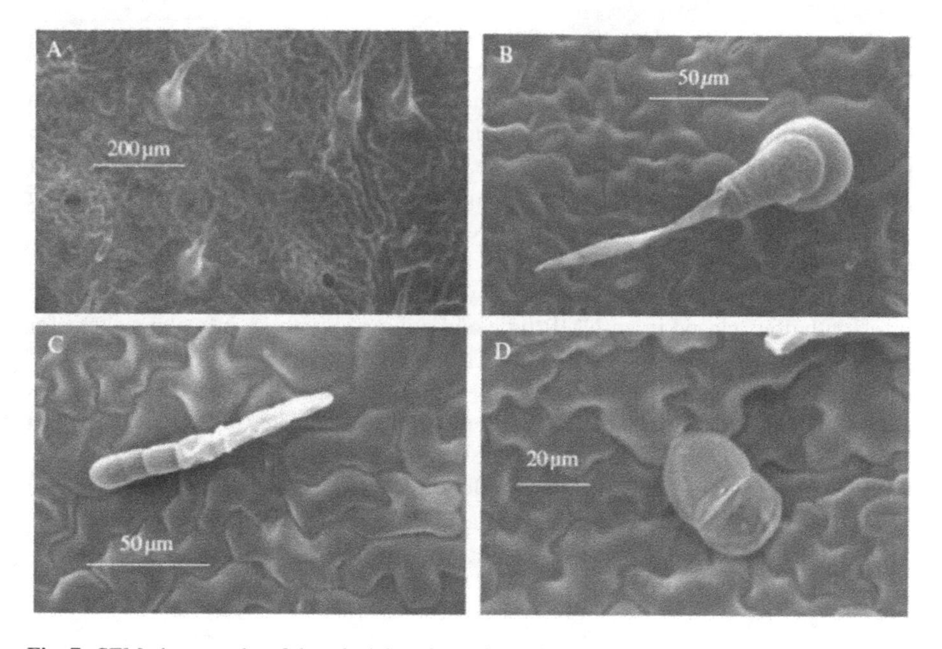

Fig. 7 SEM photographs of the adaxial surface of a leaf of *Ambrosia artemisiifolia*. (**a**) Vision of the different structures; (**b**) acute long tricha; (**c**) medium-sized tricoma; (**d**) tricomanoid. (Source: Grangeot et al. 2006)

To quantify the wax content of the leaf, 48 cm^2 of the leaf is collected and immersed in chloroform. The extract obtained is filtered and evaporated in rotavapor. The remaining solution containing the wax is transferred to pre-heavy test tubes. After complete evaporation of chloroform, the tubes are weighed again to obtain the content of epicuticular wax expressed in mg/cm^2 (Hamilton 1995). Biochemical analysis of wax is performed in a mass spectrometer by the use of gas chromatography technique. A specific solution of waxes that are injected into the apparatus is created, and as a result, there are mass spectra specific to each component, which are expressed according to the areas of the peaks obtained (Chacalis et al. 2001).

5.2 *Translocation*

Translocation is the process of movement of a herbicide from its absorption point to the place of action. The impediment of this translocation has been studied in different plants; however, there are still few confirmations of this action. The causes of the herbicide translocation, sequestration of this compound in the vacuole or cell walls are already reported causes of the impediment, as occurs in plants of *Conyza canadensis* and *Lolium multiflorum* (Ge et al. 2010, 2012; Sammons and Gaines 2014). The impediment of herbicide translocation can occur by leaf abscission, referred to in weeds as "phoenix effect" (Heap and Duke 2018) or by root exudation in plants, as occurs in plants from *Raphanus raphanistrum* to MCPA (Ghanizadeh and Harrington 2017).

Studies using Nuclear Magnetic Resonance with ^{31}P observed glyphosate in vivo entering cells and cell compartments identifying a mechanism that results in restricted translocation of this herbicide. In this technique, the herbicide is first applied, in this case glyphosate, and the tissue is collected for in vivo study. Young growing leaves in apical meristem are protected from direct exposure to the herbicide with aluminum foil. Leaf tissues are harvested at several moments after spraying, repeatedly washed with deionized water before vacuum infiltration with the infusion buffer, placed in an NMR tube and equipped with an infusion system for NMR studies (Couldwell et al. 2009). The relative partition of glyphosate (cytoplasm and vacuole) is established from signal magnitudes using signal analysis algorithms based on Bayesian probability theory and associated software (Bretthorst 1990). Therefore, after glyphosate dosage in a plant before or during NMR observation, pH changes make it possible to follow the passage of the herbicide from the apoplastic buffer (pH 5) to the cell cytoplasm (pH 6.8) and later to the vacuoles (pH 5.5) (Sammons and Gaines 2014).

5.3 Metabolization

The evaluation of herbicide metabolization can be done indirectly when metabolization occurs through the P450 enzyme. P450-specific inhibitors, such as the insecticide organophosphate (malathion), increase damage to resistant biotypes when applied in conjunction with the herbicide due to inactivation of this enzyme (Christopher et al. 1994; Yun et al. 2005; Yasuor et al. 2009). In these studies, plants in the four to six leaf stage are treated with malathion. After approximately 1.5 h, the herbicide to be tested is applied to plants. Four weeks after treatment, the percentage of survival and dry matter of plants is estimated (Dalazen et al. 2018). Torra et al. (2017) demonstrated that pretreatment of 2,4-D resistant *Papaver rhoeas* biotypes with malathion reversed the resistant to susceptible biotype in both resistant populations, showing that 2,4-D is metabolized by the plant.

6 Radiometric Techniques for the Evaluation of Herbicide Absorption, Translocation, and Metabolism by Plants

Radiolabeled herbicides with ^{14}C are used in studies of absorption, translocation and metabolism in plants because they provide some advantages compared to chemical measures, including greater sensitivity, step-by-step description of a given element in a metabolic system, herbicide position, detection by X-ray films and self-radiography imaging, in addition to liquid scintillation (Mendes et al. 2017).

Radiolabeled herbicides are also used for studies of soil herbicide dynamics (Cuervo and Fluentes 2014; Mendes et al. 2016, 2019) and phytoremediation studies (Mendes et al. 2020; Teófilo et al. 2020).

In the studies of absorption and translocation, the leaf that will receive the radio-tagged herbicide should be covered and the application of the commercial herbicide (non-radiomarked) is carried out throughout the plant. Then, a solution containing the commercial formulation of the herbicide and the radiomarked should be applied only to the leaf that has been protected. The radiolabeled product should be applied by means of a microsyringe (1 μL) to the leaf blade at the top of the leaf. The duration of this study is short; however, at least six collection times are recommended, including time zero. In each established collection time, the samples are destroyed, so the characterization of the absorption pattern in the plant is allowed, considering the appropriate statistical planning and analysis (Kniss et al. 2011). The treated leaf of each plant should be washed with appropriate solvent for each herbicide, and then the radioactivity of the washing solution is quantified by means of the Liquid Scintillation Spectrometer (LSS), quantifying the unabsorbed herbicide. Leaf absorption is calculated by the difference between applied and unabsorbed radioactivity.

The translocation of the ^{14}C-herbicide can be quantitatively analyzed by biological combustion and/or qualitatively by techniques involving self-radiography in phosphorus slide images. The usual procedure for quantifying herbicide translocation in plants is biological combustion. In this study, the sections of the plant (both the treated leaf and the part above and below it, as well as the roots) are dried in an air circulation greenhouse at 70 °C for 48 h. Subsequently, the samples are submitted to combustion in a biological oxidizer, releasing $^{14}C\text{-}CO_2$ which is fixed in scintillation solution and quantified by LSS. The radioactivity present in the parts of the plant, less of the treated leaf, are considered as the translocation of the herbicide (Ferreira and Reddy 2000). In the qualitative analysis, the treated plants, as in the absorption study, should be exposed in the phosphorus slide for 72 h, being obtained self-radiographs of the herbicide translocation in the plant using a radioscanner (Fig. 8) (Walker and Oliver 2008).

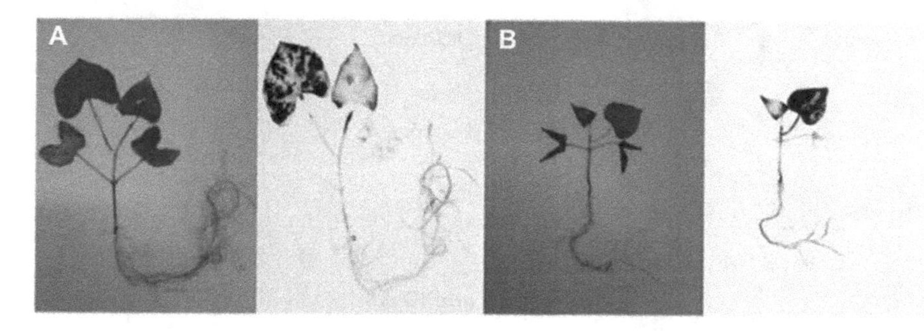

Fig. 8 Autoradiography of plants *Ipomoea purpurea* (**a**) and *Ipomoea triloba* (**b**) 48 h after treatment with ^{14}C-aminocyclopyrachlor. The pressed plant is on the right and the self-radiography of the herbicide translocation in the plant is on the left. (Source: Silva et al. 2018)

In the analysis of herbicide metabolism in plants, radiometric techniques are efficient methods and comprise three fundamental steps: plant preparation and application of treatments; extraction and separation; identification of the herbicide and its metabolites, if any (Mendes et al. 2017). The preparation of plants and application of treatments should be performed as described in the absorption and translocation study. The treated leaf is washed with non-polar solvent (usually ethanol or methanol). The plant is then frozen in liquid nitrogen and stored at −80 °C until its use. Plant tissue is macerated and homogenized with the specific solvent. This solution is then subjected to centrifugation and the supernatant is collected. The macerated plant extract is dried and then resuspended with methanol, and the liquid of this suspension is analyzed by LSS. The mass balance is expressed as the rate between the radioactivity applied at the beginning of the study and the measured total radioactivity (derived from the washing of all parts of the plant). Approximately 7 mL of the overnatant should be evaporated, resuspended at 300 mL of the solvent at 50% and centrifuged. The final sample can be analyzed by any previously described technique, such as thin-layer chromatography (TLC), high-performance liquid chromatography (HPLC), and gas chromatography (GC) (Monquero et al. 2004).

The ascending unidirectional movement in a capillary flow in the plate is the most used technique in TLC in herbicide analysis (Fried and Sharma 1999). For this, the plate on which the extract was applied will be positioned inside a glass tub containing relatively large volume of eluent (moving phase). The tank will be closed to favor the movement of samples on the plate (Mendes et al. 2016). The plates can be read in an automatic analyzer, which generate chromatographic peaks (Fig. 9). Radioactivity-sensitive plates can also be used, which will then be read in radioscanner (Fig. 10). The metabolization study is not necessarily derived from the study with radiolabeled herbicides. For example, it is possible to apply the herbicide,

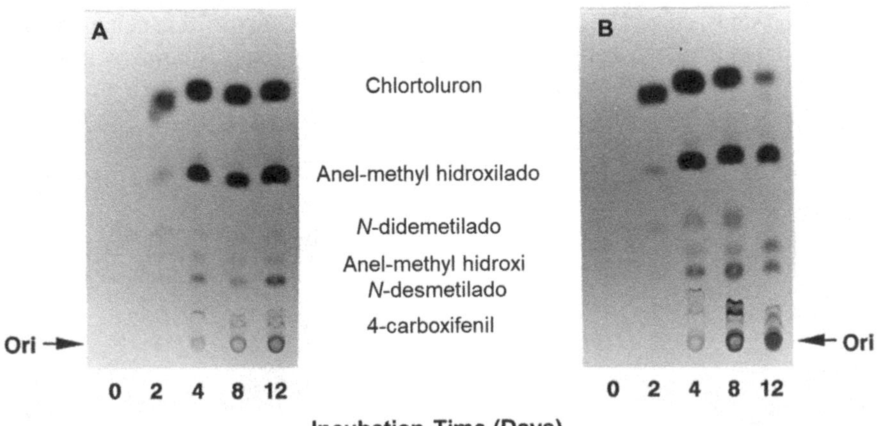

Fig. 9 TLC autoradiography of whole-plant extracts of *Nicotiana tabacum* cv. Xanthi (**a**) and Transgenic *Nicotiana tabacum* (GFC19-1) (**b**). Ori origin (chlortoluron). The signs below chlortoluron are its metabolites. (Source: Shiota et al. 1996)

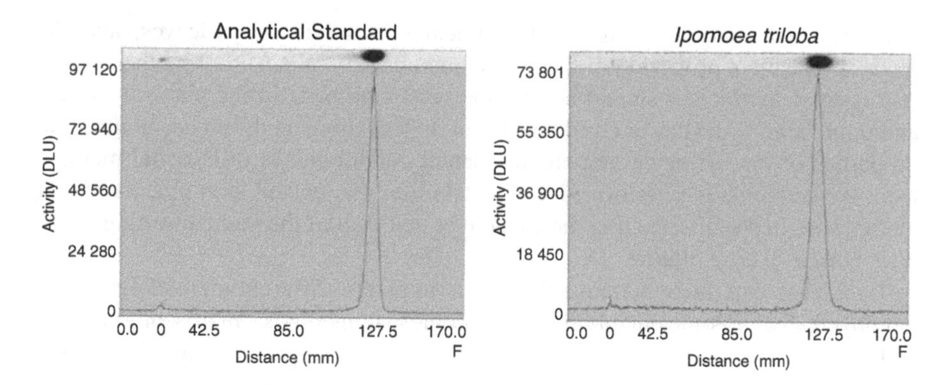

Fig. 10 Chromatograms of *Ipomoea triloba* leaf extract treated with ^{14}C-aminocyclopyrachlor at 72 h after treatment compared with analytical standard. (Source: Silva et al. 2018)

collect the tissues for them to be evaluated by HPLC, and/or mass spectrometry identifying the compounds of the route where the herbicide operates (Carvalho et al. 2012).

7 Key Concepts

- Herbicide absorption
- Leaf retention and absorption
- Absorption by the stem
- Absorption by roots
- Herbicide translocation
- Simplastic translocation
- Apoplastic translocation
- Phases of herbicide metabolization
- Methods of evaluating herbicide behavior in plants
- Radiometric techniques for herbicide studies

8 Concluding Remarks

Understanding the absorption, translocation, and metabolism of herbicides are fundamental to establish the best form of application and ensure their effectiveness of weed control. These routes are responsible for the differential selectivity of agricultural crops to herbicides and also for the resistance mechanisms not related to the place of action of these compounds in weeds.

Herbicides face different barriers until they reach the place of action and ensure the death of the plant. Absorption depends on herbicide application technology, environmental conditions, morphoanatomical characteristics of leaves, and the physical-chemical properties of the herbicide molecule. When crossing these barriers, systemic herbicides should be translocated to the parts of the plants by xylem and/or phloem, with slow or rapid translocation depending on the physical-chemical properties of the herbicide and environmental conditions. In differential metabolism, the herbicide is transformed into metabolites conjugated with glucose and/or amino acids and sequestered in the vacuole or cell wall of the plant, ensuring that it does not cause crop injuries.

In order to determine herbicide behaviors in plants, different methodologies are employed. Technologies involving chromatographic methods, microscopy images, and radiometric techniques seek to elucidate the absorption, translocation and metabolism of herbicides in weeds or crops. These techniques are widely used to determine the possible mechanisms of resistance of weeds not related to the site of herbicide action, which is a growing and worrying problem worldwide.

References

Almeida M, Almeida CV (2018) Morfologia da folha de plantas com sementes. ESALQ/USP, Piracicaba

Anderson MP, Gronwald JW (1991) Atrazine resistance in a velvetleaf (*Abutilon theophrasti*) biotype due to enhanced glutathione S-transferase activity. Plant Physiol 96(1):104–109

Anderson DM, Swanton CJ, Hall JC et al (1993) The influence of temperature and relative humidity on the efficacy of glufosinate-ammonium. Weed Res 33(2):139–147

Barroso AAM, Galeano E, Albrecht AJP et al (2015) Does sourgrass anatomy influence glyphosate resistance? Commun Sci 6(4):445–453. https://doi.org/10.14295/cs.v6i4.1124

Bretthorst GL (1990) Bayesian analysis III Applications to NMR signal detection, model selection, and parameter estimation. J Magn Reson 88(3):571–595

Bromilow RH, Chamberlain K, Evans AA (1990) Physicochemical aspects of phloem translocation of herbicides. Weed Sci 38(3):305–314

Carvalho LB (2013) Herbicida. Editado pelo autor, Lages

Carvalho SJPD, Nicolai M, Ferreira RR et al (2009) Herbicide selectivity by differential metabolism: considerations for reducing crop damages. Sci Agric 66(1):136–142. https://doi.org/10.1590/S0103-90162009000100020

Carvalho LB, Alves PLDCA, Gonzales-Torralva F et al (2012) Pool of resistance mechanisms to glyphosate in *Digitaria insularis*. J Agric Food Chem 60(2):615–622. https://doi.org/10.1021/jf204089d

Chacalis D, Reddy KN, Elmore CD, Steele ML (2001) Herbicide efficacy, leaf structure, and spray droplet contact angle among *Ipomoea* species and small flower morning glory. Weed Sci 49(5):628–634

Christoffoleti PJ, Nicolai M (2016) Aspectos de resistência de plantas daninhas a herbicidas, vol 4. HRAC-BR, Paulínia

Christopher JT, Preston C, Powles SB (1994) Malathion antagonizes metabolism-based chlorsulfuron resistance in *Lolium rigidum*. Pestic Biochem Physiol 49(3):172–182. https://doi.org/10.1017/S0043174500091748

Ciamporová LM (1992) Root structure. In: Kolek K, Kozinka V (eds) Physiology of the plant root system. Kluwer Academic Publishers, Dordrecht, pp 204–285

Cobb AH, Reade JPH (2010) Herbicide uptake and movement. In: Cobb AH, Reade JPH (eds) Herbicides and plant physiology, 2nd edn. Wiley, New York, NY, pp 50–69

Coleman J, Blake-Kalff M, Davies E (1997) Detoxification of xenobiotics by plants: chemical modification and vacuolar compartmentation. Trends Plant Sci 2(4):144–151

Couldwell DL, Dunford R, Kruger NJ et al (2009) Response of cytoplasmic pH to anoxia in plant tissues with altered activities of fermentation enzymes: application of methyl phosphonate as an NMR pH probe. Ann Bot 103(2):249–258. https://doi.org/10.1093/aob/mcn174

Cuervo JL, Fluentes CL (2014) Mineralization and sorption of ^{14}C-glyphosate in samples from three soil types collected in El Espinal, Colombia. Rev Acad Colomb Cienc Exac Fis Nat 38(148):287–297

Dalazen G, Pisoni A, Rafaeli RS et al (2018) Degradation enhancement as the mechanism of resistance to imazethapyr in barnyard grass. Planta Daninha 36:e018179504. https://doi.org/10.1590/S0100-83582018360100119

Dalazen G, Pisoni A, Menegaz C et al (2020) Hairy fleabane (*Conyza bonarienis*) response to saflufenacil in association with different formulations of glyphosate subjected to simulated rainfall. Agron Res 18(1):63–76. https://doi.org/10.15159/AR.20.033

Dimaano NG, Iwakami S (2020) Cytochrome P450-mediated herbicide metabolism in plants: current understanding and prospects. Pest Manag Sci 77:22–32. https://doi.org/10.1002/ps.6040

Dodge AD (1990) The mode of action and metabolism of herbicides. In: Hance RJ, Holly K (eds) Weed control handbook: principles, vol 8. Blackwell Scientific Publications, Oxford, pp 201–215

Donaldson TW, Bayer DE, Leonard OA (1973) Absorption of 2,4-dichlorophenoxyacetic acid and 3-(p-chlorophenyl)1, 1-dimthylurea (monuron) by barley roots. Plant Physiol 52:638

Eerd LLV, Hoagland RE, Zablotowicz RM, Hall JC (2003) Pesticide metabolism in plants and microorganisms. Weed Sci 51(4):472–495

Ferreira JFS, Reddy K (2000) Absorption and translocation of glyphosate in *Erythroxylum coca* and *E. novofranatense*. Weed Sci 48(2):193–199

Ferreira EA, Demuner AJ, Silva AA et al (2005) Chemical composition of epicuticular wax and characterization of leaf surface in sugarcane genotypes. Planta Daninha 23(4):611–619. https://doi.org/10.1590/S0100-83582005000400008

Ferreira LR, Machado AFL, Ferreira FA et al (2010) Manejo integrado de plantas daninhas na cultura do eucalipto. In: Ferreira LR et al (eds) Manejo integrado de plantas daninhas na cultura do eucalipto. Editora UFV, Viçosa, pp 39–63

Fried B, Sharma J (1999) Thin-layer chromatography, vol 4. Marcel Dekker Inc, New York, NY

Gaines TA, Lorentz L, Figge A et al (2014) RNA-Seqtranscriptome analysis to identify genes involved in metabolism-based diclofop resistance in *Lolium rigidum*. Plant J 78(5):865–876. https://doi.org/10.1111/tpj.12514

Gardin JAC, Gouzy J, Carrère S et al (2015) ALOMYbase, a resource to investigate non-target-site-based resistance to herbicides inhibiting Acetolactate-Synthase (ALS) in the major grass weed *Alopecurus myosuroides* (black-grass). BMC Genomics 16(1):590. https://doi.org/10.1186/s12864-015-1804-x

Gauvrit C (2003) Glyphosate response to calcium, ethoxylated amine surfactant, and ammonium sulfate. Weed Technol 17(4):799–804

Ge X, D'Avignon DA, Ackerman JJ et al (2010) Rapid vacuolar sequestration: the horseweed glyphosate resistance mechanism. Pest Manag Sci 66(4):345–348. https://doi.org/10.1002/ps.1911

Ge X, D'Avignon DA, Ackerman JJ et al (2012) Vacuolar glyphosate-sequestration correlates with glyphosate resistance in ryegrass (*Lolium* spp.) from Australia, South America, and Europe: a 31P NMR investigation. J Agric Food Chem 60(5):1243–1250. https://doi.org/10.1021/jf203472s

Ghanizadeh H, Harrington KC (2017) Non-target site mechanisms of resistance to herbicides. Crit Rev Plant Sci 36(1):24–34. https://doi.org/10.1080/07352689.2017.1316134

Grangeot M, Chauvel B, Gauvrit CG (2006) Spray retention, foliar uptake and translocation of glufosinate and glyphosate in *Ambrosia artemisiifolia*. Weed Res 46(2):152–162. https://doi.org/10.1111/j.1365-3180.2006.00495.x

Grover R, Cessna AJ (1991) Environmental chemistry of herbicides. CRC Press, Boca Raton, FL

Hamilton RJ (1995) Waxes: chemistry, molecular biology and functions. Orly Press, Edinburgh
Hatterman-Valenti H, Pitty A, Owen M (2011) Environmental effects on velvetleaf (*Abutilon theophrasti*) epicuticular wax deposition and herbicide absorption. Weed Sci 59(1):14–21
Heap I, Duke SO (2018) Overview of glyphosate-resistant weeds worldwide. Pest Manag Sci 74(5):1040–1049
Iwakami S, Uchino A, Kataoka Y et al (2014) Cytochrome P-450 genes induced by bispyribac-sodium treatment in a multiple-herbicide-resistant biotype of *Echinochloa phyllopogon*. Pest Manag Sci 70(4):549–558. https://doi.org/10.1002/ps.3572
Jakelaitis A, Ferreira LR, Silva AA, Miranda GV (2001) *Digitaria horizontalis* control by the herbicides glyphosate, sulfosate and potassium glifosate under different post-application rainfall intervals. Planta Daninha 19(2):279–285. https://doi.org/10.1590/S0100-83582001000200017
Johansen DA (1940) Plant microtechnique. McGrow-Hill Book, New York, NY
Jugulam M, Shyam C (2019) Non-target-site resistance to herbicides: recent developments. Plants 8(10):417. https://doi.org/10.3390/plants8100417
Kniss AR, Vassios JD, Nissen SJ et al (2011) Nonlinear regression analysis of herbicide absorption studies. Weed Sci 59(4):601–610. https://doi.org/10.1614/WS-D-11-00034.1
Kreuz K, Tommasini R, Martinoia E (1996) Old enzymes for a new job (herbicide detoxification in plants). Plant Physiol 111(2):349
Marchi G, Marchi ECS, Guimarães TG (2008) Herbicidas: mecanismos de ação e uso. Embrapa Cerrados, Planaltina
Marques RP, Martins D, Rodella RA et al (2016) Leaf anatomy of emerald grass submitted to quantitative application of herbicides. Semina: Ciências Agrárias 37(4):1767–1777
Mendes KF, Dos Reis MR, Inoue MH et al (2016) Sorption and desorption of mesotrione alone and mixed with S-metolachlor+ terbuthylazine in Brazilian soils. Geoderma 280:22–28. https://doi.org/10.1016/j.geoderma.2016.06.014
Mendes KF, Silveira RF, Inoue MH et al (2017) Procedures for detection of resistant weeds using ^{14}C-herbicide absorption, translocation and metabolism. In: Pacanoski Z (ed) Herbicide resistance in weeds and crops. IntechOpen, Rijeka, pp 159–176
Mendes KF, Alonso FG, Mertens TB et al (2019) Aminocyclopyrachlor and mesotrione sorption–desorption in municipal sewage sludge-amended soil. Bragantia 18(1):131–140. https://doi.org/10.1590/1678-4499.2017441
Mendes KF, Maset BA, Mielke KC et al (2020) Phytoremediation of quinclorac and tebuthiuron-polluted soil by green manure plants. Int J Phytoremed 23(5):474–481. https://doi.org/10.1080/15226514.2020.1825329
Miller MR, Norsworthy JK (2018) Influence of soil moisture on absorption, translocation, and metabolism of florpyrauxifen-benzyl. Weed Sci 66(4):418–423. https://doi.org/10.1017/wsc.2018.21
Monquero PA, Hirata ACS (2014) Comportamento de Herbicidas nas Plantas. In: Monquero PA (ed) Aspectos da Biologia e Manejo das Plantas Daninhas. RiMa, São Carlos, pp 145–166
Monquero PA, Christoffoleti PJ, Osuna MD et al (2004) Absorption, translocation and metabolism of glyphosate by plants tolerant and susceptible to this herbicide. Planta Daninha 22(3):445–451. https://doi.org/10.1590/S0100-83582004000300015
Nakka S, Godar AS, Thompson CR et al (2017) Rapid detoxification via Glutathione S-transferase (GST) conjugation confers a high level of atrazine resistance in Palmer amaranth (*Amaranthus palmeri*). Pest Manag Sci 73(11):2236–2243. https://doi.org/10.1002/ps.4615
Nissen SJ, Sterling TM, Namuth DM (2021a) Root absorption and xylem translocation. https://passel2.unl.edu/view/lesson/56ba9801dcbd. Accessed 20 Feb 2021
Nissen SJ, Sterling TM, Namuth DM (2021b) Foliar absorption and phloem translocation. https://passel2.unl.edu/view/lesson/c5acc4095d02. Accessed 20 Feb 2021
Oliveira Júnior RS (2001) Seletividade de herbicidas para culturas e plantas daninhas. In: Oliveira Júnior RS, Constantin J (eds) Plantas daninhas e seu manejo. Agropecuária, Guaíba, pp 291–314

Oliveira Júnior RS, Bacarin MA (2011) Absorção e translocação de herbicidas. In: Oliveira Júnior RS, Constantin J, Inoue MH (eds) Biologia e Manejo de Plantas Daninhas. Omnipax, Curitiba, pp 215–242

Palma-Bautista C, Torra J, Garcia MJ et al (2019a) Reduced absorption and impaired translocation endows glyphosate resistance in *Amaranthus palmeri* harvested in glyphosate-resistant soybean from Argentina. J Agric Food Chem 67(4):1052–1060. https://doi.org/10.1021/acs.jafc.8b06105

Palma-Bautista C, Alcántara-de la Cruz R, Rojano-Delgado AM et al (2019b) Low temperatures enhance the absorption and translocation of ^{14}C-glyphosate in glyphosate-resistant *Conyza sumatrensis*. J Plant Physiol 240:153009. https://doi.org/10.1016/j.jplph.2019.153009

Pan L, Gao H, Xia W et al (2016) Establishing a herbicide-metabolizing enzyme library in *Beckmannia syzigachne* to identify genes associated with metabolic resistance. J Exp Bot 67(6):1745–1757. https://doi.org/10.1093/jxb/erv565

Pires NM, Ferreira FA, Silva AA et al (2000) Glyphosate and sulfosate quantification in the water after rainfall simulation. Planta Daninha 18(3):491–499. https://doi.org/10.1590/S0100-83582000000300015

Plaza GA, Osuna MD, De Prado R et al (2006) Absorption and translocation of imazethapyr as a mechanism responsible for resistance of *Euphorbia heterophylla* L. biotypes to Acetolactate Synthase (ALS) inhibitors. Agron Colomb 24(2):302–305

Powles SB, Yu Q (2010) Evolution in action: resistant to herbicides plants. Annu Rev Plant Biol 61:317–347. https://doi.org/10.1146/annurev-arplant-042809-112119

Price CE (1982) A review of the factors influencing the penetration of pesticides through plant leaves. In: Cutler DF, Alvin KL, Price CE (eds) The plant cuticle. Academic Press, New York, NY, pp 237–252

Procópio SO, Silva AM, Silva AA et al (2003) Anatomia foliar de plantas daninhas do Brasil. Editora UFV, Viçosa

Reis FC, Mendes KF, Baccin L et al (2021) Seletividade, hormesis e fisiologia dos herbicidas nas plantas. In: Barroso AAM, Murata AT (eds) Matologia, Estudos sobre plantas daninhas. Editora Fábrica da Palavra, Jaboticabal, pp 295–323

Riar DS, Burke IC, Yenish JP et al (2011) Inheritance and physiological basis for 2,4-D resistance in prickly lettuce (*Lactuca serriola* L.). J Agric Food Chem 59(17):9417–9423. https://doi.org/10.1021/jf2019616

Riechers DE, Kreuz K, Zhang Q (2010) Detoxification without intoxication: herbicide safeners activate plant defense gene expression. Plant Physiol 153:3–13. https://doi.org/10.1104/pp.110.153601

Rigon CA, Ganies TA, Kupper A et al (2020) Metabolism-based herbicide resistance, the major threat among the non-target site resistance mechanisms. Outlooks Pest Manag 31(4):162–168. https://doi.org/10.1564/v31_aug_04

Robineau T, Batard Y, Nedelkina S et al (1998) The chemically inducible plant cytochrome P450 CYP76B1 actively metabolizes phenylureas and other xenobiotics. Plant Physiol 118(3):1049–1056. https://doi.org/10.1104/pp.118.3.1049

Roman ES (2005) Como funcionam os herbicidas: da biologia à aplicação. Berthier, Passo Fundo

Rosario JM, Cruz-Hipolito H, Smeda RJ et al (2011) White mustard (*Sinapis alba*) resistance to ALS-inhibiting herbicides and alternative herbicides for control in Spain. Eur J Agron 35(2):57–62. https://doi.org/10.1016/j.eja.2011.03.002

Sammons RD, Gaines TA (2014) Glyphosate resistance: state of knowledge. Pest Manag Sci 70(9):1367–1377. https://doi.org/10.1002/ps.3743

Shepherd T, Griffiths DW (2006) The effects of stress on plant cuticular waxes. New Phytol 171(3):469–499. https://doi.org/10.1111/j.1469-8137.2006.01826.x

Shimabukuro RH, Frear DS, Swanson HR et al (1971) Glutathione conjugation: an enzymatic basis for atrazine resistance in corn. Plant Physiol 47(1):10–14

Shiota N, Inui H, Ohkawa H (1996) Metabolism of the herbicide chlortoluron in transgenic tobacco plants expressing the fused enzyme between rat cytochrome P4501A1 and yeast NADPH-cytochrome P450 oxidoreductase. Pestic Biochem Physiol 54(3):190–198

Silva AFM, Silva GS, Mendes KF et al (2018) Absorption, translocation and metabolism of aminocyclopyrachlor in young plants of *Ipomoea purpurea* and *Ipomoea triloba*. Weed Res 59(1):58–66. https://doi.org/10.1111/wre.12345

Stagnari F (2007) A review of the factors influencing the absorption and efficacy of lipophilic and highly water-soluble post-emergence herbicides. Eur J Plant Sci Biotechnol 1(1):22–35

Sterling TM, Namuth DM, Nissen SJ (2021) Metabolism of herbicides or xenobiotics in plants. http://passel.unl.edu/pages/. Accessed 20 Feb 2021

Steudle E (2000) Water uptake by roots: effects of water deficit. J Exp Bot 51:1531–1542

Taiz L, Zeiger E, Moller IM et al (2017) Fisiologia e desenvolvimento vegetal, vol 6. Artmed, Porto Alegre

Tanaka FS, Hoffer BL, Shimabukuro RH et al (1990) Identification of the isomeric hydroxylated metabolites of methyl 2-[4-(2,4-dichlorophenoxy) phenoxy] propanoate (diclofop-methyl) in wheat. J Agric Food Chem 38(2):559–565

Teófilo TMS, Mendes KF, Fernandes BCC et al (2020) Phytoextraction of diuron, hexazinone, and sulfometuron-methyl from the soil by green manure species. Chemosphere 256:e127059. https://doi.org/10.1016/j.chemosphere.2020.127059

Torra J, Rojano-Delgado AM, Rey-Caballero J et al (2017) Enhanced 2,4-D metabolism in two resistant *Papaver rhoeas* populations from Spain. Front Plant Sci 8:1584. https://doi.org/10.3389/fpls.2017.01584

Trezzi MM, Teixeira SD, De Lima VA et al (2020) Relationship between the amount and composition of epicuticular wax and tolerance of *Ipomoea* biotypes to glyphosate. J Environ Sci Health B 55(11):959–967. https://doi.org/10.1080/03601234.2020.1799657

Varanasi A, Prasad PVV, Jugulam M (2016) Impact of climate change factors on weeds and herbicide efficacy. In: Sparks DL (ed) Advances in agronomy, vol 135. Elsevier, Amsterdam, pp 107–146

Walker ER, Oliver LR (2008) Translocation and absorption of glyphosate in flowering sicklepod (*Senna obtusifolia*). Weed Sci 56(3):338–343

Yamada T, Kambara Y, Imaishi H et al (2000) Molecular cloning of novel cytochrome P450 species induced by chemical treatments in cultured tobacco cells. Pestic Biochem Physiol 68(1):11–25. https://doi.org/10.1006/pest.2000.2496

Yao C, Myung K, Wang N et al (2014) Spray retention of crop protection agrochemicals on the plant surface. In: Myung K, Satchivi NM, Coleen KK (eds) Retention, uptake, and translocation of agrochemicals in plants. American Chemical Society, Washington, DC, pp 1–22

Yasuor H, Osuna MD, Ortiz A et al (2009) Mechanism of resistance to penoxsulam in late watergrass [*Echinochloa phyllopogon* (Stapf) Koss.]. J Agric Food Chem 57(9):3653–3660. https://doi.org/10.1021/jf8039999

Yu Q, Powles S (2014) Metabolism-based herbicide resistance and cross-resistance in crop weeds: a threat to herbicide sustainability and global crop production. Plant Physiol 166(3):1106–1118. https://doi.org/10.1104/pp.114.242750

Yuan JS, Tranel PJ, Stewart CN Jr (2007) Non-target-site herbicide resistance: a family business. Trends Plant Sci 12(1):6–13. https://doi.org/10.1016/j.tplants.2006.11.001

Yun MS, Yogo Y, Miura R et al (2005) Cytochrome P-450 monooxygenase activity in herbicide-resistant and-susceptible late watergrass (*Echinochloa phyllopogon*). Pestic Biochem Physiol 83:107–114. https://doi.org/10.1016/j.pestbp.2005.04.002

Zimmerlin A, Durst F (1990) Xenobiotic metabolism in plants: aryl hydroxylation of diclofop by a cytochrome P-450 enzyme from wheat. Phytochemistry 29(6):1729–1732

Induced Hormesis in Plants with Herbicide Underdoses

Kamila Cabral Mielke, Maura Gabriela da Silva Brochado, Dilma Francisca de Paula, and Kassio Ferreira Mendes

Abstract Herbicides are important tools for weed control, but both target and non-target plants are exposed to sublethal underdose composed in the field. Low herbicide doses can be used beneficially in plant growth and development. The stimulating effect of a low dose of a toxic substance is known as hormesis. The effect of hormesis is undesirable in situations where weeds are exposed to doses of hormonal herbicides, favoring their development and competition to the detriment of culture. Therefore, the importance of studies of this mechanism for professionals involved in weed management. Despite numerous theories about hormesis, few studies have evaluated its frequency, magnitude and distribution among plant chemicals. Herbicide hormesis is observed in cultivated plants and weeds, presenting several effects with potentials for exploitation in the production systems of crops of interest. However, more research on this phenomenon is needed, elucidating the negative and positive points and impacts on environmental toxicology. Therefore, this chapter brings the definition of hormesis, the hormetic dose–response of herbicides in cultivated plants and weeds and the mechanism of hormesis in the plant.

Keywords Dose–response · Pesticide · Management · Weeds · Stimulating effect · Agricultural crop

1 Introduction

Herbicides are important tools for weed control in modern agriculture, and both target and non-target plant species are often exposed to sublethal doses of these compounds under field conditions (Mobli et al. 2020). All herbicides act in crucial

K. C. Mielke (✉) · M. G. da Silva Brochado · D. F. de Paula
Federal University of Viçosa, Viçosa, Minas Gerais, Brazil
e-mail: kamilamielke@ufv.br; maura.brochado@ufv.br; dilma.paula@ufv.br

K. F. Mendes
Department of Agronomy, Federal University of Viçosa, Viçosa, Minas Gerais, Brazil
e-mail: kfmendes@ufv.br

© The Author(s), under exclusive license to Springer Nature Switzerland AG 2022

K. F. Mendes, A. Alberto da Silva (eds.), *Applied Weed and Herbicide Science*,
https://doi.org/10.1007/978-3-031-01938-8_6

pathways or processes to the plants survival, in an inhibiting or stimulating way. However, low doses of any herbicide can be used to benevolently modulate plant growth, development or composition (Velini et al. 2010). The stimulating effect of a low dose of a toxin is known as **hormesis**. This effect is common with herbicides and occurs in doses below the limit concentration capable of causing injury to plants (Belz and Duke 2014).

Originally from the English language, the term hormesis, which comes from the Greek "horme," which means "excite," has toxicological origin whose response to exposure to a certain agent considered toxic is biphasic. That is, it has two branches with different behaviors, first, stimulus or beneficial effect in low doses, and second, inhibition or toxic effect in high doses (Mattson 2008; Murado and Vázquez 2010; Marques 2019).

Initially, several herbicides were developed as plant growth regulators, proving the hypothesis of hormesis (Silva et al. 2012). Auxin-based herbicides are well-known examples of chemicals that increase growth in non-toxic concentrations, mimetizing auxin, growth hormone, but which are lethal at higher doses (Americo et al. 2017; Marques 2019). Another example is glyphosate itself, whose predecessor, glyphosine (used in Brazil as a ripener), is still used as a growth regulator in several countries (Halter 2009; Silva et al. 2012).

Despite numerous theories about the cause of hormesis, few studies have systematically evaluated its frequency, magnitude, and distribution among plant chemicals, including comparable dose–response curve studies (Cedergreen 2008; Pinheiro 2020).

Thus, this chapter will address the definition of hormesis, as well as the hormetic dose–response of herbicides in cultivated plants and mechanism of hormesis in the plant. Finally, the effect of hormesis on weeds, which is also very important for all professionals involved in the field of weed science and management.

2 Definition of Hormese

The term hormesis is widely known in the field of toxicology. This phenomenon is common in most plant species and with different types of toxic agents, including herbicides, being influenced by biological and environmental factors (Belz and Duke 2017). In general, hormesis protects plants from stress and improves crop productivity. It also improves plant tolerance against adverse conditions and provides them with a competitive advantage, allowing them to survive and maintain their biomass production (Agathokleous and Calabrese 2019).

Hormesis is defined as an adaptive response of a biphasic dose in which high doses of a toxic agent can cause inhibition and low doses of the same toxic can cause stimulation (Calabrese 2018), as previously reported. Adaptive response is induced by the initial disturbance of the state of homeostasis (Calabrese and Mattson 2017). **Homeostasis** is defined as the constant subsistence of the internal state of an organism with efficient functions and performances ensuring a physiologically stable environment is maintained in the face of a disturbance (Brito and Haddad 2017). This plant response is reproduced in a series of physiological processes, such as photosystem II (PSII) efficiency, photosynthetic rate, chlorophyll content, Hill reaction activity, signaling pathways, and antioxidant enzymes whose responses are eventually reflected in hormonal responses (Mattson 2008; Agathokleous and Calabrese 2019).

Thus, hormesis management can be a strategy to satisfy the demand of agriculture and maintain desirable yields in crops. However, doses of chemicals with hormetic effect and levels of injuries in plants are close, which makes it difficult to use toxic products to improve crop productivity.

3 Hormetic Dose–Response of Herbicides in Cultivated Plants

The first reports of hormetic effect of herbicides in plants appeared in 1950, and since then, there have been studies that have evaluated the effect of low doses of herbicides on plant growth (Reis et al. 2021). Plants may receive low doses of herbicides intentionally to regulate growth, modify the composition of plant biomass, or accumulate specific compounds (Meseldžija et al. 2020). Low herbicide doses can also reach plants through drift deposition or inadequate herbicide application conditions (wind, rain, dew, high temperatures, or humidity). Contact may also occur between treated and untreated weeds, incorrectly calibrated equipment, protection of weeds due to coverage by cultivated plants (popularly known as umbrella effect), and/or dead cover (Velini et al. 2010; Belz et al. 2018; Duke et al. 2017). Due to these factors, target and non-target plants are commonly exposed to underdoses of herbicides under field conditions (Belz et al. 2018).

To find the dose that causes the hormetic effect in plants, it is essential to expand the range of doses, since the hormonal effect can occur in very low doses compared to the commercial dose. It is necessary to consider doses ranges 100 times lower than the recommended herbicide dose in the field (Brito et al. 2018). Hormesis appears as a maximum stimulator response occurring within the response zone induced by herbicide underdoses (Fig. 1). The maximum response is commonly twice the control response and <100 times the level of adverse effect in which it causes injuries in the plant (Calabrese and Blain 2011; Agathokleous et al. 2019).

The required doses of an herbicide to stimulate hormesis are usually found in a narrow range, just before the onset of the negative effects of the herbicide (Belz and Duke 2017). Experiments with dose–response curves are commonly used to identify the relationships between the different exposure levels and the effect caused by stressor agents, in this case herbicides (Calabrese 2004). An example of a dose–response curve of glyphosate in coffee is presented in Fig. 2. The stimulating effect was observed at doses 738 g a.e./ha, 416 g a.e./ha, 620 g a.e./ha, 536 g a.e./ha,

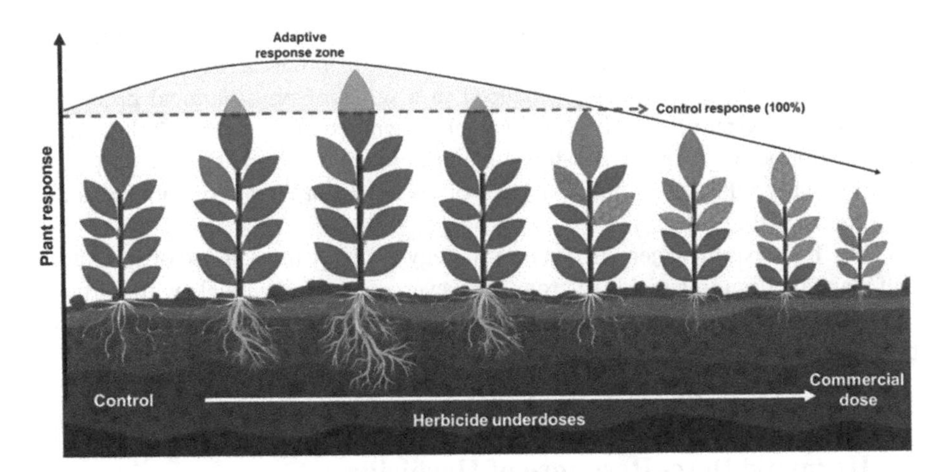

Fig. 1 Induction of hormesis in plants by herbicide underdoses. (Source: Adapted from Agathokleous et al. 2019 and Jalal et al. 2021)

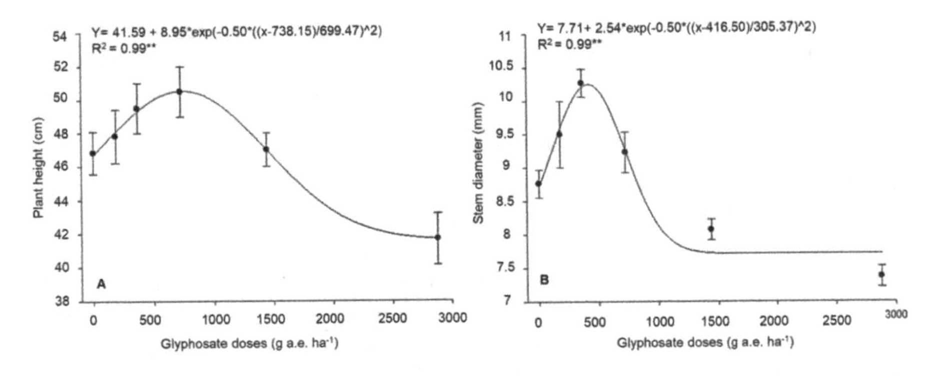

Fig. 2 Plant height (**a**) and stem diameter (**b**) of coffee, 60 days after exposure to glyphosate applied to plants at 45 days after transplantation. Vertical bars indicate the standard error of the mean and ** the significance at 1% probability. (Source: Carvalho et al. 2013)

488 g a.e./ha, 530 g a.e./ha, and 513 g a.e./ha of the herbicide, with maximum gains of 21–18% in plant height and stem diameter, respectively, in relation to the control, without glyphosate application. The hormetic effect was observed only in plants exposed to glyphosate at 45 days after transplantation (DAT) of seedling, while the exposure of younger plants (10 DAT) did not present this growth stimulus (Carvalho et al. 2013).

Herbicides that inhibit biosynthetic processes, such as lipids, carotenoids, and amino acid synthesis, are options for use in low doses, because their application can alter the concentration of many compounds in plants (Velini et al. 2008). However, the most widely used herbicide in recent years, glyphosate, has often been studied for hormetic potential in several plant species, including cultivated plants and weeds (Brito et al. 2018). The application of underdoses of different herbicides and their effects on cultivated plants are shown in Table 1. The hormesis stimulation values of herbicides vary on average between 20% and 30% above the control (Belz et al. 2011).

Glyphosate is the only product that has as its mechanism of action the inhibition of the enzyme 5-enolpyruvylshikimate-3-phosphate synthase (EPSPS) in concentrations that allow its commercial use as an herbicide (Brito et al. 2018). EPSPS is part of the shikimic acid pathway and is responsible for one of the steps in the synthesis of aromatic amino acids tryptophan, phenylalanine, and tyrosine (Oliveira 2011). Most reports of hormetic effects of glyphosate are about morphological changes in plants, such as the height and diameter of the stem; fresh and dry matter of the aerial part and/or root system; leaf area; plant tillering; mass and number of seeds or flowers (Brito et al. 2018). However, other effects were also observed with the application of glyphosate underdoses. For example, underdoses of 6.2–20.2 g a.e./ha increased the assimilation of CO_2, the stomatic conductance and transpiration of sugarcane and eucalyptus plants (Nascentes et al. 2018). On the other hand, underdose of 3.6 g a.e./ha provided increments of 203% and 170% in benzoic and salicylic acid, respectively, and these compounds are present in the shikimate route (Silva 2014).

Low doses of any herbicide can be used to modulate plant growth, composition, and physiology (Brito et al. 2018). Low doses of 2,4-D responses have been analyzed and have shown "hormonal" responses in plants (Americo et al. 2017). The underdose-induced hormesis of 2,4-D is related to increased metabolism of nucleic acids, cell division and elongation, cell wall plasticity, and RNA polymerase stimulation, while higher doses are lethal to plants (Oliveira 2011). The hormetic effects of 2,4-D were related to morphological gains of the plant, such as height, dry biomass, and reproductive structures (Table 1). The main practical use of auxinic herbicides, such as 2,4-D, is to control the ripening of citrus fruits in reducing fruit fall or in the extension of the harvest period (Almeida et al. 2004; Rufini et al. 2008). The combination of 25 mg/L of gibberellin (GA_3) plus 25 mg/L of 2,4-D reduced the natural fall of *Citrus sinensis* fruits (Orange Pear) by 78%, inhibiting fruit abscission within 3 months (Almeida et al. 2004). Treatment with 16 mg/L of 2,4-D in the form of isopropyl ester reduced the fruit drop of *Citrus sinensis* by 62% compared to untreated control (Anthony and Coggins 1999).

Herbicides are not equally effective in inducing hormesis in the plant, and some may not even provide hormetic response. The herbicides glyphosate, amicarbazone, mesotrione, imazapic, and carfentrazone were evaluated for hormetic potential in soybean plants. Only the herbicide glyphosate presented hormesis in the crop. Paraquat in the underdose of 24 g a.i./ha increased plant height and cotton seed production (Melero 2016); however, there are no reports of the hormetic effect of paraquat on other cultivated plants. Acifluorfen-sodium, glyphosate, diquat, haloxyfop-methy, MCPA, metsulfuron-methyl, pendimethalin, and terbuthylazine have been tested for hormetic effect on barley plants (*H. vulgare*) (Cedergreen 2008). The authors found that glyphosate and metsulfuron-methyl provided a 25%

Table 1 Underdoses of different herbicides and their effects on cultivated plants

Scientific name	Common name	Herbicide	Concentration (g a.i./ha or a.e./ha)	Hormetic effects	Sources
Oryza sativa	Rice	Glyphosate	11.6–32.0	It stimulated the number of total grains and the grenade panicles	Gitti et al. (2011)
Hordeum vulgare	Barley	Glyphosate	11.0–45.0	It increased photosynthetic rates	Cedergreen and Olesen (2010)
Hordeum vulgare	Barley	Glyphosate	2.5–20.0	It increased barley grain yield by 15%	Cedergreen et al. (2009)
Hordeum vulgare	Barley	Glyphosate and metsulfuron-methyl	143.0 and 1.3	It increased biomass growth and accumulation by 25%	Cedergreen (2008)
Glycine max	Soybean	Glyphosate	20.0	It increased sprout dry mass by 30%	Velini et al. (2008)
Zea mays	Corn		35.0	It increased sprout dry mass by 30%	
Eucalyptus spp.	Eucalyptus		7.2	It increased total dry mass by 70%	
Pinus caribea	Pinus		5.4	It increased total dry mass by 20%	
Zea mays	Corn	Glyphosate	0.084 and 7.0	It improved germination by 16%	Barbosa et al. (2017)
Saccharum spp.	Sugarcane	Glyphosate	1.8	The number of byes increased by 32%, the number of stems (tons per hectare), and sucrose accumulation (6.3% higher) increased by 32%	Pincelli et al. (2020)
Saccharum spp. and *Eucalyptus* spp.	Sugarcane and eucalyptus	Glyphosate	6.2–20.2	The total dry weight of sugarcane and eucalyptus increased by 28.8% and 35.3%. The transpiration rate of sugarcane increased by 80.6% and that of eucalyptus 86.1%	Nascentes et al. (2018)

(continued)

Table 1 (continued)

Scientific name	Common name	Herbicide	Concentration (g a.i./ha or a.e./ha)	Hormetic effects	Sources
Saccharum spp.	Sugarcane	Glyphosate	72.0–180.0	Increases in phenylalanine and tyrosine levels and plant stem biomass	Carbonari et al. (2014)
Schizolobium amazonicum	Parica	Glyphosate	9, 18, and 36	39.3% increase in plant height	Marques et al. (2020)
Coffea arabica	Coffee	Glyphosate	416–738	Increases of 18–39%, were observed at plant height, stem diameter, number of leaves, leaf area, and dry matter of the plant	Carvalho et al. (2013)
Phaseolus vulgaris	Bean	Glyphosate	12.0	Increase in bean vegetative mass and increases in productivity of up to 375 kg/ha	Silva et al. (2016a, b)
Solanum lycopersicum	Tomato	Glyphosate	0.2–5	It increased the growth of hypocotyl and radicle and increased rates of photosynthesis (up to 2 times)	Khan et al. (2020)
Triticum aestivum	Wheat	Glyphosate	18.0	It stimulated plant growth, and effects were maintained until maturity, including increased grain production	Abbas et al. (2016a, b)
Cicer arietinum	Chickpea	Glyphosate	18.0–72.0	Maximum stimulation in different growth and production parameters	Abbas et al. (2015)
Glycine max	Soybean	2,4-D	19.2–20.3	Number of grains and grain mass per plant reached maximum values	Silva et al. (2019)
Gossypium hirsutum	Cotton	2,4-D	1.75	It increased cotton seed productivity	Americo et al. (2017)
Caryocar brasiliense	Pequi	2,4-D	3.3	20.7% increase in leaf area	Tavares et al. (2017)

(continued)

Table 1 (continued)

Scientific name	Common name	Herbicide	Concentration (g a.i./ha or a.e./ha)	Hormetic effects	Sources
Triticum aestivum	Wheat	2,4-D	5.0–20.0	It increased chlorophyll content index and chlorophyll fluorescence values	Kaur and Kaur (2019)
Gossypium hirsutum	Cotton	2,4-D	0.855–1.71	It increase in height, number of leaves, shoot dry matter, and total dry matter	Marques et al. (2019)
Citrus reticulata	Citrus "ponkan"	2,4-D	10.0[a]	It influenced fruit texture, increased acidity, and reduced total soluble solids/fruit acidity ratio	Rufini et al. (2008)
Citrus aurantium	Citrus "tangerine"	2,4-D	50 and 100[a]	It increased fruit size, peel weight, and juice content	El-Otmani et al. (1993)
Secale cereale and *Pisum sativum*	Rye and pea	Simazine	0.05 and 0.8[b]	79% increase in plant protein content. The level of the enzyme nitrate reductase is increased	Ries et al. (1967)
Glycine max	Soybean	Sulfentrazone and lactofen	9.0 and 70.0	It increased phytoalexin production and reduction of lesions caused by *Sclerotinia sclerotiorum*	Nelson et al. (2002)
Glycine max	Soybean	Chlorimuron-ethyl and 2,4-D	0.4 and 20.0	Increases in height, stem diameter, biomass accumulation, and plant dry matter	Silva et al. (2020a, b)

Source: Adapted from Belz and Duke (2014) and Jalal et al. (2021)

[a] mg/L

[b] μM

increase in biomass. The other six herbicides tested did not induce consistent hormesis, showing that this is not a general response to the stress caused by the herbicides. Thus, the hormetic response depends on the type of herbicide, time of application, dosage, species, physiological stage, the measured characteristics, environmental conditions, and among other practices (Cedergreen and Olesen 2010).

The growth stage of the plant and physiological stage showed to have influence on the response of the plant to the underdoses of the herbicide. In soybean, a small dose of 2,4-D (about 13 g a.e./ha) during stage V4 or V6 can lead to a significant increase in crop yield (Silva et al. 2019). However, even if there are no age-related differences in the hormesis, younger ones are generally more sensitive to herbicides than older plants of the same species. Therefore, the doses that induce hormesis may be lower in younger plants (Belz and Duke 2014).

The characteristics evaluated are usually related to plant development, and the hormetic effects may be distinct. Velini et al. (2008) observed that application of glyphosate underdose (2.0–14.6 g a.e./ha) provided pine hormesis (*Pinus caribaea*) with an increase of 153% of root dry matter, while leaf dry matter increased 121% and that of the stem did not present a hormetic response in any evaluated dose. Low doses of 2,4-D (5–20 g a.e./ha) increased chlorophyll contents of wheat plants compared to control. However, no significant difference was observed in grain and straw yield at harvest, indicating that growth stimulation by underdoses of 2.4-D was not sustained over time (Kaur and Kaur 2019). Thus, the hormetic responses observed in low doses rarely lead to a general or sustainable improvement in plant fitness (Belz and Duke 2014).

Atmospheric levels of CO_2, light intensity, temperature, and nutritional availability may have effects on the plant's hormetic response. Prestressing a plant can also influence hormesis (Belz and Duke 2014). The pretreatment of soybean plants with underdoses of glyphosate (3.6 and 7.2 g a.e./ha), conditioned soybean plants to have greater growth stimulus when treated later with glyphosate, than plants without pretreatment with the herbicide (Silva et al. 2016a, b). Under water deficit conditions, low doses of glyphosate (21–37 g a.e./ha) can act as mitigating water stress in *Carthamus tinctorius* plants, allowing plants to maintain their metabolism and reach levels close to those of plants under optimal conditions (Santos et al. 2021). Low nitrogen availability showed a positive correlation with the hormesis of white oat plants (*Avena sativa*) to glyphosate, in which the dose of 14.6 g a.e./ha of the herbicide provided an increase of 43% in the plant's dry matter and increased the crop yield by ~30% under conditions of low nitrogen availability (Silva et al. 2020a, b). Hormesis is influenced by almost all tested parameters, which makes it a phenomenon that is sometimes difficult to reproduce, even with minor changes study (Belz and Duke 2017).

4 Hormesis Mechanism in Plants

Cellular and molecular mechanisms under the effect of hormesis include activation of signaling pathways, ionic channels, kinases and deacetylases, and transcription factors responsible for the production of cytoprotective proteins, such as antioxidant enzymes (superoxide dismutases and glutathione peroxidase) (Mattson 2008; Vargas-Hernandez et al. 2017). The way herbicides act in causing hormetic effects is still unclear; however, there are some proposed mechanisms. The hormetic response to a stress factor can have a specific mode of action, such as a growth stimulator (Belz and Duke 2014). Some chemicals can affect plant hormones in low doses. For example, chemicals that stimulate the hormonal responses responsible for leaf or root stretching may initiate an increase in these characteristics at low doses, while they may have deleterious effects in higher doses due to the same or other mode of action (Duke et al. 2006). Auxinic herbicides are well-known examples of chemicals that increase growth in non-toxic concentrations, mimicking growth hormone auxin, but which are lethal at higher doses (Allender 1997).

The proposal that the mechanism of hormesis is dependent on the mechanism of action of the herbicide and not on a hormonal mechanism was also analyzed (Duke et al. 2006; Belz and Duke 2014). This finding was observed when the effect of glyphosate underdoses was evaluated in soybean and corn plants resistant to glyphosate. These plants did not present hormetic effects. This indicated that the stimulating effect was dependent on the same sensitive target site, the enzyme EPSPS, responsible for plant injury (Velini et al. 2008). In this case of glyphosate, it has been speculated that low doses of the herbicide inhibit lignin synthesis by inhibiting the enzyme EPSPS, which allows the fixation of more carbon in the production and sucrose by plants (Kruse et al. 2000). Therefore, underdoses of glyphosate instill lignin synthesis that can increase cell wall plasticity and prolong the plant growth period (Duke et al. 2006).

There have been attempts to use herbicides to stimulate hormesis in plants in the past, but it has not reached commercialization, other than the use of glyphosate and sulfometuron-methyl used in maturation, to increase the production of sucrose in sugarcane (Leite and Crusciol 2008; Meschede et al. 2010). The mechanism of glyphosate and sulfometuron-methyl in sucrose accumulation in sugarcane is probably unique (Dalley and Richard 2010). It is known that glyphosate and sulfometuron-methyl are mobile in phloem accumulate preferentially in meristematic tissues, reducing metabolic activity. This should reduce growth, leading to the accumulation of sugar that would normally be used for plant growth that are less affected by herbicides (Belz and Duke 2014).

A more indirect mechanism of hormesis, **overcompensation**, is related to the induction of plant defenses. The mechanism of overcompensation occurs by induction by minimal stress caused by the herbicide in enzymatic systems, which can promote beneficial effects in plants (Belz and Duke 2014). However, in this case, the response is more specifically related to the mode of action of the herbicide, which generates reactive oxygen species (ROS) which induce defenses of the host plant (Belz and Duke 2014). ROS production includes hydrogen peroxide (H_2O_2), which at high concentrations can cause cell death, while at low concentrations can improve plant tolerance to biotic and abiotic stress (Jalal et al. 2021). Low stress levels also induce compensatory enzymatic systems that are induced by the interruption of homeostasis. This can cause a hormetic response, such as growth, although this mechanism has not been proven with herbicides (Belz and Duke 2014). In general, the physiological and biochemical mechanisms of herbicide-associated hormesis have hardly been studied. The identification of mechanisms associated with hormetic responses is limited, since there is difficulty in consistent responses in specific doses.

5 Effect of Hormesis on Weeds

The use of herbicides is an important tool in weed management in agriculture (Hamadache et al. 2016). The effect of hormesis may be undesirable in situations where weeds are unintentionally exposed to hormonal doses (Belz and Duke 2014). The phenomenon of hormesis can alter the competition relationship of plant species, altering the competition between weeds themselves and between weeds and agricultural crops. Species that have a growth gain due to hormonal underdoses may have competitive advantages when compared to those that were not exposed to underdoses or were not affected by the herbicide (Cedergreen 2008; Abbas et al. 2016a, b).

Research on herbicide underdoses is also important for herbicide-resistant weeds. The resistant biotype may have a hormetic response to herbicide doses recommended in the package leaflet, increasing its competitive potential with the culture of interest (Petersen et al. 2008). Doses of metamitron (21.0–4200 g a.i./ha) induced considerable hormesis in a white goosefoot biotype (*Chenopodium album*) resistant to photosystem II inhibitor herbicides. Metamitron doses led to higher biomass and a 32% increase in seed production, reflecting the improvement in reproductive fitness of the resistant biotype. Herbicide doses in the field can not only select resistant plants but also increase their reproductive fitness (Belz 2020). Underdoses of phenoxaprop-p-ethyl in little seed canarygrass (*Phalaris minor*) susceptible and resistant to herbicide increased seed growth and production. However, in the sensitive biotype, only 8% of the recommended dose was required, while in the resistant one, it was 20% (Farooq et al. 2019). The application of underdoses of glyphosate (0.70–45 g a.e./ha) in resistant fleabane biotype (*Conyza* spp.) provided an increase in plant dry matter (Fig. 3a). The resistant biotype had an increase in the number of chapters and number of seeds per plant of 48.2% and 114%, respectively, at a dose of 5.62 g a.e./ha in relation to the control treatment (Fig. 3b, c) (Cesco 2018).

The effects of herbicide underdoses on weeds can be variable within the same species and between different species. The application of 11.25 g a.e./ha of glyphosate promoted an increase in dry matter of the aerial part of Urochloa plants (*Urochloa decumbens*). However, the plants showed a different response regarding the application of herbicide underdoses and may promote stimulation and inhibition of plant growth (Moraes et al. 2019). An underdose of 7.2 g a.e./ha glyphosate inhibited the fresh matter of pigweed (*Amaranthus retroflexus*) by 80.8%, while in the same dose, the fresh soybean matter was reduced by 6.36% (Meseldžija et al. 2020). Therefore, even under controlled conditions, the effect of glyphosate on plants is variable and dependent on environmental, morphological, physiological conditions of plants and the applied dose of the product. This indicates that

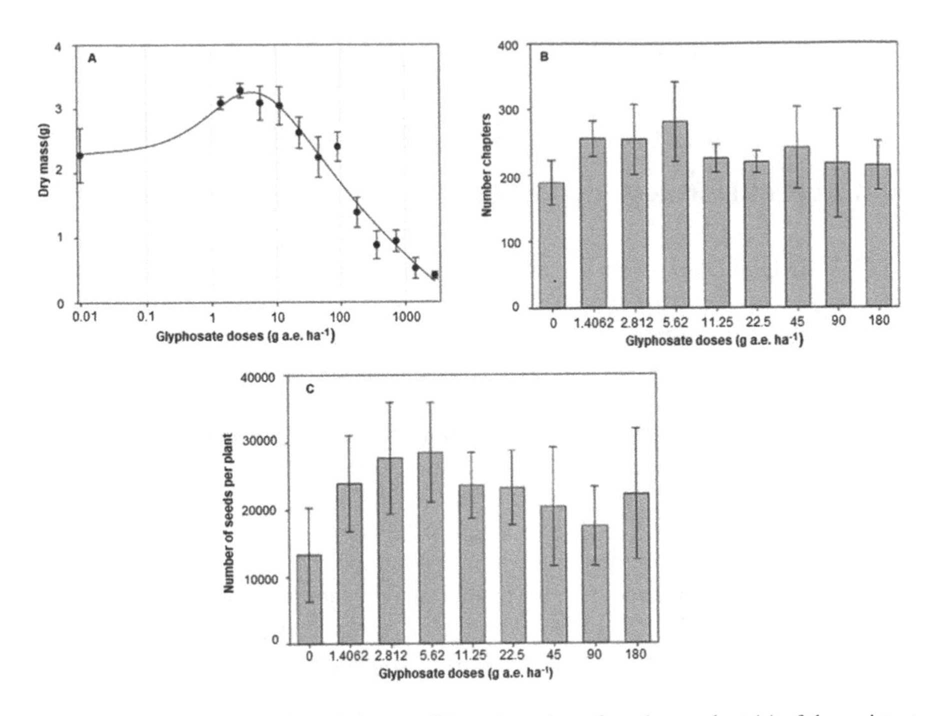

Fig. 3 Dry matter (**a**), number of chapters (**b**), and number of seeds per plant (**c**) of the resistant fleabane biotype due to glyphosate doses. Vertical bars indicate the standard error of the mean. (Source: Cesco 2018)

hormesis can play a significant role in changing crop/weed competition and can directly influence management.

Studies on weed hormesis are quite limited; however, some research has shown the effects of herbicide underdoses on hormetic response of weeds (Table 2). The stimulation in germination, growth, and production of weed seeds by herbicide underdoses can translate into greater competitive capacity under practical conditions within field crops (Ather et al. 2017). The sourgrass (*Digitaria insularis*) subjected to underdoses of glyphosate (5.62–45 g a.e./ha) showed an increase in height, biomass, seed weight, germination speed, and reduction of the time required for inflorescence emission. These effects can give a great advantage to the spread of this species, assisting in the alteration of the weed flora (Anunciato 2018). Hormesis was also reported in aquatic weeds, warning about the problem of herbicide residues in the contamination of water bodies, and the difficulty of controlling these plants (Peres et al. 2017). The aquatic plant *Lemna minor*, the microalgae *Pseudokirchneriella subcapitata*, and two terrestrial plants *Tripleurospermum inodorum* and *Stellaria media* were exposed to nine herbicides to evaluate the dose–response curve of hormesis. Hormesis was found in >70% of dose–response curves

Table 2 Underdoses of different herbicides and their effects on weeds

Scientific name	Common name	Herbicide	Concentration (g a.i./ha or a.e./ha)	Hormetic effects	Sources
Coronopus didymus	Lesser swinecress	Glyphosate	16.0	The number of seeds per plant and stem length increased by 26%. Root length increased by 30%	Ather et al. (2017)
Chenopodium album	Lambsquarters		16.0	The length of the root and stem increased by ~70%. The number of seeds per plant increased by 28%	
Rumex dentatus	Dock		16.0 and 32.0	The number of seeds per plant increased by 25%. Increased root and stem length by ~80%	
Lathyrus aphaca	Yellow pea		32.0	It increased the number of seeds per plant by 60%	
Egeria densa	Leafy elodea	2,4-D	11.2[a]	It stimulated stem tip growth by 64.8%	Peres et al. (2017)
Egeria najas	Narrowleaf anacharis		1.0[a]	Plant growth stimulus by 75%	
Ranunculus aquatilis	Whitewater crowfoot	2,4-D	10–30[a]	It increased stem length and number of roots	Belgers et al. (2007)
Phalaris minor and *Avena fatua*	Canarygrass and oat	Fenoxaprop-p-ethyl	6.0	It increased seed production by 28% and 17% for *P. minor* and *A. fatua*, respectively	Abbas et al. (2016a, b)
Echinochloa colona	Junglerice	Glyphosate	5.0	It increased total dry biomass by 64% (resistant biotypes) and 54% (susceptible biotypes)	Mollaee et al. (2020)

(continued)

Table 2 (continued)

Scientific name	Common name	Herbicide	Concentration (g a.i./ha or a.e./ha)	Hormetic effects	Sources
Sonchus oleraceus	Sowthistle	Glyphosate	5.0	Significant increase in plant height, number of leaves, in addition to increase in the number of seeds by 101%	Mobli et al. (2020)
Digitaria insularis	Sourgrass	2,4-D	100.7	It increased plant height by 39.1%	Gazola et al. (2019)
Digitaria horizontalis	Jamaican crabgrass			It increased growth by 13.5%	
Urochloa decumbens	Signal grass	Glyphosate	11.2	It increased dry biomass by 30%	Moraes et al. (2019)
Conyza bonariensis	Hairy fleabane	Glyphosate	5.6	It increased floral development by 48.2% and the number of seeds per plant by 114%	Cesco (2018)
Prymnesium parvum	Golden algae	Glyphosate	1.0	It increased plant growth rate by 45%	Dabney and Patiño (2018)
Urochloa brizantha	Palisade grass	Glyphosate	20.0	It increased dry matter and lignin production	Codognoto et al. (2021)

[a] mg/L

with glyphosate and metsulfuron-methyl and in >50% of the curves with acifluorfen and terbuthylazine (Cedergreen et al. 2007).

In view of the research, the importance of the influence of hormesis on the development of weeds is evident. Resistant biotypes submitted to herbicide underdoses were more able to reproduce, an important factor for selection and increase in the number of resistant individuals in the field (Gomes 2014; Cesco 2018). This unwanted aspect of hormesis should be considered in the implementation of effective weed management practices (Abbas et al. 2016a, b). Observing climatic conditions at the time of application, having knowledge of the herbicide and the influence of underdoses in weeds, can reduce the hormesis problems of these plants in the field. With the implementation of genetically modified crops, the challenge becomes greater, as herbicides are used more comprehensively in the crop, requiring better control and knowledge of the relationship of underdoses in weeds. The understanding of hormesis mechanisms can contribute to the use of herbicides more cautiously in the field, always aiming at increasing yield and crop quality.

6 Key Concepts

- Definition of hormesis
- Hormetic dose–response of herbicides in cultivated plants and weeds
- The mechanism of hormesis in the plant
- The importance of the study of the phenomenon of hormesis in weed management and agricultural crop

7 Concluding Remarks

Herbicide hormesis was observed in different cultivated plants and weeds, presenting a diversity of hormetic effects that provides potential approaches to explore this phenomenon in the production systems of crops of interest. Several factors influence the hormetic response of plants and the mechanism that provides this effect is still poorly understood. So far, it is observed that the results of the hormesis depends on the exposure of the plant to a underdose of herbicides that can bring different results, according to the characteristic of the vegetable to be evaluated. Therefore, herbicides do not always bring consistent responses to hormesis. This variability makes it difficult to establish the ideal underdose to provide a beneficial effect to plants. Thus, more research on this phenomenon is needed, elucidating the negative and positive points, such as the benefits provided by this technique in agricultural production and environmental toxicology, as well as the mechanism of action of hormesis in plants.

References

Abbas T, Nadeem MA, Tanveer A et al (2015) Glyphosate hormesis increases growth and yield of chickpea (*Cicer arietinum* L.). Pak J Weed Sci Res 21(4):533–542

Abbas T, Nadeem MA, Tanveer A et al (2016a) Low doses of fenoxaprop-p-ethyl cause hormese in littleseed canarygrass and wild oat. Planta Daninha 34(3):527–533. https://doi.org/10.1590/S0100-83582016340300013

Abbas T, Nadeem MA, Tanveer A et al (2016b) Glyphosate herbicide causes hormesis in wheat. Pak J Weed Sci Res 22(4):575–586

Agathokleous E, Calabrese EJ (2019) Hormesis can enhance agricultural sustainability in a changing world. Glob Food Sec 20:150–155. https://doi.org/10.1016/j.gfs.2019.02.005

Agathokleous E, Kitao M, Calabrese EJ (2019) Hormesis: a compelling platform for sophisticated plant science. Trends Plant Sci 24(4):318–327. https://doi.org/10.1016/j.tplants.2019.01.004

Allender WJ (1997) Effect of trifluoperazine and verapamil on herbicide stimulated growth of cotton. J Plant Nutr 20(1):69–80

Almeida IMLD, Rodrigues JD, Ono EO (2004) Application of plant growth regulators at preharvest for fruit development of 'PERA' oranges. Braz Arch Biol Technol 47(4):511–520. https://doi.org/10.1590/S1516-89132004000400003

Americo GHP, Americo-Pinheiro JHP, Furlani Júnior E (2017) Hormesis effect of dichlorophenoxy acetic acid sub-doses and mepiquat chloride on cotton plant. Planta Daninha 35:e017166065. https://doi.org/10.1590/S0100-83582017350100078

Anthony MF, Coggins CW Jr (1999) The efficacy of five forms of 2,4-D in controlling preharvest fruit drop in citrus. Sci Hortic 81(3):267–277

Anunciato VM (2018) Efeitos do glyphosate no crescimento e reprodução de biótipos de *Digitaria insularis* resistente ou suscetível a este herbicida. Dissertation, Universidade Estadual Paulista

Ather NM, Abbas T, Tanveer A (2017) Glyphosate hormesis in broad-leaved weeds: a challenge for weed management. Arch Agron Soil Sci 63(3):344–351. https://doi.org/10.1080/03650340.2016.1207243

Barbosa AP, Zucareli C, Freiria GH et al (2017) Low rates of glyphosate on the process germination and corn seedling development. Rev Bras Milho Sorgo 16(2):240–250. https://doi.org/10.18512/1980-6477/rbms.v16n2p240-250

Belgers JDM, Van Lieverloo RJ, Van der Pas LJ et al (2007) Effects of the herbicide 2,4-D on the growth of nine aquatic macrophytes. Aquat Bot 86(30):260–268. https://doi.org/10.1016/j.aquabot.2006.11.002

Belz RG (2020) Herbicide hormesis can act as a driver of resistance evolution in weeds–PSII-target site resistance in *Chenopodium album* L. as a case study. Pest Manag Sci 74(12):2874–2883. https://doi.org/10.1002/ps.5080

Belz RG, Cedergreen N, Duke SO (2011) Herbicide hormesis–can it be useful in crop production? Weed Res 51(4):321–332. https://doi.org/10.1111/j.1365-3180.2011.00862.x

Belz RG, Duke SO (2014) Herbicides and plant hormesis. Pest Manag Sci 70(50):698–707. https://doi.org/10.1002/ps.3726

Belz RG, Duke SO (2017) Herbicide-mediated hormesis. In: Duke SO, Kudsk P, Solomon K (eds) Pesticide dose: effects on the environment and target and non-target organisms. American Chemical Society, Washington, DC, pp 135–148

Belz RG, Farooq MB, Wagner J (2018) Does selective hormese impact herbicide resistance evolution in weeds? ACCase-resistant populations of *Alopecurus myosuroides* Huds as a case study. Pest Manag Sci 74(8):1880–1891. https://doi.org/10.1002/ps.4890

Brito I, Haddad HA (2017) A formulação do conceito de homeostase por Walter Cannon. Filos Hist Biol 12(1):99–113

Brito IP, Tropaldi L, Carbonari CA et al (2018) Hormetic effects of glyphosate on plants. Pest Manag Sci 74(5):1064–1070

Calabrese EJ (2004) Hormesis: a revolution in toxicology, risk assessment and medicine: Re-framing the dose–response relationship. EMBO Rep 5(1):37–40. https://doi.org/10.1038/sj.embor.7400222

Calabrese EJ (2018) Hormesis: path and progression to significance. Int J Mol Sci 19(10):2871. https://doi.org/10.3390/ijms19102871

Calabrese EJ, Blain RB (2011) The hormesis database: the occurrence of hormetic dose responses in the toxicological literature. Regul Toxicol Pharmacol 61(1):73–81. https://doi.org/10.1016/j.yrtph.2011.06.003

Calabrese EJ, Mattson MP (2017) How does hormesis impact biology, toxicology, and medicine? NPJ Aging Mech Dis 3(1):13

Carbonari CA, Gomes GLG, Velini ED (2014) Glyphosate effects on sugarcane metabolism and growth. Am J Bot 5(24):3585. https://doi.org/10.4236/ajps.2014.524374

Carvalho LB, Alves PL, Duke SO (2013) Hormesis with glyphosate depends on coffee growth stage. An Acad Bras Cienc 85(2):813–822. https://doi.org/10.1590/S0001-37652013005000027

Cedergreen N (2008) Herbicides can stimulate plant growth. Weed Res 48(5):429–438. https://doi.org/10.1111/j.1365-3180.2008.00646.x

Cedergreen N, Olesen CF (2010) Can glyphosate stimulate photosynthesis? Pestic Biochem Physiol 96(3):140–148. https://doi.org/10.1016/j.pestbp.2009.11.002

Cedergreen N, Streibig JC, Kudsk P (2007) The occurrence of hormesis in plants and algae. Dose-Response 5(2):6–8. https://doi.org/10.2203/dose-response.06-008.Cedergreen

Cedergreen N, Felby C, Porter J (2009) Chemical stress can increase crop yield. Field Crop Res 114(1):54–57. https://doi.org/10.1016/j.fcr.2009.07.003

Cesco VJS (2018) Hormese de glyphosate no desenvolvimento vegetativo e reprodutivo de biótipos de *Conyza* spp. Dissertation, Universidade Estadual Paulista

Codognoto LC, Conde TT, Maltoni KL (2021) Glyphosate in the production and forage quality of marandu grass. Semin Cienc Agrar 42(3):1695–1706. https://doi.org/10.5433/1679-0359.2021v42n3Supl1p1695

Dabney BL, Patiño R (2018) Low-dose stimulation of growth of the harmful alga *Prymnesium parvum* by glyphosate and glyphosate-based herbicides. Harmful Algae 80:130–139. https://doi.org/10.1016/j.hal.2018.11.004

Dalley CD, Richard EP (2010) Herbicides as ripeners for sugarcane. Weed Sci 58(3):329–333. https://doi.org/10.1614/WS-D-09-00001.1

Duke SO, Cedergreen N, Velini ED (2006) Hormesis: is it an important factor in herbicide use and allelopathy? Outlooks Pest Manag 17(1):29–33

Duke SO, Kudsk P, Solomon KR (2017) Pesticide dose: effects on the environment and target and non-target organisms. American Chemical Society, Washington, DC

El-Otmani M, Agustí M, Aznar M et al (1993) Improving the size of 'Fortune' mandarin fruits by the auxin 2,4-D. Sci Hortic 55(4):283–290

Farooq N, Abbas T, Tanveer A (2019) Differential hormetic response of fenoxaprop-p-ethyl resistant and susceptible *Phalaris minor* populations: a potential factor in resistance evolution. Planta Daninha 37:e019187554. https://doi.org/10.1590/S0100-83582019370100045

Gazola T, Dias MF, Dias RC, Carbonari CA, Velini ED (2019) Effects of 24-D herbicide on species of the *Digitaria Genus*. Planta Daninha 37:e019220694. https://doi.org/10.1590/S0100-83582019370100131

Gitti DDC, Arf O, Peron IBG et al (2011) Use of glyphosate as growth regulator in upland rice. Pesqui Agropecu Trop 41(4):500–507. https://doi.org/10.5216/pat.v41i4.10160

Gomes GLGC (2014) Caracterização bioquímica e morfofisiológica de populações de buva (*Conyza* spp.) resistentes ao glyphosate. These, Universidade Estadual Paulista

Halter S (2009) História do herbicida agrícola glyphosate. In: Velini ED, Carbonari CA, Meschede DK et al (eds) Glyphosate: uso sustentável. FEPAF, Botucatu, pp 11–16

Hamadache M, Hanini S, Benkortbi OM et al (2016) Artificial neural network-based equation to predict the toxicity of herbicides on rats. Chemom Intell Lab Syst 154:7–15. https://doi.org/10.1016/j.chemolab.2016.03.007

Jalal A, Oliveira JC, Ribeiro JS et al (2021) Hormesis in plants: physiological and biochemical responses. Ecotoxicol Environ Saf 207:111225. https://doi.org/10.1016/j.ecoenv.2020.111225

Kaur A, Kaur N (2019) Effect of sub-lethal doses of 2,4-D sodium salt on physiology and seed production potential of wheat and associated dicotyledonous weeds. Indian J Weed Sci 51(4):352–357. https://doi.org/10.5958/0974-8164.2019.00074.1

Khan S, Zhou JL, Ren L et al (2020) Effects of glyphosate on germination, photosynthesis and chloroplast morphology in tomato. Chemosphere 258:127350. https://doi.org/10.1016/j.chemosphere.2020.127350

Kruse ND, Trezzi MM, Vidal RA et al (2000) Herbicidas inibidores da EPSPS: revisão de literatura. Rev Bras Herbic 1(2):139–146. https://doi.org/10.7824/rbh.v1i2.328

Leite GHP, Crusciol CAC (2008) Growth regulators in the development and productivity of sugarcane. Pesqui Agropecu Bras 43(8):995–1001

Marques RF (2019) Hormesis de 2,4-D sal colina em algodoeiro cultivado no Cerrado. Dissertation, Universidade Federal de Goiás

Marques RF, Marchi SR, Pinheiro GH et al (2019) Hormesis of 2,4-D choline salt in biometric aspects of cotton. J Agric Sci 11(13):283–294. https://doi.org/10.5539/jas.v11n13p283

Marques KDM, Moreira WC, Silva DFJ et al (2020) Efeito hormético de glyphosate no crescimento inicial de mudas de paricá (*Schizolobium amazonicum*). Agrarian 3(47):9–16. https://doi.org/10.30612/agrarian.v13i47.8074

Mattson MP (2008) Hormesis defined. Ageing Res Rev 7(1):1–7. https://doi.org/10.1016/j.arr.2007.08.007

Melero MM (2016) Aplicação de subdoses dos herbicidas glyphosate, 2,4-D e paraquat em algodoeiro. Dissertation, Universidade Estadual Paulista "Júlio de Mesquita Filho"

Meschede DK, Velini ED, Carbonari CA (2010) Effect of glyphosate and sulfometuron-methyl on the growth and technological quality of sugarcane. Planta Daninha 28:1135–1141. https://doi.org/10.1590/S0100-83582010000500021

Meseldžija M, Lazić S, Dudić M et al (2020) Is there a possibility to involve the hormesis effect on the soybean with glyphosate sub-lethal amounts used to control weed species *Amaranthus retroflexus* L.? Agronomy 10(6):e850. https://doi.org/10.3390/agronomy10060850

Mobli A, Matloob A, Chauhan BS (2020) Glyphosate-induced hormesis: impact on seedling growth and reproductive potential of common sowthistle (*Sonchus oleraceus*). Weed Sci 68(6):605–611. https://doi.org/10.1017/wsc.2020.77

Mollaee AM, Matloob A, Mobli M, Thompson BS, Chauhan BS (2020) Response of glyphosate-resistant and susceptible biotypes of *Echinochloa colona* to low doses of glyphosate in different soil moisture conditions. PLOS ONE 15(5):e0233428. https://doi.org/10.1371/journal.pone.0233428

Moraes CP, Brito IP, Tropaldi L et al (2019) Hormetic effect of glyphosate on *Urochloa decumbens* plants. J Environ Sci Health B 55(4):376–381. https://doi.org/10.1080/03601234.2019.1705114

Murado MA, Vázquez JA (2010) Biphasic toxicodynamic features of some antimicrobial agents on microbial growth: a dynamic mathematical model and its implications on hormesis. BMC Microbiol 10(1):220. https://doi.org/10.1186/1471-2180-10-220

Nascentes RF, Carbonari CA, Simões PS et al (2018) Low doses of glyphosate enhance growth, CO_2 assimilation, stomatal conductance and transpiration in sugarcane and eucalyptus. Pest Manag Sci 74(5):1197–1205. https://doi.org/10.1002/ps.4606

Nelson KA, Renner KA, Hammerschmidt R (2002) Effects of protoporphyrinogen oxidase inhibitors on soybean (*Glycine max* L.) response, *Sclerotinia sclerotiorum* disease development, and phytoalexin production by soybean. Weed Technol 16(2):353–359. https://doi.org/10.1614/0890-037X(2002)016[0353:EOPOIO]2.0.CO;2

Oliveira RS Jr (2011) Mecanismos de ação de herbicidas. In: Oliveira RS Jr, Constantin J, Inoue MH (eds) Biologia e Manejo de Plantas Daninhas. Omnipax, Curitiba, pp 141–192

Peres LRS, Della Vechia JF, Cruz C (2017) Hormese effect of herbicides subdoses on submerged macrophytes in microassay conditions. Planta Daninha 35:e017165857. https://doi.org/10.1590/S0100-83582017350100076

Petersen J, Neser JM, Dresbach-Runkel M (2008) Resistant factors of target-site and metabolic resistant black-grass (*Alopecurus myosuroides* Huds.) biotypes against different ACCase-inhibitors. J Plant Dis Prot 21:25–29

Pincelli SRP, Bortolheiro FP, Carbonari CA et al (2020) Hormetic effect of glyphosate persists during the entire growth period and increases sugarcane yield. Pest Manag Sci 76(7):2388–2394. https://doi.org/10.1002/ps.5775

Pinheiro GHR (2020) Hormesis na cultura da soja em resposta à aplicação de 2,4-D sal colina. Dissertation, Universidade Federal de Goiás

Reis FC, Mendes KF, Baccin L et al (2021) Seletividade, hormesis e fisiologia dos herbicidas nas plantas. In: Barroso AAM, Murata AT (eds) Matologia, Estudos sobre plantas daninhas. Editora Fábrica da Palavra, Jaboticabal, pp 295–323

Ries SK, Chmiel H, Dilley DR (1967) The increase in nitrate reductase activity and protein content of plants treated with simazine. Proc Natl Acad Sci U S A 58(2):526

Rufini JCM, Ramos JD, Mendonça V et al (2008) Harvest season extent of tangerin 'Ponkan' fruits with the application of GA3 AND 2,4-D. Cienc Agrotecnol 32(3):834–839. https://doi.org/10.1590/S1413-70542008000300020

Santos JCC, Silva DMR, Amorim DJ et al (2021) Glyphosate hormesis mitigates the effect of water deficit in safflower (*Carthamus tinctorius* L.). Pest Manag Sci 77:2029–2044. https://doi.org/10.1002/ps.6231

Silva FML (2014) Hormesis de herbicidas em soja. These, Universidade Estadual Paulista "Júlio de Mesquita Filho"

Silva JCD, Arf O, Gerlach GAX (2012) Hormesis effect of glyphosate on common bean cultivars. Pesqui Agropecu Trop 42(3):295–302. https://doi.org/10.1590/S1983-40632012000300008

Silva FML, Duke SO, Dayan FE et al (2016a) Low doses of glyphosate change the responses of soyabean to subsequent glyphosate treatments. Weed Res 56(2):124–136. https://doi.org/10.1111/wre.12189

Silva JCD, Gerlach GAX, Rodrigues RAF et al (2016b) Influência de doses reduzidas e épocas de aplicação sobre o efeito hormético de glyphosate em feijoeiro. Rev Fac Agron 115(2):191–199

Silva JRO, Marques JNR, Godoy CVC et al (2019) 2,4-D hormesis effect on soybean. Planta Daninha 37:e019216022. https://doi.org/10.1590/S0100-83582019370100146

Silva AM, Lima VMM, Silva VL (2020a) Different types of herbicides in inducing the hormetic effect on soybean culture. Sci Electron Arch 13(11):30–36. https://doi.org/10.36560/13820201046

Silva DRO, Silva ÁA, Novello D et al (2020b) Nitrogen availability and glyphosate hormesis on white oat. Planta Daninha 38:e020230864. https://doi.org/10.1590/S0100-83582020380100071

Tavares CJ, Pereira LS, Araújo ACF et al (2017) Crescimento inicial de plantas de pequi após aplicação de 2,4-D. Pesqui Florest Bras 37(89):81–87. https://doi.org/10.4336/2017.pfb.37.89.1280

Vargas-Hernandez M, Macias-Bobadilla I, Guevara-Gonzalez RG et al (2017) Plant hormesis management with biostimulants of biotic origin in agriculture. Front Plant Sci 8:e1762. https://doi.org/10.3389/fpls.2017.01762

Velini ED, Alves E, Godoy MC et al (2008) Glyphosate applied at low doses can stimulate plant growth. Pest Manag Sci 64(4):489–496. https://doi.org/10.1002/ps.1562

Velini ED, Trindade ML, Barberis LRM et al (2010) Growth regulation and other secondary effects of herbicides. Weed Sci 58(3):351–354. https://doi.org/10.1614/WS-D-09-00028.1

Evolution of Weed Resistance to Herbicides

Kassio Ferreira Mendes, Kamila Cabral Mielke, Ricardo Alcántara-de La Cruz, Antonio Alberto da Silva, Evander Alves Ferreira, and Leandro Vargas

Abstract Chemical control is an efficient method and the most used in agriculture; however, resistant weeds have evolved worldwide due to the selection pressure caused by the repeated use of herbicides with the same mechanism of action. Some mechanisms of action have a greater disposition to evolve resistance, with the Acetolactate Synthase (ALS) and Photosystem II (PSII) inhibitors. Resistant biotypes seriously compromise agricultural crops, making management difficult and increasing control costs; therefore, management alternatives must be adopted to minimize the damage caused by resistant biotypes. For genetically modified varieties, such as glyphosate-resistant soybeans, other herbicides with different mechanisms of action are being included in management programs to control resistant biotypes. The use of herbicides with different mechanisms of action in annual rotations, tank mixes and sequential applications, can delay the evolution of resistance, minimizing the selection pressure imposed by a single specific mechanism of action. Integrated weed management combining different control methods is a viable alternative for controlling resistant biotypes. Thus, the chapter covers the evolution of resistance in the world and in Brazil, the different mechanisms of resistance to tar-

K. F. Mendes (✉) · A. Alberto da Silva
Department of Agronomy, Federal University of Viçosa, Viçosa, Minas Gerais, Brazil
e-mail: kfmendes@ufv.br; aasilva@ufv.br

K. C. Mielke
Federal University of Viçosa, Viçosa, Minas Gerais, Brazil
e-mail: kamilamielke@ufv.br

R. A. L. Cruz
Federal University of São Carlos, São Carlos, SP, Brazil

E. A. Ferreira
Federal University of Minas Gerais, Montes Claros, Minas Gerais, Brazil

L. Vargas
Empresa Brasileira de Pesquisa Agropecuária, Passo Fundo, RS, Brazil
e-mail: leandro.vargas@embrapa.br

© The Author(s), under exclusive license to Springer Nature Switzerland AG 2022
K. F. Mendes, A. Alberto da Silva (eds.), *Applied Weed and Herbicide Science*,
https://doi.org/10.1007/978-3-031-01938-8_7

get and non-target site, methods of identification of resistant biotypes, and alternatives for integrated management of herbicide resistant weeds.

Keywords Resistant biotypes · Resistance mechanism · Susceptibility · Chemical control · Integrated management

1 Introduction

Chemical weed control is the most widely used method in the world. This is due to the fact that this method is very efficient, has relatively low cost compared to other control methods, and is easy to use and professionally adequate. However, most producers have only an immediate and economic view of weed control, which can generate environmental problems in the medium and long term. Repeated applications of herbicides with the same mechanism of action have been common practice in various parts of the world. Due to high continuous selection pressure in most cases, many weed populations have selected resistance to herbicides in various agricultural and non-agricultural regions of the world (Heap 2014).

Herbicide resistance is the hereditary ability of weeds to grow and reproduce after exposure, under field conditions, to a herbicide dose recommended by the manufacturer that is lethal to wild (susceptible) individuals of the same species (Vencill et al. 2012). Due to the evolution of herbicide resistance, weed control has become a difficult to manage problem in several areas of intensive agriculture Peterson et al. (2018). Resistance is conferred by genetic characteristics that allow individuals from a weed population to survive the exposure of a herbicide, as already reported, giving them a strong competitive advantage over susceptible individuals, allowing them to dominate the population (WSSA 1998).

The process of evolution of herbicide resistance goes through three stages (Vila-Aiub et al. 2019):

1. Elimination of highly sensitive biotypes, leaving only the most tolerant and resistant ones
2. Elimination of all biotypes, except resistant ones, and selection of homozygous individuals within a population with high tolerance
3. Intercrossing between the surviving biotypes, generating new individuals with a higher degree of resistance, which may later undergo a new selection.

Weed resistance to herbicides is of great importance, especially when there is no or few alternative herbicides to be used to control resistant biotypes. This happens in species of great occurrence in various parts of the world making it increasingly difficult and costly to control these resistant biotypes. Currently, there are 521 single cases of resistance (species × mechanism of action) in 263 species (152 eudicotyledons and 111 monocotyledons) reported in 71 countries (Fig. 1). Weeds developed resistance to 23 of the 26 known herbicide mechanisms of action and to 164 different herbicides worldwide (Heap 2021).

Weed resistance to herbicides is undoubtedly one of the main concerns of modern agriculture and some mechanisms of action present greater resistance problems

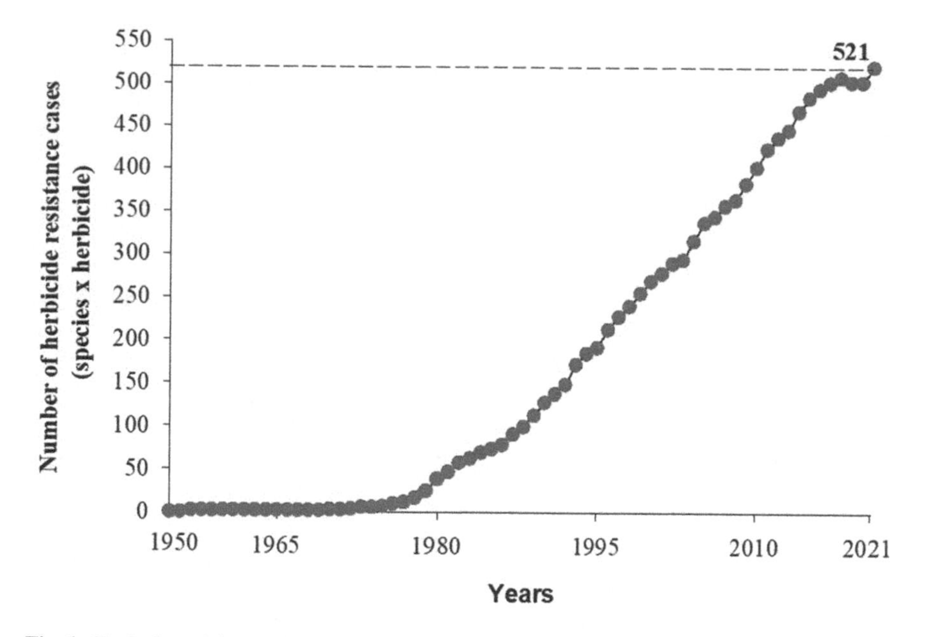

Fig. 1 Evolution of the number of unique cases (species × herbicide) of herbicide-resistant weeds worldwide. (Source: Heap 2021)

than others. The largest number of cases of resistant biotypes belongs to the mechanism of action of Acetolactate Synthase inhibitors (ALS), followed by Photosystem II inhibitors (PSII) (Fig. 2). It is understood that the number of cases is mainly related to the greater use of these herbicides in agriculture; however, some mechanisms of action tend to more easily select resistant biotypes due to the specificity of the herbicide mechanism of action (Vrbničanin et al. 2017).

The chemical control of resistant weed biotypes is seriously compromised in different agricultural crops, and it is necessary to implement other less efficient alternative methods (Silva et al. 2007). Wheat crop has the highest number of cases of resistance to herbicides (species × herbicide), followed by corn, rice, and soybean crop (Fig. 3). The management of herbicide resistance becomes difficult and costly in these crops, since resistant biotypes can develop resistance to more than one mechanism of action, reducing the number of molecules of alternative herbicides that can be used to control these weeds (Peterson et al. 2018). In Brazil, some weed genera and species, such as *Amaranthus* spp., *Bidens* spp., *Conyza* spp., *Cyperus* spp., *Digitaria* spp., *Echinochloa* spp., *Eleusine indica*, *Lolium* spp., among others, have a high facility to develop resistance to ALS-inhibiting herbicides, Acetyl Coenzyme Carboxylase (ACCase), 5-enolpyruvylshikimate-3-phosphate synthase (EPSPS), protoporphyrinogen oxidase (PPO or PROTOX), and synthetic auxins, such as 2,4-D and dicamba, especially due to their congenital genetic variability (Andres et al. 2017).

Given the above, the concepts of sensitivity, tolerance, and resistance; factors that favor the emergence of resistance; field diagnosis and resistance confirmation; mechanisms that confer resistance; practices to avoid resistance; in addition to the evolution of weed resistance in Brazil, and integrated management of resistant weeds are reported in this chapter.

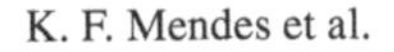

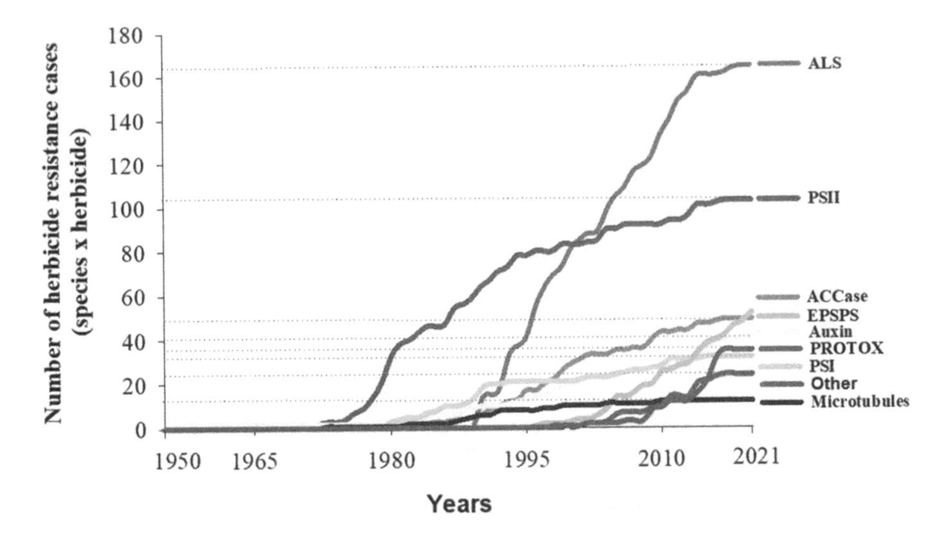

Fig. 2 Number of unique cases of resistance (species × herbicide) by herbicide action mechanisms from 1950 to 2021 worldwide. Acetolactate Synthase Inhibitors (ALS); Photosystem II (PSII); Acetyl Coenzyme Carboxylase (ACCase); 5-enolpyruvylshikimate-3-phosphate synthase (EPSPS); Auxin mimetizers (Synthetic Auxins); Protoporphyrinogen Oxidase (PPO or PROTOX); Photosystem I (PSI); Formation of Microtubules (Microtubules); Others: Inhibition of 4-hydroxy-phenyl-pyruvate-dioxygenase (HPPD), Synthesis of carotenoids and Glutamine Synthetase (GS). (Source: Heap 2021)

2 Sensitivity, Tolerance, and Resistance

First, it is important to define what is the place of action of a herbicide. The target of these compounds is a cell structure (receptor), usually an enzyme, charged with developing an essential function for plant survival, which can be activated by different substances, both internal (substrate) and external (herbicide) to the cell. Thus, the herbicide behaves as a homolog of a given substrate occupying its place in an essential process, which ends up being interrupted in whole or in part (Fig. 4).

Thus, a plant is sensitive to a herbicide when its growth and development are altered by the action of a particular herbicide in response to the recommended dose in the field. There are several degrees of susceptibility of a weed, from the effects on growth and development, to the inability to withstand the action of the herbicide.

Tolerance is an innate feature of a weed species in surviving and reproducing after exposure to a recommended field herbicide dose, even if they suffer injuries (WSSA 1998). This is due to the existing natural genetic variability in a population of plants of a species, that is, there is no selection process imposed by the herbicide (application history) to make the plant tolerant, which differentiates from the species described as resistant. As an example, we can mention the *Commelina benghalensis* and *Commelina diffusa*, *Artemisia verlotorum*, *Euphorbia heterophylla*, *Richardia brasiliensis*, *Spermacoce latifolia*, and *Ipomoea* spp. that are tolerant to glyphosate; and *Digitaria horizontalis* tolerant to atrazine that due to morphological

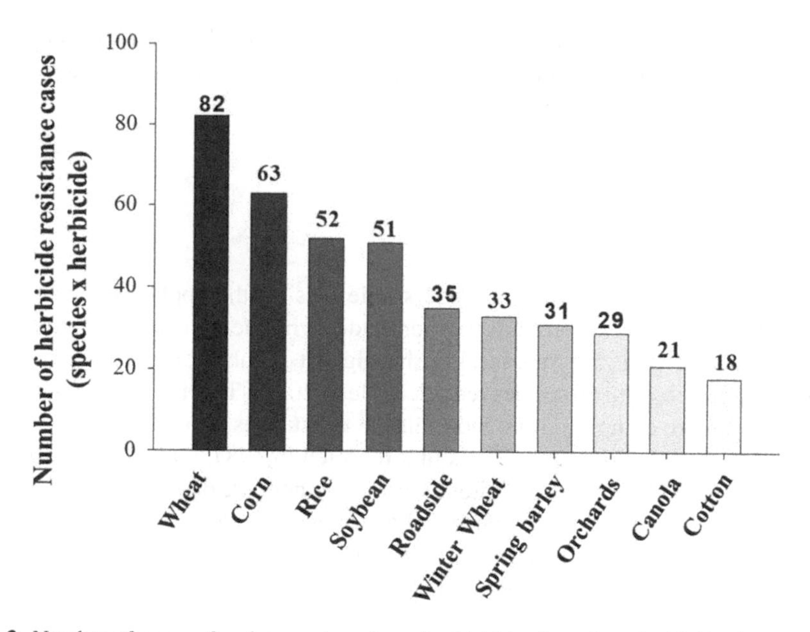

Fig. 3 Number of cases of resistance (species × herbicide) of weeds to herbicides in the main agricultural crops in the world. (Source: Heap 2021)

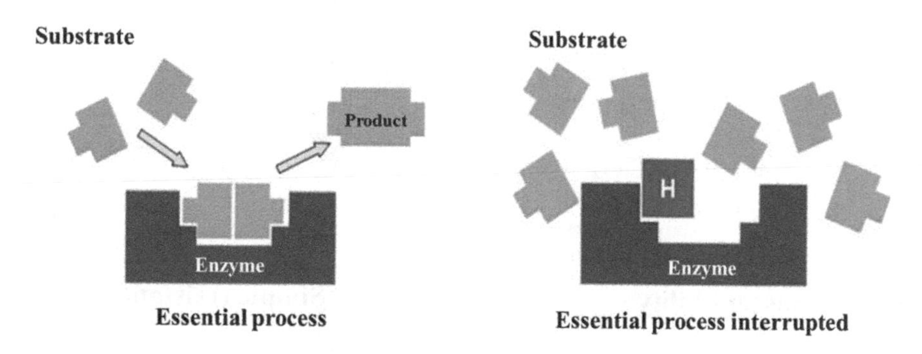

Fig. 4 Generic representation of the mechanism of action of a herbicide (H) by interrupting an essential process of plants by taking the place of the substrate in an enzymatic reaction mediated by a given enzyme

and physiological factors have difficulty in controlling the use of these herbicides (Mendes 2020). However, the continued selection pressure exerted by the same herbicide can lead to resistance evolution in tolerant plants. For example, *E. heterophylla* (Mendes et al. 2020a, b, c) and *Chloris barbata* (Bracamonte et al. 2018) considered glyphosate tolerant species, also developed resistance to this herbicide.

Resistance is the hereditary ability of some biotypes to survive and reproduce after herbicide treatments that, under normal conditions, control susceptible individuals of the same species (WSSA 1998). Resistance can occur naturally (selected in weed populations of natural occurrence in the field through the use of a herbicide) or induced by techniques such as genetic engineering or selection of variants, produced by tissue crops or mutagenesis (Heap 2014). Resistance can be simple, cross-crossed, and multiple (Fig. 5) (Moss 2017).

2.1 Simple Resistance

Simple resistance occurs when a weed species resists the application of a single herbicide but presents susceptibility to herbicides with the same or different mechanisms of action (Fig. 5). Atrazine has the highest number of weed species with simple resistance, with 66 species recorded (Heap 2021). This type of resistance can evolve into more complex cross-and-multiple resistances. For example, in zealous biotypes (*Lolium multiflorum*), it was first reported in 1987, with simple resistance to diclofop-methyl (ACCase inhibitor). Currently, there are 68 cases of resistance of *L. multiflorum* to different mechanisms of action worldwide, including inhibitors of EPSPS, ALS, ACCase, Glutamine Synthase (GS), Photosystem I (PSI) and PSII, and long-chain fatty acid inhibitors (Heap 2021).

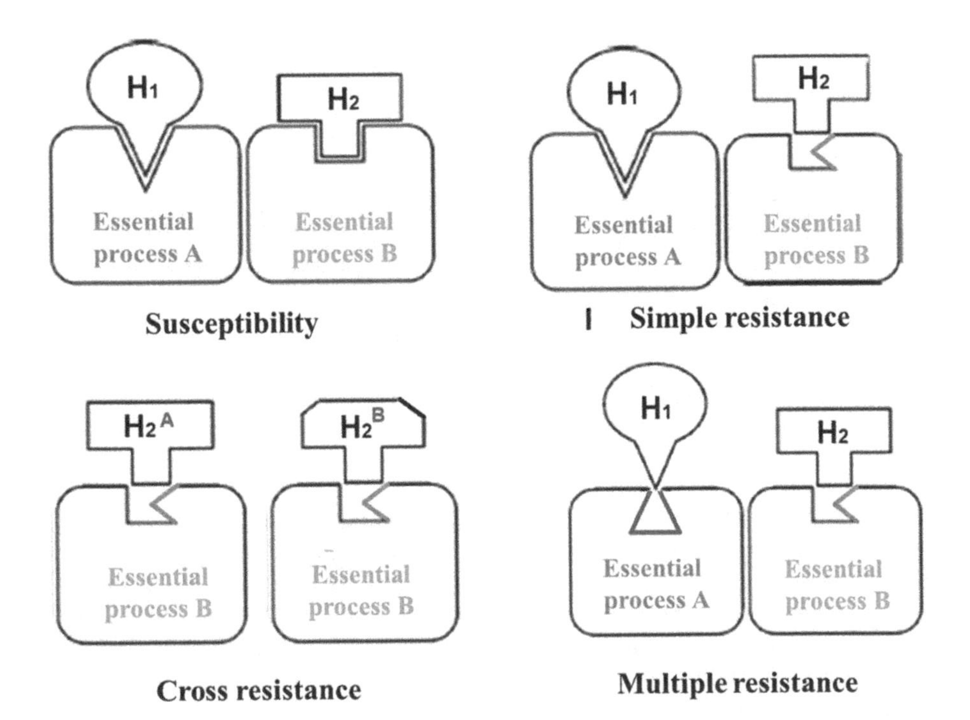

Fig. 5 Diagrammatic representation of susceptibility conditions and resistances (simple, crossed, and multiple) to weed herbicides. H1 and H2 are herbicides of different mechanism of action that instill distinct essential processes (= target sites A and B) within a weed. $H2^A$ and $H2^B$ are different herbicides that instill the same essential process. (Source: Adapted from Moss 2017)

2.2 Cross-Resistance

Cross-resistance occurs when a weed resists different herbicides to which the initial resistance has been selected (Fig. 5). This resistance pattern is perceived when the plant survives a second herbicide or subsequent herbicides, of which there is no history of application, which may belong to the same chemical group to the herbicide that selected the resistance, or distinct chemical groups, but with the same mechanism of action. Cross-resistance can be conferred by a single (monogenic) gene that contributes to a single resistance mechanism, although it is possible that more than one (polygenic) gene contributes to the same resistance mechanism (Preston and Mallory-Smith 2001). There are two types of resistance mechanisms: related to the place of action and not related to the place of action (Vrbničanin et al. 2017), which will be described in detail in the next items.

When cross-resistance is related to the site of action, i.e., a change in the structural conformation of the herbicide action site, resistance to other herbicides of different chemical groups can also be conferred, which act at the same site of action in the plant (Powles and Preston 2021). One example is the substitution of Amino Acids Trp-574-Leu in the ALS gene found in two populations of tiririca (*Cyperus iria*) after exposure to bispyribac-sodium, halosulfuron, imazamox, and penoxsulam (Riar et al. 2015). A study developed by Délye et al. (2008) suggested patterns of cross-resistance of the species *Alopecurus myosuroides* to ACCase inhibitors. The authors found that in the herbicides of the chemical group of aryloxyphenoxypropionates (FOPs) (fenoxaprop, clodinafop, and haloxyfop) are related to mutations of amino acids Leu-1781, Cys-2027, Asp-2041, and Ala-2096 of the ACCase gene. The Leu-1781 also confers resistance to cycloxydim (chemical group of cyclohexanodiones—DIMs). On the other hand, Gly-2078 gives resistance to both FOPs and DIMs (Délye et al. 2008).

There may also be variations in the level of cross-resistance of biotypes to herbicides of different groups. *Lolium rigidum* and *Kochia scoparia* biotypes that have cross-resistance to ALS-inhibiting herbicides showed varying levels of resistance to different herbicide groups. This is due to a change in the gene encoding the ALS enzyme, preventing all or part of the herbicide's binding to this target enzyme (Powles and Preston 2021). The biotype of *L. rigidum* (VLR 69) resistant to at least nine chemical groups of herbicides in five different mechanisms of action (Burnet et al. 1994) has an ACCase insensitive to FOPs, with resistance levels ranging from 4 to 29 times more resistant in relation to the susceptible biotype, but showed no resistance to IDs (sethoxydim and tralkoxydim) (Preston et al. 1996). The level of resistance may be the result of the different mutations that occur in the gene encoding the ACCase enzyme and the type of allo involved (Powles and Preston 2021).

Cross-resistance, due to other mechanisms, is also exemplified by *L. rigidum* biotypes resistant to ACCase-inhibiting herbicides, which in addition to having this insensitive enzyme, the resistant biotype shows improved diclofop-methyl metabolism. Metabolic herbicide detoxification is an essential feature that provides the basis for cross-resistance in *L. rigidum* and selectivity in wheat plants (Christopher et al. 1991). In *L. rigidum*, cross-resistance to various herbicide mechanisms of action is often endiated by herbicide metabolism through improved rates of enzymatic activity of cytochrome P450 enzyme complex (Yu et al. 2013). This species can rapidly evolve cross-resistance to various wheat selective herbicides under recurrent selection of a single mechanism of action (Busi and Powles 2013). Cross-resistance is common in *A. myosuroides*, *Echinochloa phyllopogon*, and *L. rigidum* and increasingly prevalent in other weed species, which makes it difficult to control these species (Yu and Powles 2014). In *L. rigidum* biotypes resistant to PSII-inhibiting herbicides, an increase in the rate of herbicide metabolism was detected (Powles and Preston 2021). Most cases of resistance related to herbicide metabolism in weeds were selected by ACCase, ALS, PSI, or PSII inhibitors, conferring cross-resistance to other herbicides that act in these sites of action (Beckie and Tardif 2012).

2.3 Multiple Resistance

Multiple resistance refers to biotypes that resist herbicides of different mechanisms of action without a history of application, e.g., these herbicides were not previously used in the area where the resistant biotype was found (Fig. 5). Multiple resistance is the biggest current problem and, in the future, related to the management of herbicide-resistant weeds. In the simplest cases, two or more resistance mechanisms confer resistance to only one herbicide or a herbicide mechanism of action. On the other hand, the most complex cases of multiple resistance, although they may involve the accumulation of two or more mechanisms conferring resistance to several herbicides of different chemical groups, usually only one mechanism is responsible for conferring resistance to different mechanisms of action of herbicides. The resistance mechanisms accumulated in *Lolium* spp., both target site-related and metabolic resistance, confer multiple resistance to ACCase and ALS inhibitor herbicides in Denmark and Italy (Scarabel et al. 2020). Cyp81A cytochrome P450 isoenzyme was able to confer wide multiple resistance to 18 herbicides belonging to ALS inhibitors, ACCase, phytoene desaturase (PDS), PSII, PROTOX, and 1-deoxy-D-xylulose-5-phosphate synthase (DXS) in *Echinochloa polygonum* (Dimaano et al. 2020). In Brazil, the most complex case corresponds to a population of *Conyza sumatrensis*, with multiple resistance to inhibitors of EPSPS, PSI, PSII, and PROTOX and to 2,4-D-auxin mimetic (five mechanisms of action), found in a soybean field in the municipality of Assis Chateaubriand, Paraná (Pinho et al. 2019).

The difficulties of controlling weed biotypes with multiple resistance increase even more when the mechanisms that confer resistance are related to the site of action and other mechanisms, such as metabolism and vacuolar compartmentalization. The resistance of *Conyza bonariensis* and *Conyza canadensis* to glyphosate and paraquat associated with reduced translocation is considered as a result of herbicide sequestration in vacuole (Moretti et al. 2017a, b). To control these weeds, it is necessary to employ mixtures of herbicides that do not have their activity affected by the resistance mechanisms in question. There are currently 101 reported cases of plants with multiple resistance worldwide, mainly to two and/or three mechanisms of action (Fig. 6) (Heap 2021).

The biotypes of *L. rigidum* and *A. myosuroides* constitute complex cases. In Europe, 11 *L. rigidum* biotypes showed three different cases of resistance, being simple resistance to glyphosate (EPSPS), with multiple resistance to glyphosate and ACCase inhibitors and multiple resistance to glyphosate and ALS inhibitors. Amino acid substitutions were found at position 106 of the EPSPS gene, at positions 1781, 2088, and 2096 of the ACCase gene and at positions 197 and 574 of the ALS gene (Collavo and Sattin 2014). *Lolium* spp. populations of Chile were also found with accumulation of different mutations conferring resistance to these three groups of herbicides (Vázquez-García et al. 2020). Not all populations presented amino acid substitutions, suggesting the presence of resistance mechanisms not related to the site of action (Collavo and Sattin 2014; Scarabel et al. 2020; Yanniccari et al. 2020). The biotypes of *A. myosuroides* metabolize chlorotoluron and some ACCase-inhibiting herbicides and have mutated ACCase (Powles and Preston 2021). This plant germinates in autumn in cereal crops in northwestern Europe, being the most important herbicide-resistant weed species in European agricultural systems, developing resistance to seven herbicide mechanisms, making it an ideal species to examine the presence and magnitude of resistance mechanisms (Comont et al. 2019).

In Brazil, dairy populations (*E. heterophylla*) were identified with multiple resistance to ALS and PROTOX inhibitors in 2004 (Trezzi et al. 2005). Subsequently, biotypes of *E. heterophylla* with multiple resistance to glyphosate and ALS inhibitors were found, and resistance to glyphosate and PROTOX inhibitors has recently

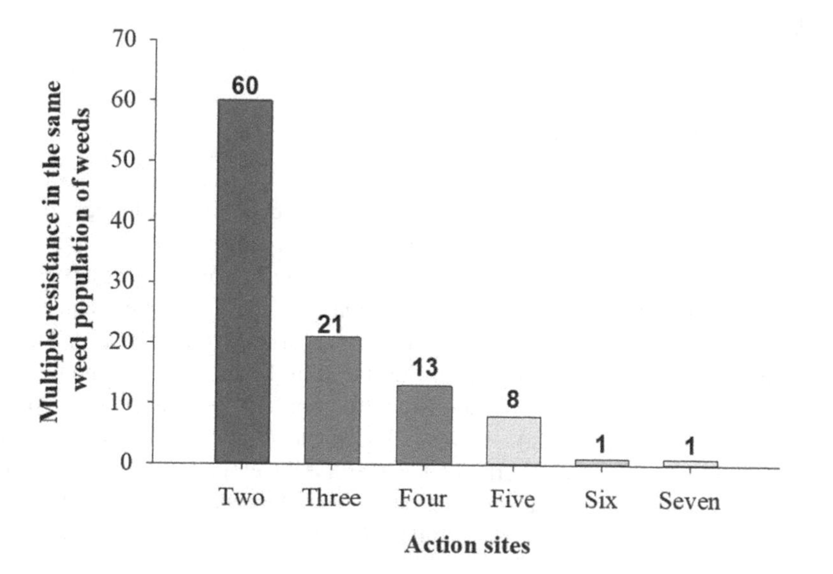

Fig. 6 Number of multiple resistance cases in the same weed population in the world. (Source: Heap 2021)

been reported (Palma-Bautista et al. 2020). Biotypes of *E. heterophylla* with multiple resistance were investigated for various aspects of the resistance mechanism, such as metabolism, detoxifying enzymatic activity, place of absorption, and site of action (Trezzi et al. 2009, 2011; Rojano-Delgado et al. 2019; Mendes et al. 2020a, b, c).

3 Factors That Favor the Emergence of Resistance

3.1 Selection Pressure

The selection pressure imposed by the continuous use of the same agricultural practice is what allows resistant individuals to survive and produce seeds, and thus occupy the niches available in the environment, left by susceptible plants that are controlled by the herbicide (Christoffoleti et al. 2016). The frequent use of herbicides provides the selection of plant species that are gradually purified (intraspecific selection) over time, until they become more adapted.

Sensitive individuals of a given species are first eliminated with herbicide application. Rarely, an individual weed develops a mutation that confers resistance to a herbicide or group of herbicides, and survives and reproduces; this happens after several generations over time and repeated selection with herbicides equal or similar, consequently the resistant biotype can become dominant in the population (Fig. 7) (Qasem 2013). It can be observed that any agricultural practice exerts selection pressure and can become problematic for farmers when applied repeatedly for a long period.

3.2 Genetic Variability

The genetic variability naturally found in weeds, associated with adequate intensity and duration of selection, makes it inevitable the emergence of resistant plants. The gene(s) that confer resistance to a given herbicide may be present(s) in a population before even this herbicide is released on the market. The greatest evidence of this was found in *A. myosuroides* from plants of a herbarium from France collected in 1888, in which an individual presented the mutation lIe-1781-Leu in its heterozygous form, widely reported as responsible for conferring resistance to ACCase inhibitors (Délye et al. 2013a). However, the presence of this mutation occurred 90 years before this group of herbicides were first developed and used. The entire natural weed population contains herbicide-resistant biotypes, which are indifferent to the application of some herbicide (HRAC 2021b). Genetically, there are two paths to the appearance of resistant plants: the occurrence of a gene or genes that

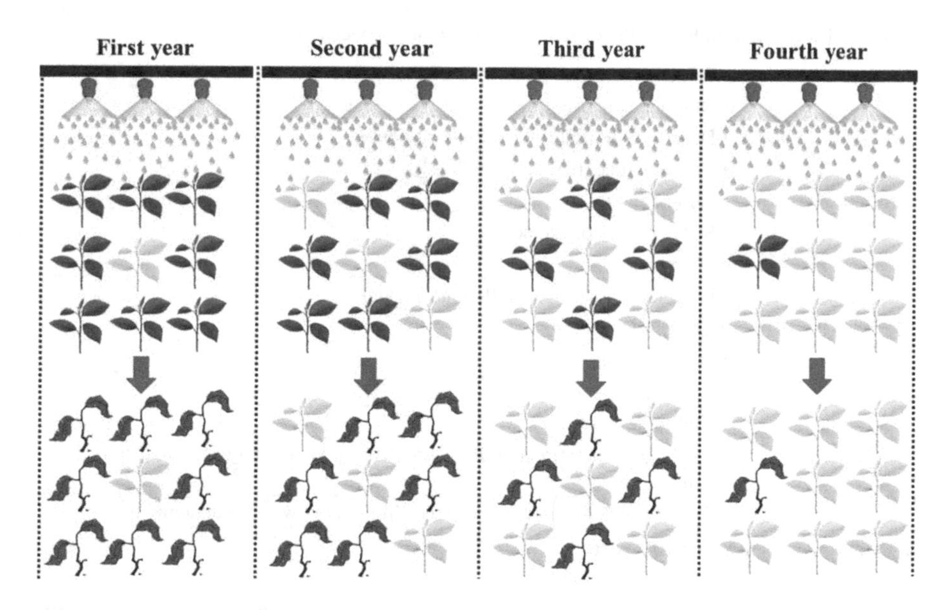

Fig. 7 Increased frequency of resistant weed biotypes over the years due to repeated applications of the same herbicide. Weeds with dark green, light green, and brown colors represent susceptible, resistant, and dead biotypes, respectively. (Source: Adapted from Christoffoleti and Nicolai 2016)

confer resistance at very low frequency in the population or through a mutation (Mortimer 1998).

Mutant gene or chromosomes are the essential source of all genetic variation (Silva et al. 2007). The selection changes the proportions between sensitive and resistant plants. The possibility of resistance in a population due to mutation is the result of the relationship between the frequency of the mutation and the size of the population. Weed characteristics such as high genetic diversity, low seed dormancy, large pollen production and propagules, combined with monoculture, and repeated use of chemical control contribute greatly to the emergence of resistant plants.

3.3 Evolution of Resistance

The evolution of weed resistance to herbicides is a dynamic process, and its impact depends on several genetic, biological, and agronomic factors that involve issues related to the herbicides used and operational (Table 1). According to Powles and Yu (2010), the key point of resistance evolution is cross fertilization among weeds.

One of the main factors involved in the evolution of resistance is the initial frequency, which refers to the number of resistant individuals present in a population that was not exposed to herbicide selection pressure. For some weed species and certain herbicides, this value is known (Christoffoleti et al. 2016). The higher the

Table 1 Factors that influence the evolution of resistance in weed populations

Genetic	• Frequency of resistance pallets
	• Number of resistant genes
	• Degree of dominance of resistant genes
	• Ecological adaptability
Biological	• Allogamy rate
	• Seed production capacity
	• Dormancy
	• Dispersal capacity of pollen grains and seeds
Herbicide used	• Chemical structure
	• Site of action
	• Residual in soil
Operating	• Herbicide dose
	• Applicator operational ability (application technology and environmental conditions)
	• Factors related to agroecosystems (integrated weed management)

Source: Powles and Yu (2010)

initial frequency of the resistant biotype, the greater the probability of increasing the proportion of resistant individuals in the population in a shorter period of time, since repeated applications of the herbicide picker may interfere in the predominance of resistant populations (Vidal and Fleck 1997).

4 Diagnosis of Field Resistance

Resistance is a phenomenon that evolves in a crop over several years. Unsatisfactory weed control does not necessarily mean that it is due to the resistance. According to the Herbicide Resistance Action Committee (HRAC 2021a), when one suspects the occurrence of resistance, initially, the following questions should be answered:

1. Was the product, dosage, time or stage of application, calibration, volume of syrup, adjuvants, type of nozzles and environmental conditions adequate?
2. Did the control failures occur in one species only?
3. Are not plants the result of reinfestation?

If the answers to these questions are affirmative, one should begin the investigation of the factors that lead to resistance.

1. Lately has repeated applications of the same herbicide or herbicides with the same mechanism of action been carried out?
2. Has the herbicide in question been losing efficiency?
3. Are there any reported cases of plants resistant to this herbicide?
4. Has the herbicide not lost efficiency over other species?

If the answer to one or more of these questions is yes, there is a possibility of resistance. In case of negative response, the causes of failure in control should be investigated. After the resistance is diagnosed, it should be confirmed. To do this, you must follow some pre-established criteria that are highlighted in the next item.

5 Resistance Confirmation

The most common method recommended by HRAC (2021a) is to harvest seeds from suspected resistance plants and sensitive plants, sowing in pots and treat with increasing doses of the herbicide in question. To be sure that the harvested plants represent the population, one must harvest around 40 plants or 1000 seeds. To serve as a sensitive pattern, seeds should be harvested from plants in places that have never received application of that herbicide. The conditions of application should follow those recommended by the manufacturer of the formulation. The doses to be applied are: half of the recommended, recommended dose, two and four times the recommended dose. After 2 and 4 weeks, the control and dry matter production of the plants should be evaluated. The results may indicate whether the resistance is due to the change in the site of action or the metabolization of the molecule. In most cases, if the control difference between resistant and sensitive biotypes is large, it suggests that the possible mechanism of resistance is related to changes in the site of action. On the other hand, if the control difference is small, it indicates that the probable mechanism involved is the reduced absorption and/or translocation of the herbicide. However, in the case of glyphosate these patterns are reversed, e.g., simple mutations provide low levels of resistance as mechanisms of the non-target site, including vacuolar compartmentalization of the herbicide, provide medium and high levels of resistance (Sammons and Gaines 2014). The level of resistance of mechanisms such as overexpression of the target enzyme, due to the multiple number of copies, and metabolism is imperishable, because each resistant population has a number of different copies and/or participation of few or many genes involved in herbicide degradation.

The differences between resistant and sensitive biotypes of a species can be quantitatively expressed by comparing the herbicide doses necessary to reduce 50% of the population (lethal dose, LD50), biomass (growth reduction, GR50), or enzyme activity (inhibition, I50) of herbicide-treated plants compared to untreated plants (Christoffoleti et al. 1994). Biochemical and molecular analyses and the use of radiometric techniques performed in the laboratory are used to identify the exact mechanism of resistance (Burgos et al. 2013; Dayan et al. 2015; Délye et al. 2015; Mendes et al. 2017; Alcántara-de la cruz et al. 2021).

There are different methodologies for studying most cases of resistance. The general steps for resistance studies are similar; however, the details of each stage vary depending on several factors, mainly the mechanism of action of the herbicide, weed species, time of application, and the objectives of the study. Rapid studies using whole plants for post-emergence herbicides, seed and seedling germination studies in petri dishes, leaf discs, pollen germination tests, DNA sequencing-based studies are independent of resistance mechanisms and are potential in identifying weed resistance (Burgos et al. 2013).

As knowledge about the various aspects of weed resistance increases, weed survey, sampling, and resistance testing techniques are also refined. Adequate field surveys, sampling of plants or seeds, seed storage, choice of studies, use of reference populations and analysis of test results are essential in planning resistance management actions or mitigation strategies. It is usually recommended after identification of the resistant biotype, immediately eradicate the remaining plants from herbicide application, aiming to reduce the addition of seeds to the soil diaspore bank. This practice is very efficient and should be the first to be used in a practical weed resistance management program.

6 Mechanisms That Confer Resistance

The effectiveness of the herbicide depends on how much of the herbicide enters a plant cell and how long it remains in its active form, available to interact with its place of action or also known as the target site. Figure 8 shows the route of a herbicide from the moment of application to the death of a sensitive plant, which is an ideal condition of chemical control of weeds. After herbicide application, the molecules penetrate the plant (leaf or root absorption), are translocated to short (contact herbicides) or long distances (systemic herbicides) to the target site, where the herbicide is connected, disrupting biosynthesis pathways or vital cellular structures and/or generating cytotoxic molecules (e.g., reactive oxygen that damages cells), and ultimately cause plant death (Délye et al. 2013b).

The mechanisms that confer resistance to herbicides will interrupt the action of the product on the plant avoiding its death. These mechanisms can be divided into two categories, called resistance mechanisms of target and non-target site. Target site related resistance usually develops by mutation within a gene encoding an enzyme from the herbicide target site or by overexpression of the target enzyme (Jablonkai 2015) (Fig. 9). Mechanisms of non-target site of action minimize the amount of active molecule of the herbicide that reaches the site of action through reduced absorption or differential translocation (reduced or increased) and increased sequestration or metabolic degradation (Fig. 9) (Gaines et al. 2020).

7 Target Site Mechanisms

These mechanisms are molecular events (single-nucleotide polymorphisms, SNPs) in the herbicide target gene and/or increased number of copies of the gene that induces overexpression of the target enzyme. These events limit the binding or

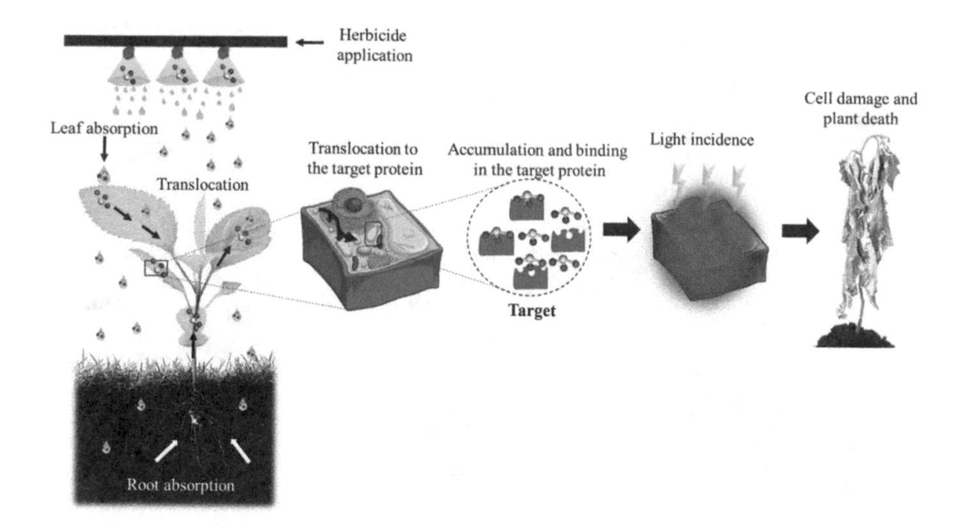

Fig. 8 Action of the herbicide after its application in a sensitive plant. (Source: Adapted from Délye et al. 2013b)

interaction of the herbicide with the site of action or require a higher herbicide concentration to inhibit them.

7.1 Changing the Site of Action

Herbicide resistance genes originate from random nucleotide substitutions (SNPs) in DNA that confer a remarkable advantage to survival and reproduction and are therefore rapidly selected and enriched in weed populations under herbicide treatment (Maxwell and Mortimer 1994). Herbicide resistance mutations may preexist or arise spontaneously in weed populations, and the rate at which they occur is very low in weeds not selected by herbicides (Casale et al. 2019). An SNP in the gene encoding a herbicide-bound protein can result in a single amino acid change, partially limiting the herbicide's ability to bind to the protein without disabling enzyme function (Gaines et al. 2020). Structural changes in enzymes are usually due to amino acid substitutions or additions in one of several possible positions in the herbicide's target protein. In some cases, the mutation may confer very high-level resistance and, in other cases, the mutation confers lower-level resistance (Délye et al. 2013b), depending on the amino acid that replaces the original amino acid. However, it is important to highlight that amino acid substitutions and/or deletions that confer some degree of resistance should occur in key positions, usually conserved from the gene encoding the target enzyme, which interact directly with the herbicide molecule.

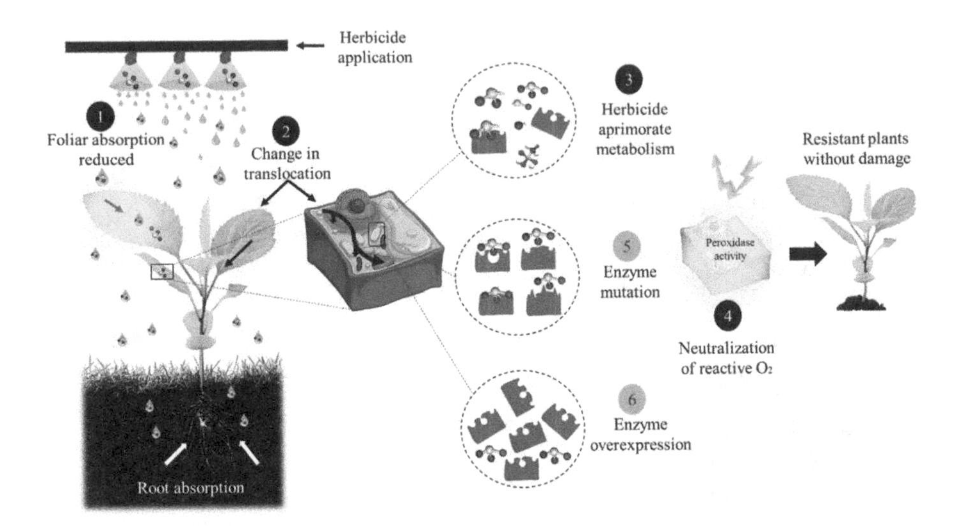

Fig. 9 Mechanisms that confer resistance to herbicides reported in weeds. Non-target site mechanisms: (1) Reduced absorption; (2) Change in translocation; (3) Herbicide metabolism and (4) Enhanced neutralization. Target site mechanisms: (5) Mutation of enzymes and (6) Enzymatic overexpression. (Source: Adapted from Délye et al. 2013b)

In the case of the EPSPS gene, an enzyme composed of more than 500 amino acids, the Glu-91-Ala mutation was reported in *Chloris truncata* (Ngo et al. 2018), the mutations Val-133-Ile and Pro-382-Leu in *Eleusine indica* (Zhang et al. 2015), the Asp-71-Met, Ala-112-Ile, and Val-201-Met mutations have been described in *Liriope platyphylla*, *Liriope spicata*, and *Ophiopogon japonicus* (Mao et al. 2016), among other cases. However, mutations that confer resistance to glyphosate must occur between positions 95 and 107 (LFLGNAGTAMRPL), as these amino acids interact directly or indirectly with the herbicide during their binding with EPSPS (Healy-Fried et al. 2007; Funke et al. 2009), that is, it is unlikely that any amino acid substitution that occurs before position 94 or after position 107, confers resistance to glyphosate. Another such case occurred in a population resistant to glufosinate of ammonium of *Lolium perenne* spp. *multiflorum* from Oregon, USA, in which resistance was initially attributed to the Asp-171-Asn mutation in the gene encoding GS (Avila-García et al. 2012). However, it has recently been proven that this mutation had no involvement in resistance to ammonium glufosinate (Brunharo et al. 2019). These clarifications have been possible, in addition to routine biochemical and molecular studies, to the use of digital tools, such as bioinformatics. The first DNA change that confers resistance to the target site was identified more than 30 years ago in trialines (PSII inhibitors). Since then, numerous mutations that confer resistance have been identified in dozens of weed species and cover eight herbicide action sites (Table 2) (Murphy and Tranel 2019).

Change in the site of action means that the herbicide molecule decreases its ability to inhibit this point due to one or more changes in the structure of this site. However, in a population of resistant biotypes occur different levels of resistance or

Table 2 Identification of mutations that confer resistance to the herbicide at the site of action

Place of action	Mechanism of action	Representative herbicide	Reference
Protein D1	PSII inhibitor	Atrazine	Hirschberg and McIntosh (1983)
Acetolactate synthase	ALS inhibitor	Chlorimuron	Guttieri et al. (1992)
Cell division	Microtubule formation inhibitors	Trifluralin	Anthony et al. (1998)
Acetyl CoA carboxylase	ACCase inhibitors	Clethodim	Zagnitko et al. (2001)
5-Enolpiruvilchiquimate-3-phosphate synthase	EPSPS inhibitors	Glyphosate	Baerson et al. (2002)
Phytoenode desaturase	Carotenoide synthesis inhibitors	Fluridone	Michel et al. (2004)
Protoporphyrinogen oxidase	PROTOX inhibitors	Lactofen	Patzoldt et al. (2006)
Auxin receiver	Auxin mimetizers	2,4-D	Le Clere et al. (2018)

Source: Murphy and Tranel (2019)

susceptibility, which may be related to the type of mutation that occurred, to the allelic forms of the gene, type of molecule, and/or the type of mechanism that is providing the resistance. The activity of the enzyme can be modified, depending on which amino acid has been altered and consequently obtains different resistance levels. For example, the site-of-action mutation of PSII-inhibiting herbicides, which compete with plastoquinone for binding to the D1 protein encoded by the psbA gene, usually amino acid substitutions of this gene confer high-level resistance to a single chemical group, but not to herbicides from other chemical groups (Gronwald 1994). The substitution of the amino acid serine by glycine at position 264 (Ser-264-Gly) confers high-level resistance to herbicides of the chemical group of trianaline in several weeds (Hirschberg and Mcintosh 1983; Ashworth et al. 2016), but moderate to no resistance to other chemical groups of PSII inhibitors, including trianalines, which are in the same group as trialines.

Several chemical groups have ALS as a place of action, including sulfonylureas and imidazolinones. To date, 21 amino acid substitutions have been reported, occurring in eight key positions (Ala-122, Pro-197, Ala-205, Asp-376, Arg-377, Trp-574, Ser-653, and Gly-654), which confer resistance to these herbicides (Tranel et al. 2021). Resistance substitutions in Pro197 have been reported more frequently, followed by mutations in Trp574 (Gaines et al. 2020). General patterns are that the Trp574 mutation confers resistance to sulfonylurea, imidazolinone, and triazolopyrimidine (Li et al. 2017), Ala-122-Tyr replacement results in resistance to imidazolinone, but not to sulfonylureas (Han et al. 2012). Pro-197-Phe substitution has been reported with resistance to sulfonylurea and sensitive to imidazolinones (depending on the specific amino acid substitution; some Pro197 mutations confer resistance to imidazolinones) (Long et al. 2019).

A possible explanation for the low frequency of herbicide resistance alleles in unselected weed populations is their selective disadvantage imposed by the associated adaptation cost. Resistance mutations in the original agroecosystem environment under no herbicide treatment would represent an adaptation cost (e.g., adequacy cost) that would limit the evolution of herbicide resistance by natural selection (Vila-Aiub 2019). Certainly, the cost of adequacy of resistant weeds corresponds to defense mechanisms against herbicides and provided that these mechanisms require a diversion of resources to operate, it would be predictable to express the costs associated with the suitability of plants carrying resistance mechanisms (Vila-Aiub 2019). Herbicides are not believed to be capable of causing mutations, since these products, before being released on the market, are evaluated for their mutagenic capacity. There is no evidence, and it is very unlikely that the mutation may occur by the action of some herbicide or other pesticide (Kissmann 1996).

7.2 Overexpression of Enzymes

Over the past decade, one of the most recent discoveries of herbicide resistance research has been gene amplification as a resistance mechanism. Although it is known that gene amplification, including genes encoding detoxifying enzymes, as well as genes encoding the site of action itself, has been resistant to insecticides for some time (Bass and Field 2011), this mechanism was not reported in relation to herbicide resistance until 2010. The involvement of gene duplication in glyphosate-resistant *Amaranthus palmeri* was the first case identified in a weed population (Gaines et al. 2010). The EPSPS gene in a population of Glyphosate-resistant *A. palmeri* was doubled 4 to more than 100 times compared to a susceptible population. The expression of EPSPS mRNA and EPSPS protein corresponded to an increase in the number of copies of genomic EPSPS. The increased expression of EPSPS, although susceptible to the herbicide, means that more of this enzyme is present at the site of action, where one fraction of the enzyme continues to develop its function in the chemical acid route and another interacts with glyphosate, e.g., the typically applied concentration of glyphosate is not sufficient to inhibit the enzymatic activity of the entire EPSPS produced by the resistant plant (Sammons and Gaines 2014).

Since then, EPSPS amplification has been found in glyphosate-resistant biotypes of several additional species, including grass and broadleaves weeds (e.g., *Lolium perenne*, *Amaranthus tuberculatus*, and *Kochia scoparia*) (Salas et al. 2012; Lorentz et al. 2014). The populations of *K. scoparia* and *A. tuberculatus* have fewer duplicate EPSPS copies than *A. palmeri*, in the range of 4–10 times (Gaines et al. 2020). Different distributions of amplified EPSPS copies in different species suggest multiple ways in which amplification arose (Tranel 2017). Currently, in addition to the duplication of the EPSPS gene, it was found that a population of mattress grass (*Digitaria sanguinalis*) had 5–7 copies of the ACCase gene resulting in 3–9 times more abundance of ACCase transcripts and conferring cross-resistance to five herbicides of this mechanism of action (Laforest et al. 2017). ALS amplification has also been recently documented in populations of *Alopecurus aequalis* and *Capsella bursa-pastoris* resistant to mesosulfuron-methyl and tribenuron-methyl, respectively (Zhao et al. 2018; Wang et al. 2019). In winter wheat in China, double mutation (Asp-376-Glu and Pro-197-Ala) was reported for ALS and also gene overexpression in *Descurainia sophia* to tribenuron-methyl (Xu et al. 2021). This is the first research that explored different functions of ALS isoenzymes in the resistance to herbicides that inhibit this mechanism of action, which may provide a worrying perspective for the management of weed resistance.

8 Non-target Site Mechanisms

Biochemical or physiological characteristic or event occurs before the herbicide reaches its place of action, preventing/decreasing a herbicide reaching its target in lethal quantity. Thus, similar to enzymatic overexpression, a fraction of the enzyme interacts with the little herbicide that has reached it, and another fraction continues to perform its essential function.

8.1 Reduced Absorption and Differential Translocation

To be effective, herbicides must be absorbed by plant cells through roots or leaves and translocated to the site of action. Thus, absorption and translocation when altered decrease the amount of herbicide that reaches the site of action, not being enough to control the plant. The procedure for detecting these resistance mechanisms in weeds may be with the use of ^{14}C radioisotopes (Mendes et al. 2017; Alcántara-de la Cruz et al. 2021). However, the reduction in herbicide absorption has been reported in resistant weeds for the main mechanisms of action of herbicides, e.g., glyphosate (Michitte et al. 2007; Vila-Aiub et al. 2012), ALS inhibitors (White et al. 2002), and ACCase inhibitors (De Prado et al. 2005). This is related to differences in the physical and chemical properties of cuticle of resistant weeds that cause a reduction in herbicide solution retention in leaves and/or a reduction in the effectiveness of herbicide penetration through cuticle (Délye 2013). Reduced absorption usually contributes to the overall resistance mechanism, e.g., glyphosate resistance in *A. tuberculatus* biotypes is due to reduced absorption and a herbicide-resistant allele of the target enzyme (EPSPS) (Nandula et al. 2013).

Differential translocation is an important mechanism of resistance and can occur through the reduction or increase of translocation (Markus et al. 2021). The reduction of herbicide translocation may involve restriction in herbicide movement within the plant and/or its compartmentalization; on the other hand, rapid translocation is mainly related to the exudation of the compounds.

Resistance due to reduced translocation occurs when the herbicide is retained in the origin leaves and has prevented its transport to growth points. Mechanisms that capture the herbicide in the leaves of origin (sequestration in vacuoles or leaf tricomas) or prevent its normal movement to the growth points through membrane barriers (altered activity of the active membrane transporters) will reduce the total amount of translocated herbicide, thus conferring resistance (Gaines et al. 2020). Ferreira et al. (2006, 2008) working with biotypes of *L. multiflorum* and *C. bonariensis* observed differences in relation to the translocation of ^{14}C-glyphosate. According to the same authors, the resistance of *L. multiflorum* and *C. bonariensis* may be related to the differential translocation of this herbicide in biotypes. Goggin et al. (2016) when comparing a susceptible population and a resistant of nabiça (*Raphanus raphanistrum*) to the 2,4-D marked in ^{14}C showed that the resistance is due to the inability of 2,4-D to transfer to other parts of the plant. Reduced absorption, translocation, and increased metabolism of atrazine were observed in a population of *Poa annua* resistant to PSII inhibitors (atrazine, diuron, and amicarbazone), where sequencing of the psbA gene confirmed the presence of a replacement of the amino acid Ser 264 for amino acid Gly (Svyantek et al. 2016).

Recently, some authors have suggested that a carbon source may be necessary to induce symptoms of rapid necrosis in resistant weeds (Moretti et al. 2017a, b; Queiroz et al. 2019), which is associated with the acute response of rapidly developing cell death in plants (Proskuryakov et al. 2003). For example, the buva biotype (*C. sumatrensis*) was reported with resistance to 2,4-D and symptoms of rapid necrosis started in approximately 2 h after herbicide application with partial wilt of leaves and mild necrotic spots (Queiroz et al. 2019). The same authors stated that rapid necrosis is a new mechanism of resistance to herbicides and characterizes the increasing complexity of weed control based only on strategies with the use of herbicides.

Other resistance mechanism not related to the target site of the herbicide much less frequent is the improved exudation of the herbicide by the roots of the plants (Ghanizadeh and Harrington 2020). With the use of ^{14}C, Jugulam et al. (2013) reported that the MCPA-resistant *Raphanus raphanistrum* biotype exuded a greater amount of herbicide into the root system than susceptible biotype. Rojano-Delgado et al. (2019) report that 65% and 38% of imazamox was exuded by the roots of resistant and susceptible dairy biotypes (*E. heterophylla*), respectively, 96 h after treatment. The authors also observed that only the parental molecule of imazamox was exuded by the roots, that is, there were no indications of imazamox metabolism. Despite the importance of root exudation potential in herbicide weed resistance and the influence on the overall performance of translocated herbicides in plants, the role of herbicide exudation from roots has rarely been reported in research involving resistant weeds and these compounds (Ghanizadeh and Harrington 2020).

These mechanisms can, alone or associated, provide tolerance or resistance to herbicides, even if they belong to different chemical groups. Thus, a weed can be sensitive, tolerant, or resistant to a herbicide.

8.2 Increased Metabolism

Plants contain a large number of genes encoding enzymes that perform biochemical reactions of synthesis of secondary metabolites and to detoxify xenobiotic compounds (e.g., herbicides) (Yuan et al. 2007). This metabolism is produced by enzymatic systems such as cytochrome P450, Glutathione S-transferases, and glucosil transferases (Nandula et al. 2019). The resistant plant in this case has the ability to degrade the herbicide molecule more quickly than sensitive plants, making it non-toxic.

Herbicide metabolism generally occurs in three phases: a conversion of the herbicide molecule to a more hydrophilic metabolite through oxidation, reduction, or hydrolysis, mediated by the cytochrome P450 monooxygenase (Phase I); followed by conjugation with biomolecules such as glutathione, mediated by glutathione-S-transferase (Phase II); and additional conjugation/dissolution/oxidation reactions with compartmentalization of herbicide metabolites in vacuole or incorporation into cell walls, where additional breakage or sequestration occurs (Phase III) (Nandula et al. 2019; Gaines et al. 2020). Improved metabolism was the main mechanism not related to target site reported in weeds resistant to HPPD inhibitors (4-hydroxyphenylpyruvate dioxygenase), as in *A. palmeri* (Nakka et al. 2017), *A. tuberculatus* (Oliveira et al. 2018), and *Echinochloa phyllopogon* (Guo et al. 2019). Resistance to mesotrione (HPPD inhibitor) in *A. palmeri* is mainly conferred by rapid detoxification (non-target site) of the herbicide and, in addition, increased expression of the HPPD gene (target site) also contributed to the resistance mechanism (Nakka et al. 2017). However, for *A. tuberculatus*, resistance was not due to duplication or overexpression of the HPPD gene before or after herbicide treatment, but rather by higher levels of mesotrione metabolism via 4-hydroxylation of the dione ring observed in resistant biotypes compared to susceptible biotype (Kaundun et al. 2017).

The rapid metabolism of synthetic auxins is an important mechanism of resistance reported in several species of eudicotyledonous weeds, in which similar to naturally tolerant monocotyledonous species, herbicide detoxification occurs by means of hydroxylation followed by conjugation, mediated by citrochrome P450 (Jugulam and Shyam 2019). This rapid detoxification of synthetic auxins was reported in *A. tuberculatus* (Figueiredo et al. 2018) and *Papaver rhoeas* (Torra et al. 2017).

In compartmentalization (Phase III of metabolism), the herbicide molecule is conjugated with plant metabolites, rendering it inactive, or removed from the metabolically active parts of the cell and stored in inactive sites (cell wall and/or vacuole) (Ghanizadeh and Harrington 2017). According to Hawkes (2014), vacuolar sequestration seems to be the main mechanism of resistance in weed species with a high level of resistance to paraquat, as it seems that other mechanisms of resistance (e.g., greater tolerance to oxidative stress) to this herbicide cause only low levels of resistance. The latter mechanism has also been reported with some frequency conferring resistance to glyphosate in species of the genera *Conyza* and *Lolium* spp. (Ge et al. 2011, 2012; D'Avignon and Ge 2018).

The metabolization of glyphosate as a resistance mechanism was reported for the first time in *D. insularis* (Carvalho et al. 2012). This finding generated controversy in the scientific community, as it had previously been concluded that weeds were unlikely to be able to degrade glyphosate (Duke 2011). In later years, metabolism was also reported as a mechanism of resistance to glyphosate eventually in some weeds (Bracamonte et al. 2016; Anagnostopoulos et al. 2020). As the results of these studies were based on analytical analyses, some researchers were still skeptical of glyphosate metabolism as a resistance mechanism because it was not fully demonstrated (Gaines et al. 2019). However, it was shown that the enzyme aldo-keto reductase was able to metabolize this herbicide into aminomethylphosphonics (AMPA) and glioxylate in *Echinochloa colona* (Pan et al. 2019). Even so, the metabolism of glyphosate mediated by aldo-keto reductase, although better accepted by researchers than previous evidence, was questioned not to be the main mechanism of resistance in *E. colona* (Duke 2019; McElroy and Hall 2020), because both the populations metabolized it. The susceptible population metabolized 44% of glyphosate, while the resistant population up to 90% of the herbicide, 72 h after application (Pan et al. 2019).

For glyphosate to be toxic to plants, it must be present in the cytoplasm and therefore mechanisms that reduce cytoplasmic glyphosate to a sublethal level may confer resistance. Pan et al. (2021) showed that an ATP-binding cassette (ABC) carrier, specifically EcABCC8, was overexpressed in a population of glyphosate-resistant *E. colona* in Western Australia. Through subcellular localization analysis, structural modeling and quantification of glyphosate levels in cells in rice plants, the researchers indicated that EcABCC8 is probably a transporter located in the plasma membrane that expels cytoplasmic glyphosate into the apoplast, decreasing the herbicide level in the cell. This discovery evidences a probable new mechanism of weed resistance to herbicides and represents a major scientific advance in the studies of resistant weeds. Further studies with other transporters, herbicides, and plants should be conducted to better understand the facts.

8.3 Neutralization of Cytotoxic Compounds

Plants have defense systems against reactive oxygen species (ROS) that, under unstressed conditions, the formation and removal of these are in balance. Most forms of biotic or abiotic stress disrupt the metabolic balance of cells, resulting in higher reactive oxygen production (Pandhair and Sekhon 2006). In case of abiotic stress (e.g., treated with herbicides), this is mainly produced in organelles (chloroplast, mitochondria, and peroxisomes) where there is intense flow of electrons during photosynthesis and respiration (Krieger-Liszkay 2005).

More than 70% of herbicide mechanisms involve the production and oxidative stress, either by direct involvement in radical production or by inhibition of biosynthetic pathways (Caverzan et al. 2019). Herbicides such as paraquat induce light-dependent oxidative damage in plants by reducing the PSI-mediated paraquat dioxide in mono cation radical, reacting with molecular oxygen to produce superoxides (O^{-2}), with subsequent production of other toxic species such as hydrogen peroxide (H_2O_2) and hydroxyl radicals (OH) (Elstner et al. 1988).

Antioxidant systems and protection against oxidative damage caused by the action of herbicides play an important role of resistance not related to the site of action (Délye 2013; Cummins et al. 2013). This system could directly eliminate EROs or complement other herbicide resistance mechanisms in a complex set of coordinated processes (Maroli et al. 2015). The different phytochemicals and enzymes produced by plants as antioxidant defense are produced in the leaves and protect the plant from damage, satiating free radicals. Plants with high antioxidant capacity are more tolerant to herbicide-induced photooxidative stress than plants with low antioxidant capacity (Arora et al. 2002). In several herbicide-resistant oat (*Avena fatua*) biotypes, enzymes with high expression related to redox potential were identified. This fact suggests that these plants exhibit a high redox maintenance capacity, since these plants are resistant to 11 herbicides of five different mechanisms of action (Keith et al. 2017). This type of resistance can cause resistance to various herbicides, posing a threat to sustainable agriculture and worries scientists, herbicide developers, and farmers.

9 Practices to Avoid Resistance

Before control failures appear in the crop, some practices can be implemented in order to minimize the risk of resistant plants (Table 3).

The main measure to prevent or reduce the risk of herbicide resistance evolution is to minimize selection pressure and control resistant individuals before they can multiply. This can be achieved by the adoption of the following practices:

1. Use herbicides with different mechanisms of action
2. Perform sequential applications of herbicides and/or herbicide mixtures with different mechanisms of action
3. Perform rotation of mechanisms of action
4. Limit the number of applications of a herbicide of the same family or mechanisms of action within a single crop
5. Apply herbicides at the registration doses and in the weed development stage recommended in the product package leaflet
6. Use herbicides with lower selection pressure (residual and efficiency)
7. Rotate crop planting
8. Rotate weed control methods
9. Track changes in flora

Table 3 Risk of resistance evolution, according to cultivation practices

Management option	Risk of resistance		
	Low	Average	High
Herbicide mechanism used	More than two	Two mechanisms	A mechanism
Herbicide mixture	More than two mechanisms	Two mechanisms	A mechanism
Control method	Cultural, mechanical, and chemical	Cultural and chemical	Chemical
Crop rotation	Complete	Limited	No
Infestation	Low	Average	High
Control in the last 3 years	Good	Declining	Bad

Source: Adapted from HRAC (2021b)

10. Use certified seeds and prevent the introduction of new weeds in agricultural areas
11. Rotate the soil preparation method
12. Clean the equipment before moving them between the plots to minimize the dispersion of weed seeds
13. Keep weed-free areas during the period without cultivation

The use of herbicides with different mechanisms of action in annual rotations, tank mixtures, and sequential applications can delay the evolution of resistance, minimizing the selection pressure imposed on weed populations by a single specific mechanism of action (Norsworthy et al. 2012). Cultural practices can be targeted to exploit crop competitiveness, thereby reducing the emergence and growth of weeds. Preventing vegetative propagation structures or seeds from being taken from infested areas to other non-infested areas should be considered important in all integrated weed management programs. Herbicide resistance management plans should begin before planting, as control options become limited when the crop is planted and weeds begin to emerge, which also favors the soil seed bank. The adoption of these practices aims to reduce selection through the diversification of weed control techniques, minimizing the dissemination of genes and resistance genotypes via pollen or propagule dispersion and eliminating additions of weed seeds to the soil seed bank (Silva et al. 2007).

10 Evolution of Weed Resistance in Brazil

Currently, 53 cases of herbicide-resistant weeds have been confirmed in Brazil (Table 4). The main groups of herbicides with resistance are als, ACCase, EPSPS, and PSII inhibitors. The crops with the highest frequency of herbicide-resistant biotypes were soybean, corn, rice, wheat, and cotton (Heap 2021). The Southern region, comprising the states of Paraná, Santa Catarina, and Rio Grande do Sul, and the Midwest region (only in Mato Grosso and Mato Grosso do Sul) have 82% of the cases, with Paraná being the State where there are more cases of resistance to herbicides (Alcántara-de la Cruz et al. 2020b).

The first officially reported cases were in *B. pilosa* and *E. heterophylla* resistant to ALS inhibitor herbicides. The ALS enzyme of the resistant biotypes proved to be less sensitive to these herbicides and thus is the most likely cause of resistance (Vargas et al. 1999; Monquero et al. 2000). In the last 5 years, the reported cases of resistance have all been multiple resistance to different mechanisms of action and among them to ALS inhibitors. ALS inhibitors have the highest risk of selecting resistance in weeds, because they have a single target site and are effective against a broad spectrum of weeds. These herbicides are used extensively in many crops and are relatively persistent—usually providing all control of propagules and germinating weed seeds (Brown et al. 1991). In addition, various mutation sites for resistance are not near the active site of the enzyme; as a result, there is no loss of conditioning due to a lower affinity for normal substrates (Christoffoleti et al. 1997).

Currently, 336 cases of glyphosate resistance have been observed in 53 weed species worldwide, in which 26 are eudicotyledonean plants and 27 are monocotyledons. In Brazil, 12 cases of glyphosate resistance are known: *Amaranthus hybridus*, *A. palmeri*, *A. viridis*, *Chloris elata*, *C. bonariensis*, *C. canadensis*, *C. sumatrensis*, *D. insularis*, *Echinochloa crus-galli* var. *crus-galli*, *E. indica*, *E. heterophylla*, and *L. multiflorum* (Mendes et al. 2020b; Alcántara-de la Cruz et al. 2020a; Heap 2021). With the technology of glyphosate-resistant soybean, quickly accepted and adopted by the producers, the use of this herbicide was expanded, reaching an average of three applications of glyphosate per soybean cycle for desiccation, and two after crop emergence. In addition, glyphosate is the main herbicide for various crops such as fruit, coffee, eucalyptus, and desiccation in no-tillage (Ferreira et al. 2009).

The chronological appearance of herbicide-resistant weeds in Brazil can be divided into two periods: the pre-glyphosate era prior to 2005, when the use of herbicides was more diverse, and the post-glyphosate era, initiated after the approval of genetically modified crops involving an almost exclusive use of glyphosate (Alcántara-de la Cruz et al. 2020b). From 1993 to 2005, 17 cases of resistant weeds were recorded, with only one case of multiple resistance reported in *E. heterophylla* in 2004. After the release of genetically modified crops resistant to glyphosate, 36 cases of biotypes resistant to different mechanisms of action were reported, and among them, 14 species with multiple resistance (Heap 2021). *C. sumatrensis* has multiple resistance to five different mechanisms of action (Pinho et al. 2019), and according to Constantin et al. (2013), the genus *Conyza* has high invasive potential due to the large seed production, the rapid and high germination capacity, causing great damage to agriculture. In addition, species of this genus present weak interspecific differentiation, which facilitates the exchange of resistant all links between species, aggravating the problem in relation to resistance management.

Some species present major problems for Brazilian agriculture, and among them, *D. insularis* and the genus *Amaranthus* are resistant to glyphosate. Biological and competitive characteristics of these species provide their ability to disseminate

Table 4 Herbicide-resistant weed species of different mechanisms of action reported in Brazil

Year	Latin name	Common name	Mechanism of action	Herbicides	Crop	Local
1993	*Bidens pilosa*	Hairy	ALS inhibitors	Imazethapyr, imazaquin, pyrithiobac-sodium, chlorimuron-ethyl, nicosulfuron	Soybeans	MS
1993	*Euphorbia heterophylla*	Poinsettia	ALS inhibitors	Imazethapyr, imazaquin, chlorimuron-ethyl, cloransulam-methyl, imazamox	Soybeans	RS
1996	*Bidens subalternans*	Hairy	ALS inhibitors	Imazethapyr, chlorimuron-ethyl, nicosulfuron	Soybeans	MS
1997	*Urochloa plantaginea*	Alexandergrass	ACCase inhibitors	Haloxyfop-methyl, diclofop-methyl, fluazifop-P-butyl, propaquizafop, quizalofop-ethyl, fenoxaprop-P-ethyl, sethoxydim, butroxydim	Soybeans	PR
1999	*Sagittaria montevidensis*	Arrowhead	ALS inhibitors	Bispyribac-sodium, pyrazosulfuron-ethyl, metsulfuron-methyl, ethoxysulfuron, cyclosulfamuron	Rice	SC
1999	*Echinochloa crus-pavonis*	Cockspur	Auxin mimetizers	Quinclorac	Rice	–
1999	*Echinochloa crus-galli* var. *crus-galli*	Barnyardgrass	Auxin mimetizers	Quinclorac	Rice	–

(continued)

Table 4 (continued)

Year	Latin name	Common name	Mechanism of action	Herbicides	Crop	Local
2000	*Cyperus difformis*	Sedge	ALS inhibitors	Pyrazosulfuron-ethyl, cyclosulfamuron	Rice	SC
2001	*Raphanus sativus*	Radish	ALS inhibitors	Imazethapyr, chlorimuron-ethyl, metsulfuron-methyl, nicosulfuron, cloransulam-methyl	Wheat	–
2001	*Fimbristylis miliacea*	Fringerush	ALS inhibitors	Pyrazosulfuron-ethyl	Rice	PR
2002	*Digitaria ciliaris*	Crabgrass	ACCase inhibitors	Haloxyfop-methyl, cyhalofop-butyl, fluazifop-P-butyl, propaquizafop, fenoxaprop-P-ethyl, sethoxydim	Soybeans	PR
2003	*Eleusine indica*	Goosegrass	ACCase inhibitors	Cyhalofop-butyl, fenoxaprop-P-ethyl, sethoxydim	Soybeans	–
2003	*Lolium perenne* ssp. *multiflorum*	Ryegrass	EPSPS inhibitors	Glyphosate	Orchards and soybeans	RS
2004	*Euphorbia heterophylla*	Poinsettia	ALS inhibitors, PROTOX inhibitors	Imazethapyr, metsulfuron-methyl, nicosulfuron, diclosulam, flumetsulam, cloransulam-methyl, fomesafen, lactofen, acifluorfen-sodium, flumiclorac-pentyl, saflufenacil	Corn and soybeans	PR

(continued)

Table 4 (continued)

Year	Latin name	Common name	Mechanism of action	Herbicides	Crop	Local
2004	*Parthenium hysterophorus*	Parthenium	ALS inhibitors	Imazethapyr, chlorimuron-ethyl, cloransulam-methyl, iodosulfuron-methyl-sodium, foramsulfuron	Soybeans	PR
2005	*Conyza bonariensis*	Fleabane	EPSPS inhibitors	Glyphosate	Corn, soybeans, wheat and fruits	RS and SP
2005	*Conyza canadensis*	Horseweed	EPSPS inhibitors	Glyphosate	Orchards, soybeans, and fruits	SP
2006	*Oryza sativa* var. *sylvatica*	Red rice	ALS inhibitors	Imazethapyr, imazapic	Rice	RS
2006	*Bidens subalternans*	Hairy	ALS inhibitors, PSII inhibitors	Atrazine, iodosulfuron-methyl-sodium, foramsulfuron	Corn	PR
2008	*Digitaria insularis*	Crabgrass	EPSPS inhibitors	Glyphosate	Corn and soybeans	PR
2009	*Sagittaria montevidensis*	Arrowhead	ALS inhibitors, PSII inhibitors	Imazethapyr, bispyribac-sodium, pyrazosulfuron-ethyl, metsulfuron-methyl, ethoxysulfuron, bentazon, penoxsulam	Rice	SC
2009	*Echinochloa crus-galli* var. *crus-galli*	Barnyardgrass	ALS inhibitors, Auxin mimetizers	Imazethapyr, bispyribac-sodium, quinclorac, penoxsulam	Rice	RS
2010	*Lolium perenne* ssp. *multiflorum*	Ryegrass	ALS inhibitors	Iodosulfuron-methyl-sodium	Wheat	RS
2010	*Lolium perenne* ssp. *multiflorum*	Ryegrass	ACCase inhibitors, EPSPS inhibitors	Clethodim, glyphosate	Corn, soybeans, and wheat	RS

(continued)

Table 4 (continued)

Year	Latin name	Common name	Mechanism of action	Herbicides	Crop	Local
2010	*Conyza sumatrensis*	Tall fleabane	EPSPS inhibitors	Glyphosate	Corn and soybeans	PR
2011	*Conyza sumatrensis*	Tall fleabane	ALS inhibitors	Chlorimuron-ethyl	Soybeans	PR
2011	*Conyza sumatrensis*	Tall fleabane	ALS inhibitors, EPSPS inhibitors	Chlorimuron-ethyl, glyphosate	Corn and soybeans	PR
2011	*Amaranthus retroflexus*	Pigweed	ALS inhibitors, PSII inhibitors	Atrazine, prometryn, trifloxysulfuron-sodium	Cotton	–
2011	*Amaranthus viridis*	Amaranth	ALS inhibitors, PSII inhibitors	Atrazine, prometryn, trifloxysulfuron-sodium	Cotton	–
2012	*Amaranthus retroflexus*	Pigweed	ALS inhibitors	Pyrithiobac-sodium, trifloxysulfuron-sodium	Cotton	–
2013	*Raphanus raphanistrum*	Radish	ALS inhibitors	Imazethapyr, chlorimuron-ethyl, metsulfuron-methyl, sulfometuron-methyl, cloransulam-methyl, iodosulfuron-methyl-sodium, imazapic	Spring barley and wheat	PR
2013	*Ageratum conyzoides*	Ageratum	ALS inhibitors	Pyrithiobac-sodium, trifloxysulfuron-sodium	Cotton and soybeans	MT
2014	*Cyperus iria*	Flatsedge	ALS inhibitors	Imazethapyr, bispyribac-sodium, pyrazosulfuron-ethyl, imazapic, penoxsulam	Rice	RS

(continued)

Table 4 (continued)

Year	Latin name	Common name	Mechanism of action	Herbicides	Crop	Local
2014	*Chloris elata*	Tall windmill grass	EPSPS inhibitors	Glyphosate	Soybeans	–
2014	*Amaranthus retroflexus*	Pigweed	PROTOX inhibitors	Fomesafen	Cotton and soybeans	MT
2015	*Echinochloa crus-galli* var. *crus-galli*	Barnyardgrass	ACCase inhibitors, ALS inhibitors, cellulose synthesis inhibitors	Cyhalofop-butyl, quinclorac, penoxsulam	Rice	Tubarão, SC
2015	*Amaranthus palmeri*	Palmer amaranth	EPSPS inhibitors	Glyphosate	Cotton	MT
2015	*Echium plantagineum*	Bugloss	ALS inhibitors	Metsulfuron-methyl	Cereals and wheat	RS
2016	*Amaranthus palmeri*	Palmer amaranth	ALS inhibitors, EPSPS inhibitors	Imazethapyr, chlorimuron-ethyl, cloransulam-methyl, glyphosate	Corn, cotton, and soybeans	MT
2016	*Digitaria insularis*	Crabgrass	ACCase inhibitors	Haloxyfop-methyl, fenoxaprop-P-ethyl	Soybeans	Midwest Region
2016	*Bidens pilosa*	Hairy	ALS inhibitors, PSII inhibitors	Imazethapyr, atrazine	Corn and soybeans	PR
2016	*Lolium perenne* ssp. *multiflorum*	Ryegrass	ACCase inhibitors, ALS inhibitors	Clethodim, iodosulfuron-methyl-sodium	Wheat	–
2016	*Conyza sumatrensis*	Tall fleabane	PSI inhibitors	Paraquat	Soybeans	PR
2016	*Eleusine indica*	Goosegrass	EPSPS inhibitors	Glyphosate	Corn, soybeans, and wheat	PR
2017	*Conyza sumatrensis*	Tall fleabane	PROTOX inhibitors	Saflufenacil	Soybeans	Palotina, PR
2017	*Conyza sumatrensis*	Tall fleabane	ALS inhibitors, EPSPS inhibitors, PSI inhibitors	Chlorimuron-ethyl, paraquat, glyphosate	Soybeans	Assis Chateau-briand, PR
2017	*Eleusine indica*	Goosegrass	ACCase inhibitors, EPSPS inhibitors	Haloxyfop-methyl, fenoxaprop-P-ethyl, glyphosate	Corn, cotton, soybeans, and bean	MT

(continued)

Table 4 (continued)

Year	Latin name	Common name	Mechanism of action	Herbicides	Crop	Local
2017	*Conyza sumatrensis*	Tall fleabane	Auxin mimetizers, EPSPS inhibitors, PROTOX inhibitors, PSI inhibitors, PSII inhibitors	Diuron, paraquat, glyphosate, 2,4-D, saflufenacil	Soybeans	Assis Chateau-briand, PR
2017	*Lolium perenne* ssp. *multiflorum*	Ryegrass	ALS inhibitors, EPSPS inhibitors	Glyphosate, iodosulfuron-methyl-sodium, pyroxsulam	Corn, soybeans, and wheat	Ronca-dor, PR
2018	*Amaranthus hybridus*	Smooth	ALS inhibitors, EPSPS inhibitors	Chlorimuron-ethyl, glyphosate	Soybeans	–
2019	*Euphorbia heterophylla*	Poinsettia	EPSPS inhibitors	Glyphosate	Soybeans	PR
2020	*Echinochloa crus-galli* var. *crus-galli*	Barnyardgrass	EPSPS inhibitors	Glyphosate	Soybeans	–
2020	*Digitaria insularis*	Crabgrass	ACCase inhibitors, EPSPS inhibitors	Haloxyfop-methyl, fenoxaprop-P-ethyl, glyphosate	Soybeans	MT and MS

Source: Heap (2021)

resistance. Martins et al. (2016) observed that the resistant biotype of *D. insularis* grew faster, was more robust and better adapted to light interception than the susceptible biotype, reaching flowering first. Dispersion of *D. insularis* populations originated in western Paraná, near the border with Paraguay, where the first case of resistance was reported. Its dispersion in the central region of Brazil, mainly in soybean producing areas, is probably a consequence of the movement of agricultural equipment and wind-mediated dispersion (Ovejero et al. 2017), but also by independent resistance selection events (Takano et al. 2018). However, this shows the importance of preventive management, such as cleaning the machinery before transport to other areas, in addition to keeping the weed-free area during the period without cultivation to prevent these plants from growing and dispersing their seeds.

A. palmeri was reported with resistance to glyphosate officially in 2015; however, the Mato-Grossense Institute of Cotton had records of the occurrence of this resistance case since 2012 in the municipalities of Ipiranga do Norte and Tapurah, in the State of Mato Grosso. This fact occurred, possibly, by the import of Argentina in 2011, via cotton collecting machines used and purchased in the USA (Bogalin et al. 2016; Ikeda et al. 2019). The multiple resistance of *A. palmeri* to EPSPS and ALS inhibitors was also observed in Mato Grosso (Heap 2021). *A. palmeri* is an aggressive weed with the ability to easily adapt to different environments, favors crossing and genetic diversity, produces large amounts of seeds, easy distribution of small seeds, and high emergence for prolonged periods (Gazziero and Silva 2017). In addition, the genus *Amaranthus* sp. can hybridism interspecifically facilitating the dispersion of resistance alleles (Nandula et al. 2014). *A. palmeri* was considered as a "pest officially controlled by the State of Mato Grosso" with the aim of eradicate it, leading to the supervision of its control in the areas of occurrence. The "zero" tolerance strategy is justified when one considers that in an area with only one female plant, being poorly managed for 3 years, it can be 95–100% infestation with *A. palmeri* at the end of this period (Norsworthy et al. 2012; Ikeda et al. 2019). Recently, *A. hybridus* with multiple resistance to EPSPS and ALS inhibitors in soybean crop in the State of Rio Grande do Sul (Heap 2021) has also been reported. The possibility that this case has been introduced from Argentina generates concern among producers and the scientific community, because in this country there have been different populations of *A. hybridus* with this resistance profile (García et al. 2020). Resistance to both herbicides involves mechanisms of the target site. The Trp-574-Lue and Asp-376-Glu mutations were found in ALS inhibitors, occurring in different individuals and conferring cross-resistance (Larran et al. 2018). In glyphosate, a triple amino acid substitution in the naturally occurring EPSPS structure was found as the main mechanism of resistance in *A. hybridus* (Perotti et al. 2019). This triple mutation is unprecedented and confers high levels of resistance to glyphosate.

The mechanism of action that involves inhibition of ACCase is also among the mechanisms with more resistant weed cases (Table 4). The species of *Urochloa plantaginea*, *D. ciliaris*, and *D. insularis* showed resistance only to ACCase inhibitors. However, the species *L. multiflorum*, *E. crus-galli*, and *E. indica* have resistance to up to three different mechanisms of action (Heap 2021). The main occurrence is in soybean crops due to the frequent use of these herbicides in grass control. Weeds resistant to PSII inhibitors have also been reported in Brazil, all of them with multiple resistance to ALS and PSII inhibitors (Table 4). The scenario faced by farmers in the coming years in relation to the management of resistant weeds will be very difficult, taking into account the complexity and evolution of resistance.

The type of management and herbicides used in an area cause changes in the type and proportion of species that make up the population of the site. This is explained by the fact that herbicides do not equally control the existing species in the area; with this, some end up benefiting and multiplying. In these situations, plants of low occurrence in the area can become a serious problem for the producer. Thus, the continuous and repeated use of the same herbicide, or herbicides with the same mechanism of action, makes the selection of species inevitable (Silva et al. 2007).

11 Integrated Management of Resistant Weeds

The integrated management considers all the factors that can provide the plant with greater and better production and allows the efficient use of the resources of the environment and represents a good alternative of the efficient control of herbicide-resistant weeds. Within this context, integrated management of resistant weeds (MIPDR) is also included. The objective of MIPDR is to include weed control strategies that consider the use of all economically available control techniques.

The premises that underheld the proposal of integrated management can be summarized in: quality assurance of the harvested product, including the exemption of pesticide residues in food; environmental sustainability, including non-degradation of soil and contamination of air and water; economic and social sustainability in production, maintaining or increasing productivity; and ensuring a better quality of life for the farmer with regard to economic return and greater safety in activities involving the use of agricultural pesticides. With the increased presence of herbicide-resistant weeds in cultivation systems, practices that impose high selection pressures, which increase gene frequencies, and that try to prevent the increase and spread of resistant weed biotypes should be discouraged. Resistance to herbicides is a very predictable result of evolution. In fact, any weed control practice will be subject to the pressures of evolution. Thus, no matter what practice, even if done consistently and long enough, weeds will evolve to survive the management practice (Heap 2014; Beckie and Harker 2017).

Integrated management strategies in different weed plant species can be divided as short or long term. Measures such as weeding or direct use of herbicides (chemical control) can be considered as short-lived, being responsible for only temporary control, with the need for new applications for each growing season. In the case of measures considered long term, the use of cultural practices and control by other biological agents have a lasting character and take into account more pronounced changes in the different agronomic practices. This results in integrated management and should integrate prevention and other control methods that promote control in the short (mechanical and chemical methods) and in the medium and long term (cultural and biological methods) (Silva et al. 2007).

Biotechnology has been developing varieties of different herbicide-resistant agronomic species, such as soybeans, corn, cotton, and glyphosate-resistant sugarcane; corn to imazaquin; from rice and soybean to ammonium-glufosinate and recently soybean and corn resistant to 2,4-D and dicamba. However, there is no doubt that the last stage of this technology is in the release of corn, soybean, and other species resistant to glyphosate and other groups of herbicides with different mechanisms of action, that is, virtually all products of this genus applied in these crops. The possibility of cultivating corn, cotton, sugarcane, and soybeans without competition with weeds attracts the Brazilian farmer, especially for the economic benefits, since the damage resulting from competition can reach the total loss of production. On the other hand, the incorrect use of this technology may, in a few years of cultivation, select weed species with tolerance and some with resistance to herbicides or, as a negative consequence, cause the elimination of several plant species, reduction of biodiversity, due to the control of almost all plant species of the planting areas. These, however, would be the direct negative effects of herbicide use (Silva et al. 2007).

It is important to remember that this negative approach is easily contestable when, associated with these crops, integrated weed management is promoted. Therefore, the importance of the study regarding the impacts of the use of transgenics in agriculture, because any and all weed management techniques will only succeed if it is applied based on detailed knowledge of the biology and ecology of weed plants in the area, mainly involving knowledge in the areas of morphology, physiology, and nutrient cycling (Silva et al. 2007).

The first method of control that must exist in crops is preventive control. Prevention strategies include sanitation and harvesting methods that do not spread seeds and vegetative propagules at all stages of production (such as seed selection, soil preparation, planting, fertilization, irrigation, weed control, harvesting, and transportation) (Christoffoleti et al. 2007). The lack of such care has contributed to the wide dissemination of several species such as *Conyza* spp. *D. insularis*, due to the lack of control of these plants in areas continuous to the property and inadequate cleaning of agricultural implements that move from infested regions to other infestation-free regions. As well as *B. pilosa* and *Cenchrus echinatus* are widely disseminated when adhering to the animal hair and *Oryza sativa* by the use of contaminated rice seeds. Other species have specific means of dissemination (Oliveira and Brighenti 2018).

Cultural control consists in the use of good agricultural practices in order to favor crop growth to the detriment of weeds. These practices help reduce the soil seed bank, reducing the levels of crop infestation in subsequent years. This control method includes the adoption of crop rotation, variation of spacing and plant population and green cover, among others, directed to the suppression of weeds. The monoculture system favors the increase of the seed bank of some weed species, due to their deficient control, due to the limitations of the herbicides used. On the other hand, the increase in the seed bank favors the emergence of biotypes more adapted to environmental conditions, that is, more competitive by resources with crops (Silva et al. 2007).

The increase in straw in soil cover can increase the control of *A. palmeri* and can reach more than 80% with 14 Mg/ha of straw (Webster et al. 2013). This is because the germination of *A. palmeri* seeds in the dark, positive photoblastic seeds, is reduced when compared to the light condition (Jha et al. 2010). In addition, straw can delay the control of *A. palmeri* in post-emergence, which may provide more flexibility in species control (Ikeda et al. 2019). Although cultural control can contribute to the control of *A. palmeri*, it is necessary to association with chemical control, mainly due to the need for eradication of the species.

A. palmeri control alternatives in soybean-cotton succession systems are: in soybean crop, the combination of pendimethalin application in pre-emergence with faminesafen or lactofen in post-emergence (control greater than 90%); in cotton culture, both the application of S-metolachlor and trifluralin in pre-emergence with the application of post-emergence glufosinate ammonium (control greater than 90%) (Ikeda et al. 2019).

The practice of soil cover has been highly indicated in Brazil in the management of other weeds that are difficult to control, such as *Conyza* ssp. and *D. insularis* (Petter et al. 2015), mainly when associated with chemical control. *Urochloa ruziziensis* contorted with corn in autumn-winter without the use of mixtures and herbicide rotation was not effective in the management of Glyphosate-resistant *C. sumatrensis*, suggesting that this cultivation system requires complementation of weed control with rotation and herbicide mixing. On the other hand, the rotation and mixture of herbicides without the cover of *U. ruziziensis* in autumn-winter were also not sufficient for the effective control of this species. Higher infestations by *C. sumatrensis* were observed in treatments without herbicide rotation and without the use of straw (Marochi et al. 2018).

It is necessary to modify some agricultural practices in order to prevent or delay the establishment of resistance in some weed biotypes (Christoffoleti et al. 1994). Crop rotation helps maintain high productivity by reducing pesticide use and fertilizer intake and can reduce the incidence of pests and pathogens, weed infestation, and weed resistance selection pressure to herbicides. However, in regions where genetically modified crops are widely adopted, there is a clear tendency toward monoculture and not crop rotation and diversification (Mortensen et al. 2012).

Resistance management using crop rotation is interesting because each agricultural crop is usually infested by weed species that have similar crop requirements or have the same growth habits; examples: *Echinochloa* spp. in rice crops; *Alternanthera tenella* in corn crops; mustard in wheat crops; and *Amaranthus deflexus* in sugarcane. Some problems that can be easily solved with the practice of crop rotation: infestation of *Solanum americanum* and *Nicandra physaloides* in tomato and potato crops treated with the herbicide metribuzim. In all these cases, crop rotation breaks the life cycle of weed species, preventing their dominance in the area.

In rice cultivation systems, problems are increasing with plant resistance to ALS inhibitors, mainly due to the intensive use of these herbicides provided by Clearfield® (imidazolinone-tolerant crops, a chemical group of ALS inhibitors) (Merotto Júnior et al. 2016) and resistance to graminicides (ACCase inhibitors). The use of crop rotation has been shown to be advantageous in controlling some herbicide-resistant weeds. Study evaluating the effects of two rotation systems of rice-soybean and rice-sorgo crops on the suppression of *Echinochloa* spp. and *U. plantaginea* showed that rice-sorgo rotation is a good option to reduce the seed bank of these weeds. In addition, the rice-soybean rotation reduced the seed bank of these species, and the reduction was dependent on the soil tillage system and the inclusion of herbicides (Scherner et al. 2018).

When the same cultural techniques are applied then, year after year, in the same soil, the interference of these weeds increases greatly. When the main objective is weed control, the choice of crop in rotation should be on plants with growth habits and very contrasting cultural characteristics (Silva et al. 2007). Thus, using crops with different physiological needs occurs the change of weed species from one crop to another, and it is necessary to use herbicides, these being with different mechanisms of action. Rotation is an efficient method both in preventing the emergence of resistant biotypes and in the management of installed resistance.

It is extremely important to rotate the mechanisms of action of herbicides to avoid the selection of resistant weed biotypes and this strategy is within the chemical management of weed resistance. Chemical management should be carried out appropriately to prevent resistance. Herbicide mixtures and sequential applications are also effective resistance management strategies. In fact, there is growing evidence that herbicide mixtures may be more effective than herbicide rotations in delaying resistance (Beckie and Reboud 2009). Ideally, each herbicide in the mixture should be active in different mechanisms of action and have a high level of effectiveness in major weeds with problems (Heap 2014). The use of herbicides in pre-emergence will continue to increase as a way to rotate the mechanisms of action of herbicides in the cultivation system and prolong control with the residual effect on the soil. To avoid the selection of low-level polygenic resistance, it is also important to use the recommended herbicide dose and proper application at the stage of development of the weed species most difficult to control in the field (Heap 2014).

Other alternatives in resistance prevention and management may be: cultivation of more competitive crops, more dense spacing and biological control. It is interesting to maintain a history of each area of the property to identify the evolution of the population of certain species, because, usually, the occurrence of resistant biotypes cannot be detected during the first years of application of the picker. This fact is only observed when the lack of control of a species that was traditionally controlled by a certain herbicide is perceived, and several years have already occurred since the beginning of the selection of the resistant biotype. Thus, the main forms of resistance management and prevention are all related to integrated weed management (Monquero and Christoffoleti 2001).

The measures mentioned above seem relatively easy to implement; however, the adoption of some of them on a medium and large scale, and some cases also on a small scale, are economically and logistically unfeasible by several factors, or are simply not a definitive solution to prevent the emergence of herbicide resistance (Lamichhane et al. 2017). The implementation of non-chemical measures is less preferred by farmers, as in addition to requiring more time and investment, they are often less efficient in the short term when compared to herbicides (Heap 2014). In herbicide-based strategies used in tank mixtures, sequential applications, herbicide rotations with different mechanism of action, as well as reduced dose applications, although they can maintain effective weed control and sustain economically acceptable harvest yields, they only delay resistance selection (Kudsk 2014; Lamichhane et al. 2017).

Crop rotation is one of the best weed management strategies. In Brazil, this practice is widespread in grain-producing agricultural regions; however, most of the time rotation is restricted to monocultures (grains per grain), involving the use of the same technology (crops resistant to glyphosate) (Alcántara-de la Cruz et al. 2020b). This is due to the high competitiveness of global commodity markets, which has led farmers to specialize in the production of one or a few closely related crops, allowing the use of the same existing agricultural implements, infrastructure, and marketing networks (Lamichhane et al. 2017). The rotation of crops that need similar weed management interventions has increased the selection pressure, because the number of herbicide applications is doubled per agricultural year (6–12 applications), because each crop requires 3–5 herbicide applications per cycle (main crop or second crop) (Alcántara-de la Cruz et al. 2020b). More complex crop rotations (grains per legume or annual crop per biennial/perennial crop) requires specific training or hiring new professionals for the second crop, investment in implements, and search for markets, which ends up discouraging producers (Lamichhane et al. 2017).

The implementation of integrated weed management measures are usually expensive at first, but in the medium and long term, they are sustainable and profitable (Livingston et al. 2016). However, a large percentage of producers lease agricultural land, and, consequently, they generally make decisions based on profits only in the short term (Lamichhane et al. 2017). In addition, the adoption or not of a particular method of weed control, chemical or non-chemical, by farmers is generally influenced by the margin of benefit, and not by the lack of available research, technologies, or management programs (Moss 2019).

12 Key Concepts

- Sensitivity, tolerance, and resistance
- Factors that favor the emergence of resistance
- Diagnosis of field resistance
- Resistance confirmation
- Mechanisms that confer resistance
- Target-site mechanisms
- Non-target site mechanisms
- Practices to avoid resistance
- Evolution of weed resistance in Brazil
- Integrated management of resistance weed

13 Concluding Remarks

Weed resistance to herbicides is a fait accompli in Brazil. It is known that its evolution in an area is dependent on the selection pressure, the genetic variability of the weed species, the number of genes involved, the inheritance pattern, the gene flow, and the dispersion of propagules. However, the problem is not related to the cultivation of genetically modified crops, but rather in the inappropriate use of these technologies as a whole, restricting to herbicide applications of the same mechanism of action, frequency of applications, and high doses of herbicides. In the case of genetically modified varieties, such as glyphosate-resistant soybeans, other herbicides with different mechanisms of action aimed at controlling resistant biotypes are being included in management programs, which is the cost of weed control. Integrated weed management combining control methods such as cultural (crop rotation, succession, green fertilization), biological, preventive, physical, and mechanical control can be a viable alternative in preventing and combating resistant weed biotypes. Resistant weed management strategies should be carried out according to the specific diagnosis of each reality. The best way to minimize the pressure of weed resistance evolution is to challenge them with diversity of control, so that any practice is not often used to select resistant biotypes. However, the biggest challenge is to show producers that the adoption of diversified weed management measures results in greater ecological and economic benefits in the medium and long term, without affecting productivity in the short term.

References

Alcántara-de la Cruz R, Amaral GDS, Oliveira GM et al (2020a) Glyphosate resistance in *Amaranthus viridis* in Brazilian citrus orchards. Agriculture 10:304. https://doi.org/10.3390/agriculture10070304

Alcántara-de la Cruz R, Oliveira GM, Carvalho LB et al (2020b) Herbicide resistance in Brazil: status, impacts, and future challenges. In: Kontogiannatos D, Kourti A, Mendes KF (eds) Pests, weeds and diseases in agricultural crop and animal husbandry production. IntechOpen, London, pp 1–25

Alcántara-de la Cruz R, Silva AG, Mendes KF et al (2021) Absorption, translocation, and metabolism studies of herbicides in weeds and crops. In: Mendes KF (ed) Radioisotopes in weed research. CRC Press, Boca Raton, FL, pp 127–154

Anagnostopoulos C, Stasinopoulou P, Kanatas P, Travlos I (2020) Differences in metabolism of three *Conyza* species to herbicides glyphosate and triclopyr revealed by LC-MS/MS. Chil J Agric Res 80(1):100–107

Andres A, Concenço G, Schreiber F (2017) Predictions for weed resistance to herbicides in Brazil: a botanical approach. In: Pacanoski Z (ed) Herbicide resistance in weeds and crops. IntechOpen, London, pp 133–157

Anthony RG, Waldin TR, Ray JA et al (1998) Herbicide resistance caused by spontaneous mutation of the cytoskeletal protein tubulin. Nature 393(6682):260

Arora A, Sairam RK, Srivastava GC (2002) Oxidative stress and antioxidative system in plants. Curr Sci 82(10):1227–1238

Ashworth MB, Han H, Knell G, Powles SB (2016) Identification of triazine-resistant *Vulpia bromoides*. Weed Technol 30(2):456–463. https://doi.org/10.1614/WT-D-15-00127.1

Avila-García WV, Sanchez-Olguin E, Hulting AG, Mallory-Smith C et al (2012) Target-site mutation associated with glufosinate resistance in Italian ryegrass (*Lolium perenne* L. ssp. *multiflorum*). Pest Manag Sci 68(9):1248–1254. https://doi.org/10.1002/ps.3286

Baerson SR, Rodriguez DJ, Tran M et al (2002) Glyphosate-resistant goosegrass. Identification of a mutation in the target enzyme 5-enolpyruvylshikimate-3-phosphate synthase. Plant Physiol 129(3):1265–1275. https://doi.org/10.1104/pp.001560

Bass C, Field LM (2011) Gene amplification and insecticide resistance. Pest Manag Sci 67(8):886–890. https://doi.org/10.1002/ps.2189

Beckie HJ, Rebound X (2009) Selecting for weed resistance: Herbicide rotation and mixture. Weed Technol 23(3):363–370

Beckie HJ, Harker KN (2017) Our top 10 herbicide-resistant weed management practices. Pest Manag Sci 73:1045–1052. https://doi.org/10.1002/ps.4543

Beckie HJ, Tardif FJ (2012) Herbicide cross resistance in weeds. J Crop Prot 35:15–28. https://doi.org/10.1016/j.cropro.2011.12.018

Bogalin DC, Silveira OR, Gazziero DLP (2016) *Amaranthus palmeri*: Programa de erradicação da espécie no Estado de Mato Grosso. In: Meschede DK, Gazziero DLP (eds) A era glyphosate: agricultura, meio ambiente e homem. Midiograf, Londrina, pp 321–336

Bracamonte E, Fernández-Moreno PT, Barro F et al (2016) Glyphosate-resistant *Parthenium hysterophorus* in the Caribbean Islands: non target site resistance and target site resistance in relation to resistance levels. Front Plant Sci 7:1845. https://doi.org/10.3389/fpls.2016.01845

Bracamonte E, Silveira HMD, Alcántara-de la Cruz R et al (2018) From tolerance to resistance: mechanisms governing the differential response to glyphosate in *Chloris barbata*. Pest Manag Sci 74:1118–1124. https://doi.org/10.1002/ps.4874

Brown HM, Fuesler TP, Ray TB et al (1991) Role of plant metabolism in crop selectivity of herbicides. In: Frehse H (ed) Pesticide chemistry: advances in international research, development and legislation. Wiley-VCH, Reinheim, pp 257–266

Brunharo CACG, Takano HK, Mallory-Smith CA et al (2019) Role of Glutamine synthetase isogenes and herbicide metabolism in the mechanism of resistance to glufosinate in *Lolium perenne* L. spp. *multiflorum* biotypes from Oregon. J Agric Food Chem 67:8431–8440. https://doi.org/10.1021/acs.jafc.9b01392

Burgos NR, Tranel PJ, Streibig JC et al (2013) Confirmation of resistance to herbicides and evaluation of resistance levels. Weed Sci 61(1):4–20. https://doi.org/10.1614/WS-D-12-00032.1

Burnet MWM, Holtum JAM, Powles SB (1994) Resistance to nine herbicide classes in a *Lolium rigidum* biotype. Weed Sci 42:369–377

Busi R, Powles SB (2013) Cross-resistance to prosulfocarb and triallate in pyroxasulfone-resistant *Lolium rigidum*. Pest Manag Sci 69(12):1379–1384

Carvalho LB, Alves PLDCA, Gonzalez-Torralva F et al (2012) Pool of resistance mechanisms to glyphosate in *Digitaria insularis*. J Agric Food Chem 60(2):615–622. https://doi.org/10.1021/jf204089d

Casale FA, Giacomini DA, Tranel PJ (2019) Empirical investigation of mutation rate for herbicide resistance. Weed Sci 67(4):361–368. https://doi.org/10.1017/wsc.2019.19

Caverzan A, Piasecki C, Chavarria G et al (2019) Defenses against ROS in crops and weeds: the effects of interference and herbicides. Int J Mol Sci 20(5):1086. https://doi.org/10.3390/ijms20051086

Christoffoleti PJ, Nicolai M (2016) Aspectos de resistência de plantas daninhas à herbicidas, 4th edn. Associação Brasileira de Ação à Resistência de Plantas Daninhas ao Herbicidas (HRAC-BR), Piracicaba

Christoffoleti PJ, Victoria Filho R, Silva CB (1994) Resistência de plantas daninhas à herbicidas. Planta Daninha 12(1):13–20

Christoffoleti PJ, Westra P, Moore F III (1997) Growth analysis of sulfonylurea-resistant and-susceptible kochia (*Kochia scoparia*). Weed Sci 45(5):691–695

Christoffoleti PJ, Carvalho SJP, Nicolai M et al (2007) Prevention strategies in weed management. In: Upadhyaya MK, Blackshaw RE (eds) Non-chemical weed management: principles, concepts and technology. CABI, Oxfordshire, pp 1–16

Christoffoleti PJ, Nicolai M, López-Ovejero RF et al (2016) Resistência de plantas daninhas à herbicidas: termos e definições importantes. In: Christoffoleti PJ, Nicolai M (eds) Aspectos de Resistência de Plantas Daninhas à herbicidas, 4th edn. Associação Brasileira de Ação à Resistência de Plantas Daninhas ao Herbicidas (HRAC-BR), Piracicaba, pp 11–31

Christopher JT, Powles SB, Liljegren DR et al (1991) Cross resistance to herbicides in annual ryegrass (*Lolium rigidum*) chlorsulfuron resistance involves a wheat-like detoxification system. Plant Physiol 94(4):1036–1043

Collavo A, Sattin M (2014) First glyphosate-resistant *Lolium* spp. biotypes found in European annual arable cropping system also affected by ACCase and ALS resistance. Weed Sci 54:325–334. https://doi.org/10.1111/wre.12082

Comont D, Knight C, Crook L et al (2019) Alterations in life-history associated with non-target-site herbicide resistance in *Alopecurus myosuroides*. Front Plant Sci 10:837. https://doi.org/10.3389/fpls.2019.00837

Constantin J, Oliveira Junior RS, Oliveira Neto AM (2013) Buva: Fundamentos e recomendações para manejo, 1st edn. Omnipax, Curitiba

Cummins I, Wortley DJ, Sabbadin F et al (2013) Key role for a glutathione transferase in multiple-herbicide resistance in grass weeds. Proc Natl Acad Sci 110(15):5812–5817. https://doi.org/10.1073/pnas.1221179110

D'Avignon DA, Ge X (2018) In vivo NMR investigations of glyphosate influences on plant metabolism. J Magn Reson 292:59–72. https://doi.org/10.1016/j.jmr.2018.03.008

Dayan FE, Owens DK, Corniani N et al (2015) Biochemical markers and enzyme assays for herbicide mode of action and resistance studies. Weed Sci 63:23–63. https://doi.org/10.1614/WS-D-13-00063.1

De Prado JL, Osuna MD, Heredia A, De Prado R (2005) *Lolium rigidum*, a pool of resistance mechanisms to ACCase inhibitor herbicides. J Agric Food Chem 53(6):2185–2191. https://doi.org/10.1021/jf049481m

Délye C (2013) Unravelling the genetic bases of non-target-site-based resistance (NTSR) to herbicides: a major challenge for weed science in the forthcoming decade. Pest Manag Sci 69(2):176–187. https://doi.org/10.1002/ps.3318

Délye C, Matéjicek A, Michel S (2008) Cross-resistance patterns to ACCase-inhibiting herbicides conferred by mutant ACCase isoforms in *Alopecurus myosuroides* Huds. (black-grass) re-examined at the recommended herbicide field rate. Pest Manag Sci 64(11):1179–1186. https://doi.org/10.1002/ps.1614

Délye C, Deulvot C, Chauvel B (2013a) DNA analysis of herbarium specimens of the grass weed *Alopecurus myosuroides* reveals herbicide resistance pre-dated herbicides. PLoS One 8:75117. https://doi.org/10.1371/journal.pone.0075117

Délye C, Jasieniuk M, Le Corre V (2013b) Deciphering the evolution of herbicide resistance in weeds. Trends Genet 29(11):649–658. https://doi.org/10.1016/j.tig.2013.06.001

Délye C, Duhoux A, Pernin F et al (2015) Molecular mechanisms of herbicide resistance. Weed Sci 63:91–115. https://doi.org/10.1614/WS-D-13-00096.1

Dimaano NG, Yamaguchi T, Fukunishi K et al (2020) Functional characterization of cytochrome P450 CYP81A subfamily to disclose the pattern of cross-resistance in *Echinochloa phyllopogon*. Plant Mol Biol 102:403–416. https://doi.org/10.1007/s11103-019-00954-3

Duke SO (2011) Glyphosate degradation in glyphosate-resistant and -susceptible crops and weeds. J Agric Food Chem 59(11):5835–5841

Duke SO (2019) Enhanced metabolic degradation: the last evolved glyphosate resistance mechanism of weeds? Plant Physiol 181(4):1401–1403. https://doi.org/10.1104/pp.19.01245

Elstner EF, Wagner GA, Schutz W (1988) Activated oxygen in green plants in relation to stress situations. Pestic Biochem Physiol 7:159–187

Ferreira EA, Santos JB, Silva AA et al (2006) Translocação do glyphosate em biótipos de azevém (*Lolium multiflorum*). Planta Daninha 24(2):365–370. https://doi.org/10.1590/S0100-83582006000200021

Ferreira EA, Galon L, Aspiazú I et al (2008) Glyphosate translocation in hairy fleabane (*Conyza bonariensis*) biotypes. Planta Daninha 26(3):637–643. https://doi.org/10.1590/S0100-83582008000300020

Ferreira EA, Germani C, Vargas L et al (2009) Resistência de *Lolium multiflorum* ao Glyphosate. In: Agostinetto D, Vargas L (eds) Resistência de plantas daninhas no Brasil. Gráfica Berthier, Passo Fundo, pp 271–289

Figueiredo MRA, Leibhart LJ, Reicher ZJ et al (2018) Metabolism of 2,4-dichlorophenoxyacetic acid contributes to resistance in a common waterhemp (*Amaranthus tuberculatus*) population. Pest Manag Sci 74(10):2356–2362. https://doi.org/10.1002/ps.4811

Funke T, Yang Y, Han H et al (2009) Structural basis of glyphosate resistance resulting from the double mutation Thr(97) → Ile and Pro(101) → Ser in 5-enolpyruvylshikimate-3-phosphate synthase from *Escherichia coli*. J Biol Chem 284:9854–9860. https://doi.org/10.1074/jbc.M809771200

Gaines TA, Zhang W, Wang D et al (2010) Gene amplification confers glyphosate resistance in *Amaranthus palmeri*. Proc Natl Acad Sci U S A 107(3):1029–1034. https://doi.org/10.1073/pnas.0906649107

Gaines TA, Patterson EL, Neve P (2019) Molecular mechanisms of adaptive evolution revealed by global selection for glyphosate resistance. New Phytol 223(4):1770–1775. https://doi.org/10.1111/nph.15858

Gaines TA, Duke SO, Morran S et al (2020) Mechanisms of evolved herbicide resistance. J Biol Chem 295:10307–10330. https://doi.org/10.1074/jbc.REV120.013572

García MJ, Palma-Bautista C, Vazquez-Garcia JG et al (2020) Multiple mutations in the EPSPS and ALS genes of *Amaranthus hybridus* underlie resistance to glyphosate and ALS inhibitors. Sci Rep 10:17681. https://doi.org/10.1038/s41598-020-74430-0

Gazziero DL, Silva AF (2017) Characterization and management of *Amaranthus palmeri*. Embrapa Soja, Londrina

Ge X, D'Avignon DA, Ackerman J (2011) Rapid vacuolar sequestration: the horseweed glyphosate resistance mechanism. Pest Manag Sci 66:345–348. https://doi.org/10.1002/ps.1911

Ge X, D'Avignon DA, Ackerman JJ et al (2012) Vacuolar glyphosate-sequestration correlates with glyphosate resistance in ryegrass (*Lolium* spp.) from Australia, South America, and Europe: a ^{31}P NMR investigation. J Agric Food Chem 60:12432–12450. https://doi.org/10.1021/jf203472s

Ghanizadeh H, Harrington KC (2017) Non-target site mechanisms of resistance to herbicides. Crit Rev Plant Sci 36:24–34. https://doi.org/10.1080/07352689.2017.1316134

Ghanizadeh H, Harrington KC (2020) Perspective: root exudation of herbicides as a novel mode of herbicide resistance in weeds. Pest Manag Sci 76(8):2543–2547. https://doi.org/10.1002/ps.5850

Goggin DE, Cawthray GR, Powles SB (2016) 2,4-D resistance in wild radish: reduced herbicide translocation via inhibition of cellular transport. J Exp Bot 67(11):3223–3235. https://doi.org/10.1093/jxb/erw120

Gronwald JW (1994) Resistance to photosystem II inhibiting herbicides. In: Powles SB, Holtum JAM (eds) Herbicide resistance in plants: biology and biochemistry. Lewis Publishers, Boca Raton, FL, pp 83–139

Guo F, Iwakami S, Yamaguchi T et al (2019) Role of CYP81A cytochrome P450s in clomazone metabolism in *Echinochloa phyllopogon*. Plant Sci 283:321–328. https://doi.org/10.1016/j.plantsci.2019.02.010

Guttieri MJ, Eberlein CV, Mallory-Smith CA et al (1992) DNA sequence variation in domain A of the acetolactate synthase genes of herbicide-resistant and -susceptible weed biotypes. Weed Sci 40:670–677

Han H, Yu Q, Purba E et al (2012) A novel amino acid substitution Ala-122-Tyr in ALS confers high-level and broad resistance across ALS-inhibiting herbicides. Pest Manag Sci 68(8):1164–1170. https://doi.org/10.1002/ps.3278

Hawkes TR (2014) Mechanisms of resistance to paraquat in plants. Pest Manag Sci 70(9):1316–1323. https://doi.org/10.1002/ps.3699

Healy-Fried ML, Funke T, Priestman MA et al (2007) Structural basis of glyphosate tolerance resulting from mutations of Pro101 in 5-enolpyruvylshikimate-3-phosphate synthase. J Biol Chem 282:32949–32955. https://doi.org/10.1074/jbc.M705624200

Heap IA (2014) Global perspective of herbicide-resistant weeds. Pest Manag Sci 70:1306–1315. https://doi.org/10.1002/ps.3696

Heap IA (2021) The international survey of herbicide resistant weeds. http://www.weedscience.org/Pages/filter.aspx. Accessed 27 Mar 2021

Herbicide Resistance Action Committee HRAC (2021a) Guideline to the management of herbicide resistance. http://ipmwww.ncsu.edu/orgs/hrac/guideline.html. Accessed 27 Mar 2021

Herbicide Resistance Action Committee HRAC (2021b) How to minimize resistance risks and how to respond to cases of suspected and confirmed resistance. http://ipmwww.ncsu.edu/orgs/hrac/hoetomil.html. Accessed 27 Mar 2021

Hirschberg J, Mcintosh L (1983) Molecular basis of herbicide resistance in *Amaranthus hybridus*. Science 222(4630):1346–1349

Ikeda FS, Avalieri S, Lima Junior FDM et al (2019) Estratégias de controle de *Amaranthus palmeri* resistente à herbicidas inibidores de EPSPS e ALS. Embrapa Agrossilvipastoril, Brasília

Jablonkai I (2015) Herbicide metabolism in weeds - selectivity and herbicide resistance. In: Price A, Kelton J, Sarunaite L (eds) Herbicides, physiology of action and safety. IntechOpen, London, pp 1609–2600

Jha P, Norsworthy JK, Riley MB, Bridges W (2010) Annual changes in temperature and light requirements for germination of palmer amaranth (*Amaranthus palmeri*) seeds retrieved from soil. Weed Sci 58(4):426–432. https://doi.org/10.1614/WS-D-09-00038.1

Jugulam M, Shyam C (2019) Non-target-site resistance to herbicides: recent developments. Plants 8(10):417. https://doi.org/10.3390/plants8100417

Jugulam M, DiMeo N, Veldhuis LJ et al (2013) Investigation of MCPA (4-chloro-2-ethylphenoxyacetate) resistance in wild radish (*Raphanus raphanistrum* L.). J Agric Food Chem 61:12516–12521. https://doi.org/10.1021/jf404095h

Kaundun SS, Hutchings SJ, Dale RP et al (2017) Mechanism of resistance to mesotrione in an *Amaranthus tuberculatus* population from Nebraska, USA. PLoS One 12(6):e0180095

Keith BK, Burns EE, Bothner B et al (2017) Intensive herbicide use has selected for constitutively elevated levels of stress-responsive mRNAs and proteins in multiple herbicide-resistant *Avena fatua* L. Pest Manag Sci 73(11):2267–2281. https://doi.org/10.1002/ps.4605

Kissmann KG (1996) Resistência de plantas à herbicidas. Basf Brasileira S.A, São Paulo

Krieger-Liszkay A (2005) Singlet oxygen production in photosynthesis. J Exp Bot 56(411):337–346. https://doi.org/10.1093/jxb/erh237

Kudsk P (2014) Reduced herbicide rates: present and future. Julius-Kühn-Archiv 443:37–44. https://doi.org/10.5073/jka.2014.443.003

Laforest M, Soufiane B, Simard MJ et al (2017) Acetyl-CoA carboxylase overexpression in herbicide resistant large crabgrass (*Digitaria sanguinalis*). Pest Manag Sci 73(11):2227–2235. https://doi.org/10.1002/ps.4675

Lamichhane JR, Devos Y, Beckie HJ et al (2017) Integrated weed management systems with herbicide-tolerant crops in the European Union: lessons learnt from home and abroad. Crit Rev Biotechnol 37:459–475. https://doi.org/10.1080/07388551.2016.1180588

Larran AS, Lorenzetti F, Tuesca D et al (2018) Molecular mechanisms endowing cross-resistance to ALS-inhibiting herbicides in *Amaranthus hybridus* from Argentina. Plant Mol Biol Rep 36:907–912. https://doi.org/10.1007/s11105-018-1122-y

Le Clere S, Wu C, Westra P, Sammons RD (2018) Cross-resistance to dicamba, 2,4-D, and fluroxypyr in *Kochia scoparia* is endowed by a mutation in an AUX/IAA gene. Proc Natl Acad Sci 115:2911–2920. https://doi.org/10.1073/pnas.1712372115

Li J, Li M, Gao X, Fang F (2017) A novel amino acid substitution Trp574Arg in acetolactate synthase (ALS) confers broad resistance to ALS-inhibiting herbicides in crabgrass (*Digitaria sanguinalis*). Pest Manag Sci 73:2538–2543. https://doi.org/10.1002/ps.4651

Livingston M, Fernandez-cornejo J, Frisvold GB (2016) Economic returns to herbicide resistance management in the short and long run: the role of neighbor effects. Weed Sci 64:595–608

Long W, Malone J, Boutsalis P, Preston C (2019) Diversity and extent of mutations endowing resistance to the acetolactate synthase (AHAS)-inhibiting herbicides in Indian hedge mustard (*Sisymbrium orientale*) populations in Australia. Pestic Biochem Physiol 157:53–59. https://doi.org/10.1016/j.pestbp.2019.03.004

Lorentz L, Gaines TA, Nissen SJ et al (2014) Characterization of glyphosate resistance in *Amaranthus tuberculatus* populations. J Agric Food Chem 62(32):8134–8142. https://doi.org/10.1021/jf501040x

Mao C, Xie H, Chen S et al (2016) Multiple mechanism confers natural tolerance of three lilyturf species to glyphosate. Planta 243:321–335. https://doi.org/10.1007/s00425-015-

Markus C, Barroso AAM, Dalazen G et al (2021) Resistência de plantas daninhas aos herbicidas. In: Barroso AAM, Murata AT (eds) Matologia: Estudos sobre plantas daninhas. Editora Fábrica da Palavra, Jaboticabal, pp 324–364

Marochi A, Ferreira A, Takano HK et al (2018) Managing glyphosate-resistant weeds with cover crop associated with herbicide rotation and mixture. Cienc Agrotec 42(4):381–394. https://doi.org/10.1590/1413-70542018424017918

Maroli AS, Nandula VK, Dayan FE et al (2015) Metabolic profiling and enzyme analyses indicate a potential role of antioxidant systems in complementing glyphosate resistance in an *Amaranthus palmeri* biotype. J Agric Food Chem 63(41):9199–9209. https://doi.org/10.1021/acs.jafc.5b04223

Martins JF, Barroso AAM, Carvalho LB et al (2016) Plant growth and genetic polymorphism in glyphosate-resistant sourgrass (*Digitaria insularis* L. Fedde). Aust. J Crop Sci 10(10):e1466. https://doi.org/10.21475/ajcs.2016.10.10.p7761

Maxwell BD, Mortimer AM (1994) Selection for resistance herbicide. In: Powles SB, Holtum JAM (eds) Herbicide resistance in plants: biology and biochemistry. Lewis Publishers, Boca Raton, FL, pp 1–25

Mcelroy JS, Hall ND (2020) *Echinochloa colona* with reported resistance to glyphosate conferred by aldo-keto reductase also contains a Pro-106-Thr EPSPS target site mutation. Plant Physiol 183(2):447–450. https://doi.org/10.1104/pp.20.00064

Mendes KF (2020) Plantas daninhas resistentes ao glyphosate no Brasil: biologia, mecanismo de resistência e manejo, 1st edn. Brazil Publishing, Curitiba

Mendes KF, Silveira RF, Inoue MH, Tornisielo VL (2017) Procedures for detection of resistant weeds using ^{14}C-herbicide absorption, translocation, and metabolism. In: Pacanoski Z (ed) Herbicide resistance in weeds and crops. IntechOpen, London, pp 161–176

Mendes RR, Takano HK, Oliveira RS et al (2020a) A Trp574Leu target site mutation confers imazamox-resistance in multiple-resistant wild poinsettia populations from Brazil. Agronomy 10:1057

Mendes RR, Takano HK, Leal JF et al (2020b) Evolution of EPSPS double mutation imparting glyphosate resistance in wild poinsettia (*Euphorbia heterophylla* L.). PLoS One 15:e0238818. https://doi.org/10.1371/journal.pone.0238818

Mendes RR, Takano HK, Adegas FS et al (2020c) Target site mutation in PPO2 evolves in wild poinsettia (*Euphorbia heterophylla*) with cross-resistance to PPO-inhibiting herbicides. Weed Sci 68:437–444

Merotto Júnior A, Goulart IC, Nunes AL et al (2016) Evolutionary and social consequences of introgression of non-transgenic herbicide resistance from rice to weedy rice in Brazil. Evol Appl 9(7):837–846. https://doi.org/10.1111/eva.12387

Michel A, Arias RS, Scheffler BE et al (2004) Somatic mutation-mediated evolution of herbicide resistance in the nonindigenous invasive plant hydrilla (*Hydrilla verticillata*). Mol Ecol 13(10):3229–3237. https://doi.org/10.1111/j.1365-294X.2004.02280.x

Michitte P, De Prado R, Espinoza N et al (2007) Mechanisms of resistance to glyphosate in a ryegrass (*Lolium multiflorum*) biotype from Chile. Weed Sci 55(5):435–440. https://doi.org/10.1614/WS-06-167.1

Monquero PA, Christoffoleti PJ (2001) Manejo de populações de plantas daninhas resistentes à herbicidas inibidores da acetolactato sintase. Planta Daninha 19(1):67–74

Monquero PA, Christoffoleti PJ, Dias CTS (2000) Weed resistance to ALS-inhibiting herbicides in soybean (*Glycine max*) crop. Planta Daninha 8(3):419–425

Moretti ML, Van Horn CR, Robertson R et al (2017a) Glyphosate resistance in *Ambrosia trifida*. Part 2. Rapid response physiology and non-target-site resistance. Pest Manag Sci 74:1079–1088. https://doi.org/10.1002/ps.4569

Moretti ML, Alárcon-Reverte R, Pearce S et al (2017b) Transcription of putative tonoplast transporters in response to glyphosate and paraquat stress in *Conyza bonariensis* and *Conyza canadensis* and selection of reference genes for qRT-PCR. PLoS One 12(7):e0180794. https://doi.org/10.1371/journal.pone.0180794

Mortensen DA, Egan JF, Maxwell BD et al (2012) Navigating a critical juncture for sustainable weed management. Biosci J 62(1):75–84. https://doi.org/10.1525/bio.2012.62.1.12

Mortimer AM (1998) Review of graminicide resistance. http://ipmwww.ncsu.edu/orgs/hrac/monograph1.htm. Accessed 27 Mar 2021

Moss S (2017) Herbicide resistance weeds. In: Hatcher PE, Froud-Williams RJ (eds) Weed research: expanding horizons, 1st edn. John Wiley & Sons Ltd., London, pp 181–214

Moss S (2019) Integrated Weed Management (IWM): why are farmers reluctant to adopt non-chemical alternatives to herbicides? Pest Manag Sci 75:1205–1211. https://doi.org/10.1002/ps.5267

Murphy BP, Tranel PJ (2019) Target-site mutations conferring herbicide resistance. Plants 8(10):382. https://doi.org/10.3390/plants8100382

Nakka S, Godar AS, Wani PS et al (2017) Physiological and molecular characterization of hydroxyphenylpyruvate dioxygenase (HPPD)-inhibitor resistance in Palmer amaranth (*Amaranthus palmeri* S. Wats.). Front Plant Sci 8:555. https://doi.org/10.3389/fpls.2017.00555

Nandula VK, Ray JD, Ribeiro DN et al (2013) Glyphosate resistance in tall waterhemp (*Amaranthus tuberculatus*) from Mississippi is due to both altered target-site and nontarget-site mechanisms. Weed Sci 61(3):374–383. https://doi.org/10.1614/WS-D-12-00155.1

Nandula VK, Wright AA, Bond JA et al (2014) EPSPS amplification in glyphosate-resistant spiny amaranth (*Amaranthus spinosus*): a case of gene transfer via interspecific hybridization from glyphosate-resistant Palmer amaranth (*Amaranthus palmeri*). Pest Manag Sci 70(12):1902–1909. https://doi.org/10.1002/ps.3754

Nandula VK, Riechers DE, Ferhatoglu Y et al (2019) Herbicide metabolism: crop selectivity, bioactivation, weed resistance, and regulation. Weed Sci 67(2):149–175. https://doi.org/10.1017/wsc.2018.88

Ngo TD, Malone JM, Boutsalis P et al (2018) EPSPS gene amplification conferring resistance to glyphosate in windmill grass (*Chloris truncata*) in Australia. Pest Manag Sci 7(4):1101–1108. https://doi.org/10.1002/ps.4573

Norsworthy JK, Ward SM, Shaw DR et al (2012) Reducing the risks of herbicide resistance: best management practices and recommendations. Weed Sci 60(1):31–62. https://doi.org/10.1614/WS-D-11-00155.1

Oliveira MF, Brighenti AM (2018) Controle de Plantas Daninhas: Métodos físico, mecânico, cultural, biológico e alelopatia. Embrapa Milho e Sorgo, Sete Lagoas

Oliveira MC, Gaines TA, Dayan FE et al (2018) Reversing resistance to tembotrione in an *Amaranthus tuberculatus* (var. *rudis*) population from Nebraska, USA with cytochrome P450 inhibitors. Pest Manag Sci 74(10):2296–2305. https://doi.org/10.1002/ps.4697

Ovejero RFL, Takano HK, Nicolai M et al (2017) Frequency and dispersal of glyphosate-resistant sourgrass (*Digitaria insularis*) populations across Brazilian agricultural production areas. Weed Sci 65(2):285–294. https://doi.org/10.1017/wsc.2016.31

Palma-Bautista C, Rojano-Delgado AM, Vázquez-García JG et al (2020) Resistance to fomesafen, imazamox and glyphosate in *Euphorbia heterophylla* from Brazil. Agronomy 10:1573. https://doi.org/10.3390/agronomy10101573

Pan L, Yu Q, Han H et al (2019) Aldo-keto reductase metabolizes glyphosate and confers glyphosate resistance in *Echinochloa colona*. Plant Physiol 181(4):1519–1534. https://doi.org/10.1104/pp.19.00979

Pan L, Yu Q, Wang J et al (2021) An ABCC-type transporter endowing glyphosate resistance in plants. Proc Natl Acad Sci 118(16):e2100136118. https://doi.org/10.1073/pnas.2100136118

Pandhair V, Sekhon BS (2006) Reactive oxygen species and antioxidants in plants: an overview. J Plant Biochem Biotechnol 15(2):71–78. https://doi.org/10.1007/BF03321907

Patzoldt WL, Hager AG, McCormick JS et al (2006) A codon deletion confers resistance to herbicides inhibiting protoporphyrinogen oxidase. Proc Natl Acad Sci 103(33):12329–12334. https://doi.org/10.1073/pnas.0603137103

Perotti VE, Larran AS, Palmieri VE et al (2019) A novel triple amino acid substitution in the EPSPS found in a high-level glyphosate-resistant *Amaranthus hybridus* population from Argentina. Pest Manag Sci 75(5):1242–1251. https://doi.org/10.1002/ps.5303

Peterson MA, Collavo A, Ovejero R et al (2018) The challenge of herbicide resistance around the world: a current summary. Pest Manag Sci 74:2246–2259. https://doi.org/10.1002/ps.4821

Petter FA, Sulzbacher AM, Silva AF et al (2015) Use of cover crops as a tool in the management strategy of sourgrass. Rev Bras Herbicid 14(3):200–209

Pinho CF, Leal JFL, Santos SA et al (2019) First evidence of multiple resistance of Sumatran Fleabane (*Conyza sumatrensis* (Retz.) E. Walker) to five- mode-of-action herbicides. Aust J Crop Sci 13:1688–1697. https://doi.org/10.21475/ajcs.19.13.10.p1981

Powles SB, Preston C (2021) Herbicide cross resistance and multiple resistance in plants. http://ipmwww.ncsu.edu/orgs/hrac/mono2.htm. Accessed 27 Mar 2021

Powles SB, Yu K (2010) Evolution in action: plants resistant to herbicides. Annu Rev Plant Biol 61:317–347. https://doi.org/10.1146/annurev-arplant-042809-112119

Preston C, Mallory-Smith CA (2001) Biochemical mechanisms, inheritance, and molecular genetics of herbicide resistance in weeds. In: Powles SB, Shaner DL (eds) Herbicide resistance and world grains. CRC Press, New York, NY, pp 23–60

Preston C, Tardif FJ, Christopher JT, Powles SB (1996) Multiple resistance to dissimilar herbicide chemistries in a biotype of *Lolium rigidum* due to enhanced activity of several herbicide degrading enzymes. Pestic Biochem Physiol 54(2):123–134

Proskuryakov SY, Konoplyannikov AG, Gabai VL (2003) Necrosis: a specific form of programmed cell death? Exp Cell Res 283:1–16

Qasem JR (2013) Herbicide resistant weeds: the technology and weed management. In: Price A, Kelton J (eds) Herbicides-current research and case studies in use. IntechOpen, London, pp 445–471

Queiroz ARS, Delatorre CA, Lucio FR et al (2019) Rapid necrosis: a novel plant resistance mechanism to 2,4-D. Weed Sci 68(1):6–18. https://doi.org/10.1017/wsc.2019.65

Riar DS, Tehranchian P, Norsworthy JK et al (2015) Acetolactate synthase-inhibiting, herbicide-resistant rice flatsedge (*Cyperus iria*): cross-resistance and molecular mechanism of resistance. Weed Sci 63(4):748–757. https://doi.org/10.1614/WS-D-15-00014.1

Rojano-Delgado AM, Portugal JM, Palma-Bautista C et al (2019) Target site as the main mechanism of resistance to imazamox in a *Euphorbia heterophylla* biotype. Sci Rep 9:e15423. https://doi.org/10.1038/s41598-019-51682-z

Salas RA, Dayan FE, Pan Z et al (2012) EPSPS gene amplification in glyphosate-resistant Italian ryegrass (*Lolium perenne* ssp. *multiflorum*) from Arkansas. Pest Manag Sci 68(9):1223–1230. https://doi.org/10.1002/ps.3342

Sammons RD, Gaines TA (2014) Glyphosate resistance: state of knowledge. Pest Manag Sci 70(9):1367–1377. https://doi.org/10.1002/ps.3743

Scarabel L, Panozzo S, Loddo D et al (2020) Diversified resistance mechanisms in multi-resistant *Lolium* spp. in three European countries. Front Plant Sci 11:608845. https://doi.org/10.3389/fpls.2020.608845

Scherner A, Schreiber F, Andres A et al (2018) Rice crop rotation: a solution for weed management. In: Shah F, Khan ZH, Iqbal A (eds) Rice crop current developments. IntechOpen, London, pp 25–44

Silva AA, Vargas LL, Ferreira EA (2007) Herbicidas: Resistência de Plantas. In: Silva AA, Silva JF (eds) Tópicos em manejo de plantas daninhas. Editora UFV, Viçosa, pp 280–324

Svyantek A, Aldahir P, Chen S et al (2016) Target and nontarget resistance mechanisms induce annual bluegrass (*Poa annua*) resistance to atrazine, amicarbazone, and diuron. Weed Technol 30:773–782. https://doi.org/10.1614/WT-D-15-00173.1

Takano HK, Oliveira Júnior RS, Constantin J et al (2018) Spread of glyphosate-resistant sourgrass (*Digitaria insularis*): independent selections or merely propagule dissemination? Weed Biol Manag 18:50–59. https://doi.org/10.1111/wbm.12143

Torra J, Rojano-Delgado AM, Rey-Caballero J et al (2017) Enhanced 2,4-D metabolism in two resistant *Papaver rhoeas* populations from Spain. Front Plant Sci 8:1584. https://doi.org/10.3389/fpls.2017.01584

Tranel PJ (2017) Herbicide-resistance mechanisms: gene amplification is not just for glyphosate. Pest Manag Sci 73(11):2225–2226. https://doi.org/10.1002/ps.4679

Tranel PJ, Wright TR, Heap IM (2021) Mutations in herbicide-resistant weeds for herbicide inhibitors ALS. http://www.weedscience.com. Accessed 27 Mar 2021

Trezzi MM, Felippi CL, Mattei D et al (2005) Multiple resistance of acetolactate synthase and protoporphyrinogen oxidase inhibitors in *Euphorbia heterophylla* biotypes. J Environ Sci Health B 40(1):101–109. https://doi.org/10.1081/pfc-200034254

Trezzi MM, Vidal RA, Kruse ND et al (2009) Fomesafen absorption site as a mechanism of resistance in an *Euphorbia heterophylla* biotype resistant to PROTOX inhibitors. Planta Daninha 27(1):139–148. https://doi.org/10.1590/S0100-83582009000100018

Trezzi MM, Vidal RA, Kruse ND et al (2011) Eletrolite leakage as a technique to diagnose *Euphorbia heterophylla* biotypes resistant to PPO-inhibitors herbicides. Planta Daninha 29(3):655–662. https://doi.org/10.1590/S0100-83582011000300020

Vargas L, Silva AA, Borém A et al (1999) Resistência de plantas daninhas à herbicidas. JARD Prod. Gráficas, Viçosa

Vázquez-García JG, Alcántara-de la Cruz R, Palma-Bautista C et al (2020) Accumulation of target gene mutations confers multiple resistance to ALS, ACCase, and EPSPS inhibitors in *Lolium* species in Chile. Front Plant Sci 11:553948. https://doi.org/10.3389/fpls.2020.553948

Vencill WK, Nichols RL, Webster TM et al (2012) Herbicide resistance: toward an understanding of resistance development and the impact of herbicide-resistant crops. Weed Sci 60:2–30. https://doi.org/10.1614/WS-D-11-00206.1

Vidal RA, Fleck NG (1997) The risk of finding herbicide resistant weed biotypes. Planta Daninha 15(2):152–161. https://doi.org/10.1590/S0100-83581997000200008

Vila-Aiub MM (2019) Fitness of herbicide-resistant weeds: current knowledge and implications for management. Plants 8(11):469. https://doi.org/10.3390/plants8110469

Vila-Aiub MM, Balbi MC, Distéfano AJ et al (2012) Glyphosate resistance in perennial *Sorghum halepense* (Johnsongrass), endowed by reduced glyphosate translocation and leaf uptake. Pest Manag Sci 68(3):430–436. https://doi.org/10.1002/ps.2286

Vila-Aiub MM, Yu Q, Powles SB (2019) Do plants pay a fitness cost to be resistant to glyphosate? New Phytol 223(2):532–547. https://doi.org/10.1111/nph.15733

Vrbničanin S, Pavlović D, Božić D (2017) Weed resistance to herbicides. In: Pacanoski Z (ed) Herbicide resistance in weeds and crops. IntechOpen, London, pp 7–35

Wang H, Zhang L, Li W et al (2019) Isolation and expression of acetolactate synthase genes that have a rare mutation in shepherd's purse (*Capsella bursa-pastoris* (L.) Medik.). Pestic Biochem Physiol 155:119–125. https://doi.org/10.1016/j.pestbp.2019.01.013

Webster TM, Scully BT, Grey TL, Culpepper AS (2013) Winter cover crops influence *Amaranthus palmeri* establishment. Crop Prot 52:130–135. https://doi.org/10.1016/j.cropro.2013.05.015

Weed Science Society of America (1998) Herbicide resistance and herbicide tolerance definitions. Weed Technol 12(4):789

White AD, Owen MD, Hartzler RG et al (2002) Common sunflower resistance to acetolactate synthase-inhibiting herbicides. Weed Sci 50:432–437. https://doi.org/10.1614/0043-1745(2002)050[0432:CSRTAS]2.0.CO;2

WSSA – Weed Science Society of America (1998) "Herbicide resistance" and "herbicide tolerance" defined. Weed Technol 12:789

Xu Y, Xu L, Shen J et al (2021) Effects of a novel combination of two mutated acetolactate synthase (ALS) isozymes on resistance to ALS-inhibiting herbicides in flixweed (*Descurainia sophia*). Weed Sci 69:430–438. https://doi.org/10.1017/wsc.2021.26

Yanniccari M, Gigón R, Larsen A (2020) Cytochrome P450 herbicide metabolism as the main mechanism of cross-resistance to ACCase- and ALS-inhibitors in *Lolium* spp. populations from Argentina: a molecular approach in characterization and detection. Front Plant Sci 11:600301. https://doi.org/10.3389/fpls.2020.600301

Yu Q, Powles SB (2014) Metabolism-based herbicide resistance and cross-resistance in crop weeds: a threat to herbicide sustainability and global crop production. Plant Physiol 166(3):1106–1118. https://doi.org/10.1104/pp.114.242750

Yu Q, Han H, Cawthray GR et al (2013) Enhanced rates of herbicide metabolism in low herbicide-dose selected resistant *Lolium rigidum*. Plant Cell Environ 36:818–827. https://doi.org/10.1111/pce.12017

Yuan JS, Tranel PJ, Stewart C Jr (2007) Non-target-site herbicide resistance: a family business. Trends Plant Sci 12(1):6–13

Zagnitko O, Jelenska J, Tevzadze G et al (2001) An isoleucine/leucine residue in the carboxyl-transferase domain of acetyl-CoA carboxylase is critical for interaction with aryloxyphenoxy-propionate and cyclohexanedione inhibitors. Proc Natl Acad Sci 98:6617–6622. https://doi.org/10.1073/pnas.121172798

Zhang C, Li FENG, He TT et al (2015) Investigating the mechanisms of glyphosate resistance in goosegrass (*Eleusine indica*) population from South China. J Integr Agric 14:909–918. https://doi.org/10.1016/S2095-3119(14)60890-X

Zhao N, Yan Y, Wang H et al (2018) Acetolactate synthase overexpression in mesosulfuron-methyl-resistant shortawn foxtail (*Alopecurus aequalis* Sobol.): reference gene selection and herbicide target gene expression analysis. J Agric Food Chem 66:9624–9634. https://doi.org/10.1021/acs.jafc.8b03054

Genetically Modified Crops Resistant to Herbicides and Weed Control

Adalin Cezar Moraes de Aguiar, Antonio Alberto da Silva, Kassio Ferreira Mendes, and Alessandro da Costa Lima

Abstract The adoption of genetically modified (GM) herbicide-resistant crops was an evolution in the management of weeds in agricultural crops. With these technologies, many non-selective herbicides began to be applied mainly in post-emergence of crops, without causing damage. All this has occurred through the improvement of some techniques such as the use of biotechnology, allowing the introduction of genes of interest in the genome of cultivated plants. In Brazil, the main crops with this technology are soybeans, cotton, and corn, cultivated on a large scale in the country. GM crops initially began to be developed with resistance to broad-spectrum herbicides, such as glyphosate and ammonium-glufosinate, and in some cases, the possibility of introducing more than one gene in the same cultivar conferring resistance to different herbicides. GM crops are important in weed management, especially in resistance management, due to the possibility of rotation of herbicides. However, these technologies will require greater knowledge of technicians and farmers, related to the different transgenic events, in addition to the management of resistant volunteer plants. The GM crops are and will be extremely important in weed management; however, the adoption of integrated management is essential to maintain the sustainability of chemical control and the longevity of these technologies.

Keywords Transgenic crops · Resistance · Gene introduction · Chemical control · Selectivity · Volunteer plants · Integrated management

---

A. C. M. de Aguiar · A. da Costa Lima
Department of Agronomy, Federal University of Viçosa, Viçosa, Minas Gerais, Brazil

A. Alberto da Silva · K. F. Mendes (✉)
Department of Agronomy, Federal University of Viçosa, Viçosa, Minas Gerais, Brazil
e-mail: aasilva@ufv.br; kfmendes@ufv.br

© The Author(s), under exclusive license to Springer Nature Switzerland AG 2022
K. F. Mendes, A. Alberto da Silva (eds.), *Applied Weed and Herbicide Science*,
https://doi.org/10.1007/978-3-031-01938-8_8

1 Introduction

Chemical control is the main tool used in weed management in large areas. Characteristics such as flexibility in application, control efficiency, and labor reduction make the use of herbicides essential in the production of agricultural crops. Currently in Brazil, due to its high importance, the use of herbicides reaches close to 50% of the total volume of pesticides marketed in the country (SINDIVEG 2020).

Among the factors responsible for the increase in herbicide use compared to other pesticides are the emergence of resistant weed biotypes and the introduction of genetically modified (GM) crops. In Brazil, about 51.3 million ha of GM crops were planted in 2018, mainly involving soybean, corn, and cotton crops. These large GM biotech crops had an average adoption of 93% (Isaaa 2019).

The adoption of these technologies has increased the use of certain herbicides worldwide. In the case of insecticides or fungicides, the GM crop itself carries out pest management, as with technologies with *Bacillus thuringiensis* (Bt) and Viptera (VIP) bacteria. On the other hand, the application of herbicides is necessary in resistant GM plants, thus maintaining the volume of applied product, or even increasing the use of some classes of herbicides. In some cases, the possibility of introducing more than one gene in the same cultivar gives the possibility of herbicide application of different chemical groups or even different mechanisms of action, in which it allows the use of several selective herbicides in the same crop cycle. One of the major limitations in herbicide use is the non-selectivity of certain crops or the botanical similarity between crop and weeds. In this sense, GM crops with specific resistance to certain herbicides, propitiated the use of these molecules in POST-emergence of plants of interest, without causing injuries to the crop, in addition to facilitating weed control, compared with conventional cultivars. Allied to this, and the emergence of resistant weed biotypes, herbicide-resistant GM crops have raised the possibility of herbicide rotation of different mechanisms of action, as already reported, thus allowing a more efficient management of weeds.

One of the main stimuli for the development of herbicide-resistant GM crops was through the improvement of some techniques with the use of biotechnologies, allowing the introduction of genes of interest in the genome of cultivated plants. In addition, research involving the development of GM crops has increased, mainly due to the difficulty and high cost of developing new herbicide molecules.

It is important to highlight that the term herbicide resistance is usually used to refer to the evolution of resistance to these products over a period of time. In addition, the term can be used for resistance characteristics introduced in a plant (transgenic) (Rao 2014). Some researchers also use the term tolerance; however, tolerance is considered an inherent ability of a species to survive and reproduce after treatment with herbicide, this implies in the non-selection or genetic manipulation to make the plant tolerant, that is, it is naturally tolerant. With this, the term resistance will be used in this chapter, to refer to crops that have been genetically modified in order to support herbicide applications.

Therefore, this chapter addresses the main herbicide-resistant GM crops in Brazil, highlighting the benefits and also implications of the adoption of these technologies. In addition to reported weed control options allied to these technologies.

2 Impacts of the Use of Genetically Modified Crops Resistant to Herbicides

The possibility of using biotechnology to produce herbicide-resistant GM crops was a scientific advance that helped revolutionize weed management. However, some precautions should be taken in the adoption of technologies in agricultural systems. The following are the positive and negative impacts of the adoption of herbicide-resistant GM crops (López-Ovejero et al. 2014; Krishnan and Preston 2018).

2.1 Positive Impacts

Among the main benefits of the use of herbicide-resistant GM crops are:

- Possibility of herbicide use within the crop, particularly in situations where weed species have a certain morphophysiological similarity with the crop of interest.
- Use of herbicides with broad spectrum of action, such as glyphosate and ammonium-glufosinate in POST-emergence of crops.
- Increased selectivity of herbicides, reducing the damage caused by injuries in cultivated plants.
- Possibility of herbicide application of different chemical groups or different mechanisms of action, helping in the management of resistant weed biotypes.
- Reduction of herbicide use with long residual effect on the soil, reducing the problems of carryover and contamination of natural resources.
- Simplified management of weeds in the short term and lower cost.
- Reduction of the preparation of the cultivation areas, facilitating the adoption and maintenance of the no-tillage system.

2.2 Negative Impacts

Some precautions that should be taken in the adoption of herbicide-resistant GM crops:

- The use of an herbicide as the only control tool can promote the selection of resistant weed biotypes.

- The appearance of voluntary weeds from harvest remains in subsequent crops. These voluntary plants, besides being difficult to control because they present resistance to certain herbicides, can serve as hosts of pests and diseases.
- In some cases, where the technology allows the use of more than one herbicide of different mechanisms of action in POST, the mixture of herbicides in tank can generate antagonistic effects on the control.
- Changes in the species of the infesting community, which may promote changes in the balance of the ecosystem.

3 Genetically Modified Crops Resistant to Herbicides Approved for Commercialization in Brazil

3.1 Soybean

The commercial release of herbicide-resistant GM soybean cultivars in Brazil by the *Comissão Técnica Nacional de Biossegurança* (CTNbio) began in 1998 (Table 1). The technology developed by Monsanto® was named *Roundup Ready®* (RR). The soybean cultivars presented resistance to glyphosate, a non-selective herbicide that was basically used in the management of weeds in desiccation and pre-sowing of the crops. However, only from 2003, the planting of RR soybeans began to be allowed in commercial areas in Brazil.

After the introduction of the first herbicide-resistant soybean cultivar in the country, 6 years passed to permit a new GM cultivar resistant to herbicide in soybean. The technology was developed by EMBRAPA in partnership with Basf under the name Cultivance®, being the first herbicide-resistant GM soybean developed in Brazil. This technology combined the use of herbicide-resistant soybean cultivars of the imidazolinone group (acetolactate synthase inhibitors, ALS) allowing the management of narrow and wide leaf weeds.

A year later, in 2010, CTNbio approved yet another herbicide-resistant soybean event, a technology developed by Bayer company named Liberty Link® (LL), that allowed applications of ammonium-glufosinate in POST soybeans. One of the positive points of the use of ammonium-glufosinate was the control of biotypes of tolerant and resistant weeds to glyphosate; however, the difficulty of controlling weeds with more developed stages, such as grasses above two tillers and dicotyledons above four true leaves (Raimondi et al. 2012; Albrecht et al. 2018a, b), made it a little difficult to use this technology.

In 2015, there was a large number of approvals of herbicide resistance events for soybean crops. This fact was very important due to the development of cultivars with more than one resistance event in the same cultivar, allowing the application in some cases of up to three herbicides of distinct mechanisms of action in the same soybean cycle. Highlight for the soybean developed by the company Dow AgroSciences®, which allowed the application of 2,4-D in POST of the crop.

Table 1 Genetically modified soybean cultivars resistant to herbicides approved for commercialization in Brazil by CTNBio

Commercial name	Donor organism	Herbicides	Protein	Company	Year
Roundup Ready	*Agrobacterium tumefaciens*	Glyphosate	*CP4-EPSPS*	Monsanto	1998
Cultivance	*Arabidopsis thaliana*	Imidazolinone chemical group (ALS)	*CSR-1-2*	Basf and EMBRAPA	2009
Liberty Link TM	*Streptomyces viridochromogenes*	Ammonium-glufosinate	*PAT*	Bayer	2010
Intacta RR2 PRO	*A. tumefaciens*	Glyphosate	*CP4-EPSPS*	Monsanto	2010
Enlist™	*Delftia acidovorans*; *S. viridochromogenes*	2,4-D and ammonium-glufosinate	*AAD-12*; *PAT*	Dow	2015
–	*Pseudomonas fluorescens*; *Zea mays*	Isoxaflutole and glyphosate	*HPPD*; *2MEPSPS*	Bayer	2015
Enlist E3	*D. acidovorans*; *S. viridochromogenes*; *Z. mays*	2,4-D; ammonium-glufosinate and glyphosate	*AAD-12*; *PAT*; *2MEPSPS*	Dow	2015
–	*P. fluorescens*; *Z. mays*; *S. viridochromogenes*	Isoxaflutole; glyphosate and ammonium-glufosinate	*HPPD*; *2MEPSPS*; *PAT*	Bayer	2015
Conkesta	*S. viridochromogenes*	Ammonium-glufosinate	*PAT*	Dow	2016
–	*Stenotrophomonas maltophilia*	Dicamba	*DMO*	Monsanto	2016
Conkesta Enlist E3	*D. acidovorans*; *S. viridochromogenes*; *Z. mays*	2,4-D; ammonium-glufosinate and glyphosate	*AAD-12*; *PAT*; *2MEPSPS*	Dow	2017
Xtend	*Agrobacterium* spp.; *S. maltophilia*	Glyphosate and dicamba	*CP4-EPSPS*; *DMO*	Monsanto	2017
Plenish™ RR1	*A. tumefaciens*	Glyphosate	*CP4-EPSPS*	Du Pont	2018
–	*S. maltophilia*; *A. tumefaciens*	Dicamba and glyphosate	*DMO*; *CP4-EPSPS*	Monsanto	2018
–	*A. tumefaciens*; *S. viridochromogenes*	Glyphosate and ammonium-glufosinate	*CP4-EPSPS*; *PAT*	TMG	2019

(–) Awaiting denomination
Source: CTNBio (2015)

Another important technology approved in 2015 for soybeans was that developed by Bayer, which allowed applications of isoxaflutole, a 4-hydroxyphenylpyruvate dioxygenase (HPPD, carotenoid synthesis inhibitor).

The last GM soybean cultivar resistant to the herbicide approved in Brazil was the Xtend technology, developed by Monsanto® in 2016. Soybean plants with this technology allow the application of dicamba in POST of the crop, being a major advance in the management of biotypes of dicotyledons weeds tolerant and resistant to other herbicides in POST of soybean.

3.2 Cotton

The first herbicide-resistant GM cotton cultivars were approved in Brazil by CTNbio in 2008. RR technology allowed the application of glyphosate in the POST of cotton and LL with the use of ammonium-glufosinate (Table 2). The possibility of the use of these herbicides was a major advance in weed control in the cultivation of the crop, due to the limited number of selective herbicides available for this crop.

The selectivity of cotton to glyphosate was improved in 2011 with the release of the second generation with resistance to glyphosate. The cultivar MON88913 allowed the application of glyphosate on the GM cotton crop until the later stages of plant development. This allowed safer management of weeds during cultivation, without causing damage to cotton crop. In 2012, herbicide-resistant cotton of two mechanisms of action in the same cultivar was approved, allowing the application of both glyphosate and ammonium-glufosinate in POST.

The launch of a new technology in cotton with resistance to a new herbicide molecule occurred only in 2017. Thus, dicamba-resistant cotton was developed, a herbicide belonging to the group of synthetic auxins, being an important alternative in the management of dicot weeds resistant and tolerant to glyphosate in cotton (Spaunhorst and Bradley 2013; Cahoon et al. 2015).

In 2018, in addition to resistance to dicamba and ammonium-glufosinate, the glyphosate resistance gene was included in the same cultivar, which allows the application of these three herbicides of different mechanisms of action in the POST of cotton. Also in 2018, there was the release of technologies that give cotton resistance to ammonium-glufosinate and 2,4-D (Enlist™), the latter being an important herbicide in the management of broad leaf weeds.

Finally, in 2019, glyphosate, ammonium-glufosinate and isoxaflutole, an inhibitory herbicide of the HPPD enzyme, were approved, used in PRE-emergence control of grass and dicotyledons weeds.

3.3 Corn

The first herbicide-resistant GM corn cultivar was released in Brazil in 2007 by CTNbio. This technology allows the application of ammonium-glufosinate in a total area after corn emergence, without causing crop damage (Table 3). Soon after 2008, glyphosate resistance technology, an important herbicide in weed management, was approved.

Table 2 Genetically modified cotton cultivars resistant to herbicides approved for commercialization in Brazil by CTNBio

Commercial name	Donor organism	Herbicide	Protein	Company	Year
Roundup Ready	*Agrobacterium tumefaciens*	Glyphosate	*CP4-EPSPS*	Monsanto	2008
Liberty Link	*Streptomyces viridochromogenes*	Ammonium-glufosinate	*PAT*	Bayer	2008
BGRR	*Agrobacterium tumefaciens*	Glyphosate	*CP4-EPSPS*	Monsanto	2009
Widestrike	*S. viridochromogenes*	Ammonium-glufosinate	*PAT*	Dow	2009
GlyTol	*Zea mays*	Glyphosate	*2MEPSPS*	Bayer	2010
TwinLink	*S. hygroscopicus*	Ammonium-glufosinate	*PAT*	Bayer	2011
MON88913	*A. tumefaciens*	Glyphosate	*CP4-EPSPS*	Monsanto	2011
GlytolxTwinLink	*Z. mays*	Glyphosate	*2MEPSPS*	Bayer	2012
GTxLL	*Z. mays, S. viridochromogenes*	Glyphosate and ammonium-glufosinate	*2MEPSPS, PAT*	Bayer	2012
BGIIFlex	*A. tumefaciens*	Glyphosate	*CP4-EPSPS*	Monsanto	2012
BGIIIRRFlex	*A. tumefaciens*	Glyphosate	*CP4 EPSPS*	Monsanto	2016
–	*Z. mays*	Glyphosate	*2MEPSPS*	Bayer	2017
DGT	*Stenotrophomonas maltophilia, S. hygroscopicus*	Dicamba and ammonium-glufosinate	*DMO, PAT*	Monsanto	2017
Enlist	*S. viridochromogenes, Delftia acidovoran*	Ammonium-glufosinate and 2,4-D	*PAT, AAD-12*	Dow	2018
–	*S. hygroscopicus*	Ammonium-glufosinate	*PAT*	Basf	2018
RRFlexDGT	*S. maltophilia, A. tumefaciens, S. hygroscopicus*	Dicamba, glyphosate, and ammonium-glufosinate	*DMO, CP4 EPSPS, PAT*	Monsanto	2018
BGIIIRRFlexDGT	*S. maltophilia, A. tumefaciens, S. hygroscopicus*	Dicamba, glyphosate, and ammonium-glufosinate	*DMO, CP4 EPSPS, PAT*	Monsanto	2018
–	*Zea mays, S. higroscopicus, Pseudomonas fluorescens*	Glyphosate, ammonium-glufosinate, and isoxaflutole	*2MEPSPS, PAT, HPPD W336*	Basf	2019
–	*A. tumefaciens, Delftia acidovorans*	Ammonium-glufosinate and 2,4-D	*PAT, AAD-12*	Dow	2019

(–) Awaiting denomination
Source: CTNBio (2015)

Table 3 Genetically modified corn cultivars resistant to herbicides approved for commercialization in Brazil by CTNBio

Commercial name	Donor organism	Herbicides	Protein	Company	Year
Liberty Link	*Streptomyces viridochromogenes*	Ammonium-glufosinate	*PAT*	Bayer	2007
TL	*Streptomyces viridochromogenes*	Ammonium-glufosinate	*PAT*	Syngenta	2007
Roundup Ready 2	*Agrobacterium tumefaciens*	Glyphosate	*CP4-EPSPS*	Monsanto	2008
TG	*Zea mays*	Glyphosate	*2MEPSPS*	Syngenta	2008
Herculex	*Streptomyces viridochromogenes*	Ammonium-glufosinate	*PAT*	Du Pont and Dow	2008
YGRR2	*A. tumefaciens*	Glyphosate	*CP4-EPSPS*	Monsanto	2009
TL/TG	*S. viridochromogenes*, *Z. mays*	Ammonium-glufosinate and glyphosate	*PAT*, *2MEPSPS*	Syngenta	2009
HR Herculex/RR2	*S. viridrochromogenes*, *A. tumefaciens*	Ammonium-glufosinate and glyphosate	*PAT*, *CP4-EPSPS*	Du Pont	2009
TL TG Viptera	*S. viridochromogenes*, *Z. mays*	Ammonium-glufosinate and glyphosate	*PAT*, *2MEPSPS*	Syngenta	2010
PRO2	*A. tumefaciens*	Glyphosate	*CP4-EPSPS*	Monsanto	2010
Yield Gard VT	*A. tumefaciens*	Glyphosate	*CP4-EPSPS*	Monsanto	2010
Power Core PW/Dow	*S. viridochromogenes*, *A. tumefaciens*	Ammonium-glufosinate and glyphosate	*PAT*, *CP4-EPSPS*	Monsanto and Dow	2010
Optimum Intrasect	*S. viridochromogenes*, *A. tumefaciens*	Ammonium-glufosinate and glyphosate	*PAT*, *CP4-EPSPS*	Du Pont	2011
TC1507xMON810	*S. viridochromogenes*	Ammonium-glufosinate	*PAT*	Du Pont	2011
MON89034 × MON88017	*A. tumefaciens*	Glyphosate	*CP4-EPSPS*	Monsanto	2011
Herculex XTRA™ corn	*S. viridochromogenes*	Ammonium-glufosinate	*PAT*	Du Pont and Dow	2013
Viptera4	*S. viridochromogenes*, *Z. mays*	Glyphosate	*PAT*, *2MEPSPS*	Syngenta	2014

(continued)

Table 3 (continued)

Commercial name	Donor organism	Herbicides	Protein	Company	Year
Enlist™	*Sphingobium herbicidorovans*	2,4-D and chemical group of FOPs (ACCase)	*AAD-1*	Dow	2015
–	*A. tumefaciens, S. viridocromogenes*	Glyphosate and ammonium-glufosinate	*CP4-EPSPS, PAT*	Monsanto	2015
Leptra	*S. viridochromogenes, A. tumefaciens*	Ammonium-glufosinate and glyphosate	*PAT, CP4-EPSPS*	Du Pont	2015
–	*S. viridochromogenes, A. tumefaciens*	Ammonium-glufosinate and glyphosate	*PAT, CP4-EPSPS*	Du Pont	2015
–	*S. viridochromogenes*	Ammonium-glufosinate	*PAT*	Du Pont	2015
–	*A. tumefaciens*	Glyphosate	*CP4-EPSPS*	Du Pont	2015
–	*S. viridochromogenes*	Ammonium-glufosinate	*PAT*	Du Pont	2015
Enlist™ RR	*S. herbicidorovans, A. tumefaciens*	2,4-D, chemical group of FOPs and glyphosate	*AAD-1, CP4-EPSPS*	Dow	2015
Agrisure Duracade 5222	*S. viridochromogenes, Z. mays*	Ammonium-glufosinate and glyphosate	*PAT, 2MEPSPS*	Syngenta	2015
VIP2	*S. viridochromogenes*	Ammonium-glufosinate	*PAT*	Syngenta	2015
PowerCore Enlist	*S. viridochromogenes, A. tumefaciens, S. herbicidovorans*	2,4-D, chemical group of FOPs, glyphosate, and ammonium-glufosinate	*AAD-1, CP4-EPSPS, PAT*	Dow	2016
SmartStax™	*S. viridochromogenes; A. tumefaciens*	Ammonium-glufosinate and glyphosate	*PAT, CP4 EPSPS*	Dow	2016

(continued)

Table 3 (continued)

Commercial name	Donor organism	Herbicides	Protein	Company	Year
–	*A. tumefaciens*	Glyphosate	*CP4-EPSPS*	Monsanto	2016
–	*A. tumefaciens*	Glyphosate	*CP4-EPSPS*	Monsanto	2016
VIP4TG	*S. viridochromogenes*, *Z. mays*	Ammonium-glufosinate and glyphosate	*PAT*, *2MEPSP*	Syngenta	2017
VIP4	*S. viridochromogenes*	Ammonium-glufosinate	*PAT*	Syengenta	2017
PowerCore Ultra	*S. viridochromogenes*, *A. tumefaciens*	Ammonium-glufosinate and glyphosate	*PAT*, *CP4 EPSPS*	Dow	2017
PowerCore Ultra Enlist	*A. tumefaciens*, *S. viridrochromogenes*, *S. herbicidovorans*	2,4-D, chemical group of FOPs, glyphosate, and ammonium-glufosinate	*AAD-1*, *CP4 EPSPS* and *PAT*	Dow	2018
–	*A. tumefaciens*, *S. viridochromogenes*, *S. herbicidovorans*	2,4-D, chemical group of FOPs, glyphosate, and ammonium-glufosinate	*AAD-1*, *CP4 EPSPS*, *PAT*	Dow	2019
–	*Agrobacterium* spp., *S. maltophilia*, *S. viridocromogenes*	Glyphosate, dicamba, and ammonium-glufosinate	*CP4 EPSPS*, *DMO* and *PAT*	Monsanto	2019
–	*A. tumefaciens*	Glyphosate	*CP4 EPSPS*	Monsanto	2019
–	*Agrobacterium* spp., *S. viridochromogenes*, and *S. herbicidovorans*	2,4-D, chemical group of FOPs, glyphosate, and ammonium-glufosinate	*AAD-1*; *CP4 EPSPS* and *PAT*	Du Pont	2020

(continued)

Table 3 (continued)

Commercial name	Donor organism	Herbicides	Protein	Company	Year
	Agrobacterium spp., *S. viridochromogenes*, and *S. herbicidovorans*	2,4-D, chemical group of FOPs, glyphosate, and ammonium-glufosinate	*AAD-1*, *CP4 EPSPS* and *PAT*	Dow	2020

(–) Awaiting denomination
Source: CTNBio (2015)

In 2009, the first glyphosate resistance technology and ammonium-glufosinate in corn were approved, allowing the application of the two herbicides in the same crop cycle. The use of these herbicides with a broad spectrum of action in POST greatly favored the management of weeds in corn.

After 6 years, in 2015, the commercialization of 2,4-D-resistant corn and herbicides of the aryloxyphenoxypropionate group FOPs (Enlist™) was approved. Important technology in the management of resistant weeds in corn POST. In addition, in the same year, 2,4-D-resistant corn, FOPs, and glyphosate (Enlist™ RR) were approved. In 2016, the technology that gave corn resistant to four herbicides of different mechanisms of action was approved for commercialization: the 2,4-D (synthetic auxins); herbicides of the FOPs group (Acetyl Coenzyme A carboxylase inhibitors—ACCase); glyphosate (5-enolpiruvilshikimate-3-phosphate synthase inhibitor—EPSPs); and also, ammonium-glufosinate (Glutamine Sintetase inhibitor—GS).

The latest approved technology with resistance to a new herbicide molecule was dicamba-resistant corn in 2019. In addition to dicamba, resistance to glyphosate and glufosinate ammonium is linked to the technology. Even though it is selective for grasses, applications of dicamba in stages not recommended in traditional corn cultivars can cause crop injuries. With this resistance technology, dicamba can be applied in the PRE and POST of corn without causing injury.

4 Herbicide Resistance Mechanism in Genetically Modified Crops and Weed Control

4.1 Resistance to Glyphosate

Glyphosate acts in the blocking of the enzyme EPSPs through competition with the phosphoenolpyruvate substrate (PEP), which is part of the biosynthesis pathway of essential aromatic amino acids, used in the synthesis of proteins and some

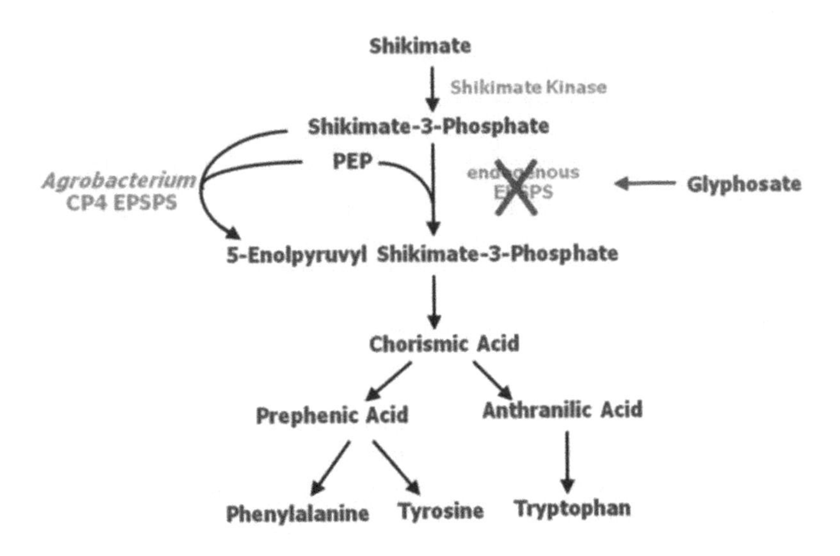

Fig. 1 Schematic representation of glyphosate mechanism of action and resistance mechanism of GM cultures mediated by CP4-EPSPs. (**Source:** CERA 2011)

secondary metabolites. When the enzyme EPSPs is blocked, the metabolic pathway is disrupted, and due to the lack of amino acids, the plant is driven to death.

In the case of soybean, cotton, and GM corn resistant to glyphosate, the *cp4 epsps* gene was introduced, isolated from the bacterium *Agrobacterium tumefaciens* that encodes the protein CP4-EPSPs in plants. The CP4-EPSPs protein has a higher affinity for PEP than glyphosate affinity; therefore, CP4-EPSPs preferentially binds to PEP even in the presence of glyphosate, and catalysis occurs exactly as in the absence of the herbicide. With this, plants continue the biosynthesis of aromatic amino acids phenylalanine, tyrosine, and tryptophan and other metabolites necessary for growth and development (Fig. 1).

Another way to promote resistance of cultures to glyphosate was by inserting the *2mepsps* gene originating from the *epsps* gene of corn plants (*Zea mays*), with alteration by site-directed mutation in two amino acids in the original peptide sequence. The mutation results in the 2mEPSPS protein with lower affinity of binding with glyphosate, maintaining the normal route of shikimate and then the synthesis of aromatic amino acids.

The adoption of this technology allows the application of glyphosate in POST of soybean, cotton, and corn crops. Glyphosate is a non-selective herbicide with a broad spectrum of systemic control, capable of causing the death of treated plants in a period of 7–14 days. The control efficiency combined with flexibility in application makes glyphosate an important herbicide in the concept of practicality. However, glyphosate is an herbicide with effect only in POST of weeds, that is, without residual effect on the soil, in this sense a single application may not be sufficient to manage weeds within GM crops.

Even though it is a systemic herbicide, capable of controlling young and adult weeds, the possibility of glyphosate application in POST of crops does not dispense with a good desiccation in pre-sowing and the use of herbicides in PRE. Desiccation of sowing areas reduces the infestation pressure of weeds from seeds and eliminates weed residues with vegetative propagation capacity. The use of glyphosate, associated with other herbicides such as 2,4-D and saflufenacil and graminicide herbicides, such as ACCase inhibitors, are important alternatives in pre-sowing crop desiccation (Quadros et al. 2020).

In many cases where weeds are in more developed stages, there may be a need for a second application of herbicide in POST of the crop, sequentially the first application or close to sowing of the crop. The herbicides used in the second application are contact herbicides, with no residual effect, such as diquat or ammonium-glufosinate.

Certain weeds have natural tolerance to the action of glyphosate, such as dayflower (*Commelina benghalensis*), spreading (*Commelina diffusa*), morning glory (*Ipomoea grandifolia*), poinsettia (*Euphorbia heterophylla*), false-buttonweed (*Spermacoce latifolia*), among others. Under these conditions, the association of glyphosate with other herbicides is necessary for efficient management of these weeds. In soybean, the associations of glyphosate with fomesafen, lactofen, bentazon, acifluorfen, imazethapyr, chlorimuron-ethyl, flumiclorac-pentyl, and chloransulam-methyl were efficient in the management of broad leaf weeds tolerant to glyphosate (Gazziero et al. 2004).

In POST cotton applications, glyphosate had a good effect on the control of pigweed (*Amaranthus hybridus*) and amaranth (*Amaranthus lividus*) (Braz et al. 2012). Glyphosate was also efficient in the control of weeds resistant to ALS-inhibiting herbicides, such as beggarticks (*Bidens pilosa*) and *E. heterophylla* in cotton (Braz et al. 2011a).

The weeds of the family Solanaceae such as apple-of-Peru (*Nicandra physalodes*) and nightshade (*Solanum americanum*) are often found in cotton crops. Glyphosate applications showed satisfactory control of these weeds in the six-leaf stage (Braz et al. 2011b).

In corn crop, glyphosate presents good control of radish (*Raphanus raphanistrum*) and alexandergrass (*Urochloa plantaginea*) plants in POST (Krenchinski et al. 2019). On the other hand, the same authors found that the isolated application did not promote efficient control of nutsedge (*Cyperus rotundus*) and pusley (*Richardia brasiliensis*).

Even though it is a herbicide with a broad spectrum of action, glyphosate has been losing effect in the control of some tolerant and resistant weed species, in this sense, the adoption of chemical control associated with other herbicides and other control methods becomes essential for the punctual control of weeds and to maintain the longevity of the efficient use of glyphosate.

4.2 Resistance to the Chemical Group of Imidazolinones

Imidazolinone-resistant soybean is a variety derived from a single processing event, called BPS-CV127-9 (CV127 soybean). The event was developed using the biobal-listic transformation technique, resulting in the introduction of the *csr1-2* gene of *Arabidopsis thaliana* into the soybean genome. The transformation was performed in the embryonic axis taken from the apical meristem of soybean seeds of the cultivar Conquista. The *csr1-2* gene encodes the main subunit of the ALS enzyme, responsible for the resistance of herbicides of the chemical group imidazolinones (CTNBio 2015). With this, the synthesis of branched amino acids usually occurs in plants in the presence of the herbicide (Fig. 2).

The technology was created by EMBRAPA and Basf, being the first GM soybean cultivar developed in Brazil. The herbicides of the imidazolinone group instill the ALS enzyme, disrupting the synthesis of branched amino acids (leucine, isoleucine, and valine), leading plants to death.

The use of herbicides from the imidazolinone group is an important tool in the management of broad and narrow leaf weeds in soybean. The herbicides of this chemical group have a broad spectrum of action and can be used as an alternative association or in the rotation system with glyphosate. The herbicides used in this technology are imazapyr and imazapic, which must be applied in the early stage of soybean so that no injuries and reduction in productivity occur.

These herbicides are efficient in the initial POST of weeds, up to two tillers in grasses and six true leaves in dicots. In the trade association of imazapyr + imazapic, the control of some weed species such as: amaranth (*Amaranthus viridis*),

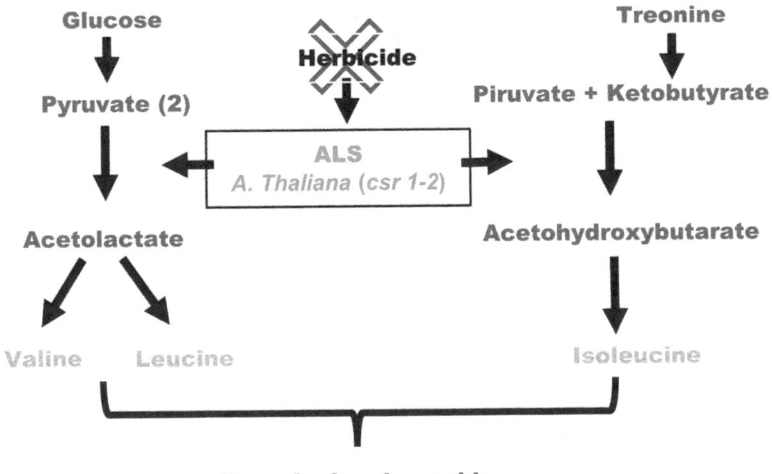

Fig. 2 Schematic representation of the mode of action of ALS-inhibiting herbicides and the resistance mechanism mediated by the csr 1-2 gene from *Arabidopsis thaliana* plants

U. plantaginea, *C. benghalensis*, Jamaican crabgrass (*Digitaria horizontalis*), sourgrass (*Digitaria insularis*), *I. grandifolia*, among others (Albrecht et al. 2018a, b).

The use of these herbicides from the chemical group of imidazolinones is an interesting alternative for rotation with glyphosate; however, this technology did not present sequence, despite its potential.

4.3 Resistance to Ammonium-Glufosinate

Soybean, cotton, and corn plants with this technology have the ability to metabolize ammonium-glufosinate by expressing the phosphinothricin acetyltransferase (*pat*) gene, also known as the bialaphos resistance gene (*bar*). Plants with this gene express phosphinothricin protein N-acetyltransferase (PAT) that rapidly converts L-phosphinothricin into *N*-acetyl-L-phosphinothricin, a non-toxic metabolite of ammonium-glufosinate. The expression of the pat or *bar* gene at high levels allows the application of ammonium-glufosinate in POST of cultures, without causing injury (Fig. 3).

The *pat* gene used is a modified version of the gene isolated from the natural soil bacterium *Streptomyces viridochromogenes* inserted into plant cells using the bio-ballistic transformation process. Ammonium-glufosinate acts in the balance between the generation and elimination of reactive oxygen species (ROS). Hydrogen peroxide (H_2O_2) is produced partially by the activity of glycolate oxidase in peroxisome. Inhibition of Glutamine Synthesis (GS) disrupts the photorespiration pathway and the linear flow of electrons in photosynthesis light reactions. Under high light intensity, the antioxidant system is overloaded, and electrons are then accepted by singlet oxygen (O_2) from water breakage in photosystem II (PSII). Subsequent accumulation of ROS leads to lipid peroxidation and forms the basis for the rapid action of ammonium-glufosinate (Takano et al. 2020).

The possibility of application of ammonium-glufosinate in POST of crops becomes an alternative in weed control, mainly of biotypes resistant and tolerant to glyphosate. The use of ammonium-glufosinate is efficient in the management of *E. heterophylla* and *B. pilosa*. However, according to Raimondi et al. (2012), no good results were observed in the control of *C. benghalensis*, redroot (*Amaranthus retroflexus*), and *I. grandifolia* in isolated applications.

In relation to horseweed and fleabane species (*Conyza* spp.), ammonium-glufosinate has good efficiency, provided that the managed plants do not exceed 10 cm (Oliveira Neto et al. 2010). On the other hand, in *D. insularis* (plants with up to 1–2 tillers), ammonium-glufosinate showed good control efficiency (Albrecht et al. 2018a, b).

Ammonium-glufosinate has a contact action, this characteristic causes the herbicide not to present satisfactory control when applied in higher stages of weeds. Everman et al. (2007) emphasized that in order to have good weed control in the POST of ammonium-glufosinate-resistant crops, applications should be timely in young plants at the beginning of the growing season.

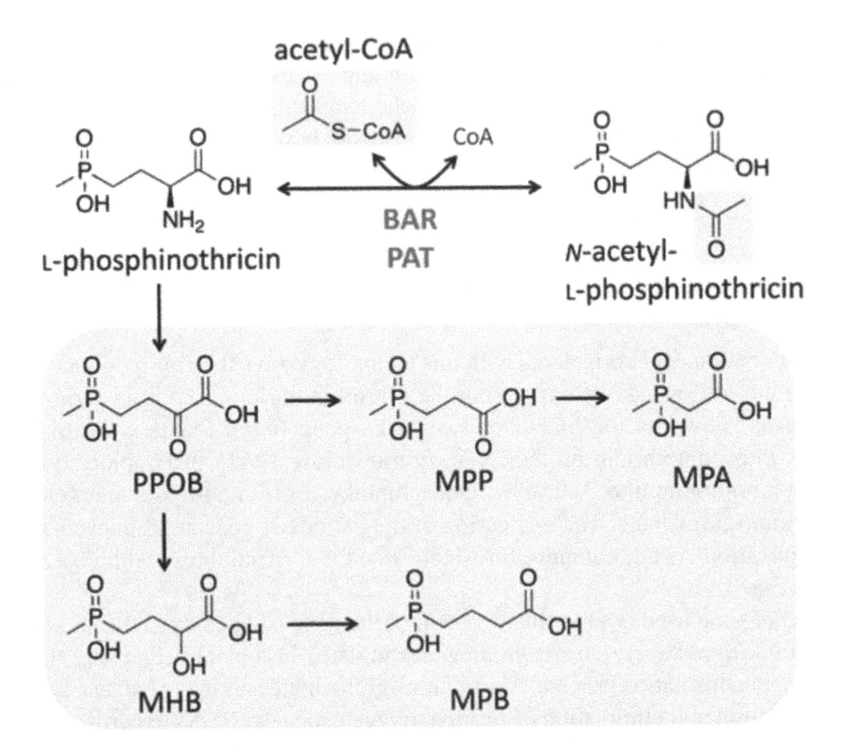

Fig. 3 Metabolism of L-phosphinothricin in N-acetyl-L-phosphinothricin by phosphinothricin acetyltransferase (pat) or bialaphos (bar) cultures in ammonium-glufosinate-resistant cultures (green). Metabolism of L-phosphinothricin in non-transgenic plants (blue), PPOB: 4-methylphosp hinico-2-oxo-butanoic acid; MPP: 3-(hydroxymethylphosphinyl)propionic acid; MPA: 2-methylphosphinicoacetic acid; MHB: 4-methylphosphinico-2-hydroxybutanoic acid; and MPB: 4-methylphosphinicobutanoic acid. (**Source:** Takano and Dayan 2020)

For weeds to be at a stage where ammonium-glufosinate has a good control, good pre-sowing desiccation and the use of herbicides with PRE effect of weeds are essential. In soybean, herbicides applied in PRE such as diclosulam, flumioxazin, and sulfentrazone are some of the alternatives mainly in the management of broad leaves, and S-metolachlor, trifluralin, and clomazone can be used in the management of narrow leaves, all of which are selective soybean (Gazziero et al. 2004).

In cotton, for the control of *D. insularis*, herbicides such as isoxaflutole, clomazone, diuron, and flumioxazin are efficient (Melo et al. 2017). In addition, they facilitate the moment of application and posterior control in POST using ammonium-glufosinate. S-metolachlor provided efficient control (above 90%) of *A. tenella* and *C. benghalensis* (Freitas et al. 2006).

In corn, the use of isolated ammonium-glufosinate also showed control failures. Krenchinski et al. (2019) reported that ammonium-glufosinate was not enough to control the plants of *R. raphanistrum*, *C. rotundus*, *U. plantaginea*, and pusley. On the other hand, when associated with bentazone, wild *R. raphanistrum* and nut grass control was satisfactory. For *U. plantaginea* control, the association of ammonium-glufosinate with nicosulfuron proved to be efficient. On the other hand, for the white-eye, there was a need for the association of ammonium-glufosinate with 2,4-D. In the case of herbicides applied in PRE, selective for corn and with residual effect on weed control, atrazine, S-metolachlor, pendimenthalin, among others can be used.

The use of ammonium-glufosinate in POST of crops is an important option in weed control; however, care should be taken with the positioning of this herbicide, especially in relation to the stage of weed development at the time of application.

4.4 Resistance to 4-Hydroxyphenylpyruvate Dioxygenase (HPPD) Inhibitor Herbicides

This technology gives soybean and cotton plants resistance to isoxaflutole. This herbicide acts by inhibiting the 4-hydroxyphenylpyruvate dioxygenase (HPPD) enzyme in sensitive plants, interrupting the conversion of 4-hydroxyphenylpyruvic acid (HPPA) into homogentisic acid (HGA). This inhibition affects the synthesis of carotenoids causing the appearance of bleaching symptoms in the leaves that emerge after application. Bleaching is caused by the oxidation of chlorophyll by light, due to lack of protection in the absence of carotenoids.

The technology has the *hppdPfw336* gene, which encodes the HPPD enzyme based on the *hppd* gene of *Pseudomonas fluorescens*. To reduce the sensitivity of this enzyme to the herbicide, HPPD was altered by replacing the amino acid tryptophan with a glycine at position 336 in the original sequence, resulting in HPPDW336. The mutated enzyme HPPDW336 is encoded by the mutant gene *hppdPfw336*, resulting in decreased affinity for isoxaflutole, allowing the survival of plants that express it when treated with this herbicide (CTNBio 2015).

This technology allows the application of isoxaflutole after sowing, in the PRE of weeds and soybean and cotton crops. Isoxaflutole acts in the control of both grasses and some dicots. The applications of isoxaflutole were efficient in the management of *D. horizontalis*, *U. plantaginea*, galinsoga (*Galinsoga parviflora*), *R. raphanistrum*, and *A. viridis*, presenting a good residual effect until 42 days after application (Adoryan et al. 2002).

When studying the efficacy of herbicides in PRE, Gonçalves Netto et al. (2019) found good control results (above 80%) with the use of isoxaflutole, suppressing populations of amaranth (*Amaranthus palmeri*) until the last evaluation (60 DAA) in sandy soils. Isoxaflutole also showed efficient control of chloris (*Chloris elata*) when applied in PRE, in which the control was higher than 96% in all populations, at 45 days after application (Correia and Resende 2018).

Resistance to isoxaflutole is the only technology that gives a crop resistance to a PRE herbicide. In this sense, based on the importance that residual herbicides have in weed management, isoxaflutole can be used as an important tool in the PRE control of grass and dicot weed species.

4.5 Resistance to 2,4-D

The resistance of soybean and cotton to 2,4-Dichlorophenoxyacetic acid (2,4-D) occurs by the insertion of the aryloxyalkanoate dioxygenase-12 gene (*add-12*) derived from the soil bacterium *Delftia acidovorans*. The *aad-12* gene encodes the protein aryloxyalkanoate dioxygenase-12 (AAD-12), this enzyme is a α-ketoglutarate-dependent dioxygenase that degrades 2,4-D by converting catalysis of 2,4-D into 2,4-Dichlorophenol (DCP), a compound without herbicide activity (Figs. 4, 5, and 6). The enzyme AAD-12 also degrades aquiratic phenoxyacetate herbicides such as 2-methyl-4-chlorophenoxyacetic acid (MCPA) and pyridyloxy-acetate herbicides such as triclopyr and fluroxypyr in their corresponding inactive phenols (Wright et al. 2010).

The possibility of application of 2,4-D in POST of soybean and cotton is an advance in the control of broadleaf weeds. 2,4-D is one of the most efficient herbicides in the management of dicot weeds; in addition, it performs well in *Conyza*

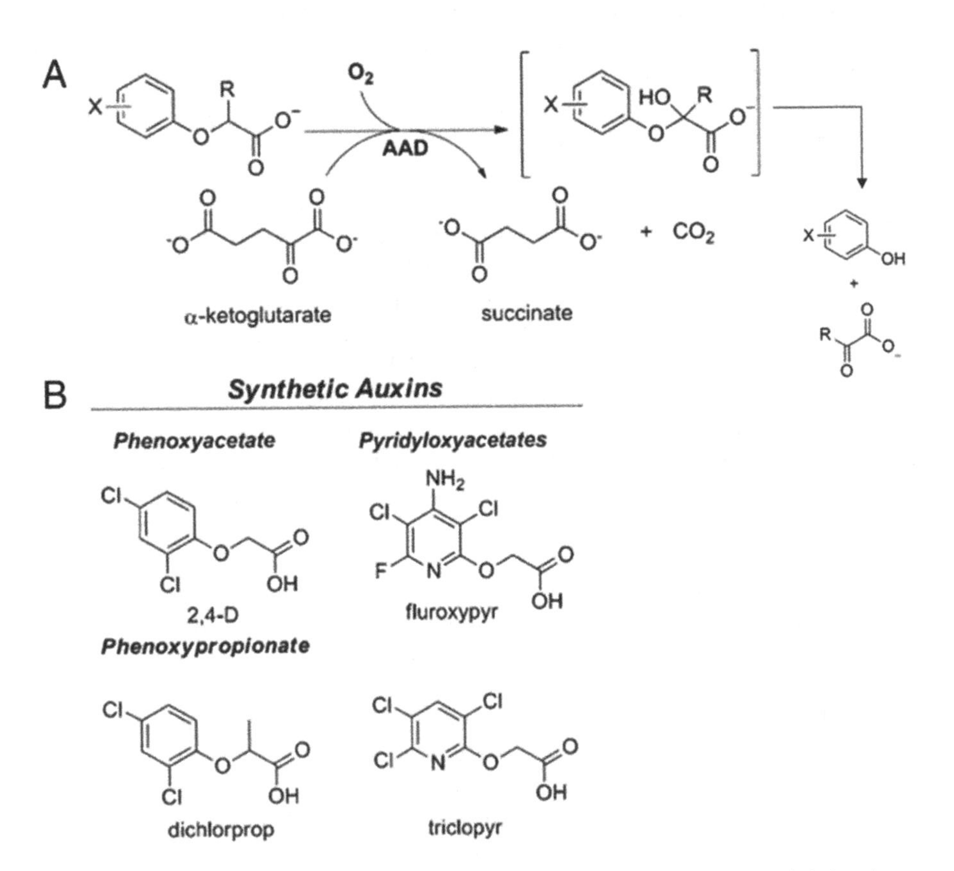

Fig. 4 General reaction catalyzed by enzyme AAD-12 (**a**) and herbicide compounds belonging to synthetic auxins that are substrates for AAD-12 (**b**). (**Source:** Adapted from Wright et al. 2010)

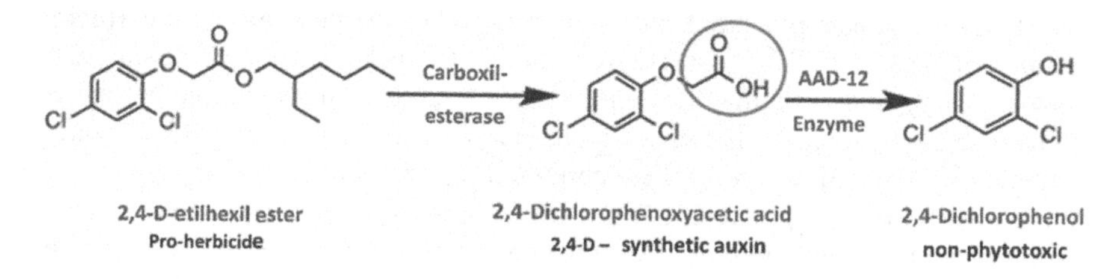

Fig. 5 Rapid metabolism of 2,4-Dichlorophenoxyacetic acid (2,4-D) into non-phytotoxic 2,4-Dichlorophenol (DCP) by the enzyme AAD-12. (**Source**: Skelton et al. 2017)

Fig. 6 Response of GM soybean plants resistant and not resistant to 2,4-D in the field. (**Source:** Wright et al. 2010)

spp. weeds, plants with a history of resistance to herbicides of different mechanisms of action. In the case of glyphosate-resistant *A. palmeri*, an efficient option is the association of 2,4-D with ammonium-glufosinate (Merchant et al. 2013).

2,4-D is an efficient herbicide in the management of weeds considered difficult to control, such as *R. brasiliensis*, *E. heterophylla*, *S. latifolia*, and *I. grandifolia* (Takano et al. 2013). Another weed susceptible to 2,4-D is *C. benghalensis*, a weed that presents natural tolerance to glyphosate, besides not obtaining satisfactory control using ammonium-glufosinate (Constantin et al. 2011; Takano et al. 2013). In the case of cotton, cultivated plants continue to vegetate after fruiting and their permanence in the field serves as host of pests and diseases in the off-season, so it is important to eliminate live plants after harvest. Today, the most efficient herbicide in cotton clumps control is 2,4-D (Ferreira et al. 2018). Thus, cotton cultivars with this technology will not allow the control of the clump with this herbicide, which can be a negative point. In this sense, there is a need for further research in search of alternative herbicides to control the cotton clump resistant to 2,4-D.

The possibility of applying 2,4-D in POST crops such as soybean and cotton is today one of the main tools of GMs created for weed management, being only behind the resistance to glyphosate. However, care with repetitive applications of this herbicide can generate the selection of resistant biotypes. In addition, especially in cotton, the farmer will lose an important tool in the destruction of clump. With this, it is necessary for the farmer to make a good planning of the crop, so that in addition to a good control of weeds, this technology does not cause future problems.

4.6 Resistance to 2,4-D and Chemical Group of Aryloxyphenoxypropionates (FOPs)

This technology was developed for corn crop, conferring resistance to 2,4-D and herbicides of the aryloxyphenoxypropionate group (FOPs). For this, the aad-1 gene, derived from the soil bacterium *Sphingobium herbicidovorans*, was introduced into corn; this gene, is able to encode the protein AAD-1 (aryloxialcanoate dioxygenase). This protein is capable of degrading the herbicide acid 2,4-D in 2,4-DCP, a compound that does not present herbicide activity. The Protein AAD-1 is also able to efficiently catalyze the degradation of FOPs herbicides, such as haloxyfop and quizalofop, to their corresponding inactive phenols, which gives the plant resistance to these herbicides (Figs. 7 and 8).

The herbicides haloxyfop and quizalofop, from the chemical group of FOPs, act as inhibitors of the lipid biosynthesis pathway in grasses, inhibiting the ACCase enzyme present in chloroplast. The possibility of application of graminicides in POST corn is a major advance in the control of grassy weeds in the crop. The AAD-1 enzyme, in addition to promoting resistance to FOPs herbicides, provides resistance to 2,4-D simultaneously. The possibility of using 2,4-D allows minimizing the ancient problems of auxinic herbicides in corn, such as bedtime, leaf folding, fragility of the stem, and deformed roots, which restricted the application of 2,4-D in certain stages of corn (Wright et al. 2010).

It is important to note that 2,4-D-resistant corn and the chemical group of FOPs do not support applications of ACCase-inhibiting herbicides from the cyclohexanodione group (DIM), such as sethoxydim and clethodim, and also the phenylpyrazolin (DEN) group, such as pinoxaden. The applications of sethoxydim and clethodim were able to generate injuries in corn ranging from 66% to 89% (Fig. 9); on the other hand, pinoxaden caused injuries of up to 61%. Regarding corn yield, herbicides caused losses between 77% and 97% (Striegel et al. 2020). This high capacity to cause injury with these herbicides reinforces the need to bring this information to

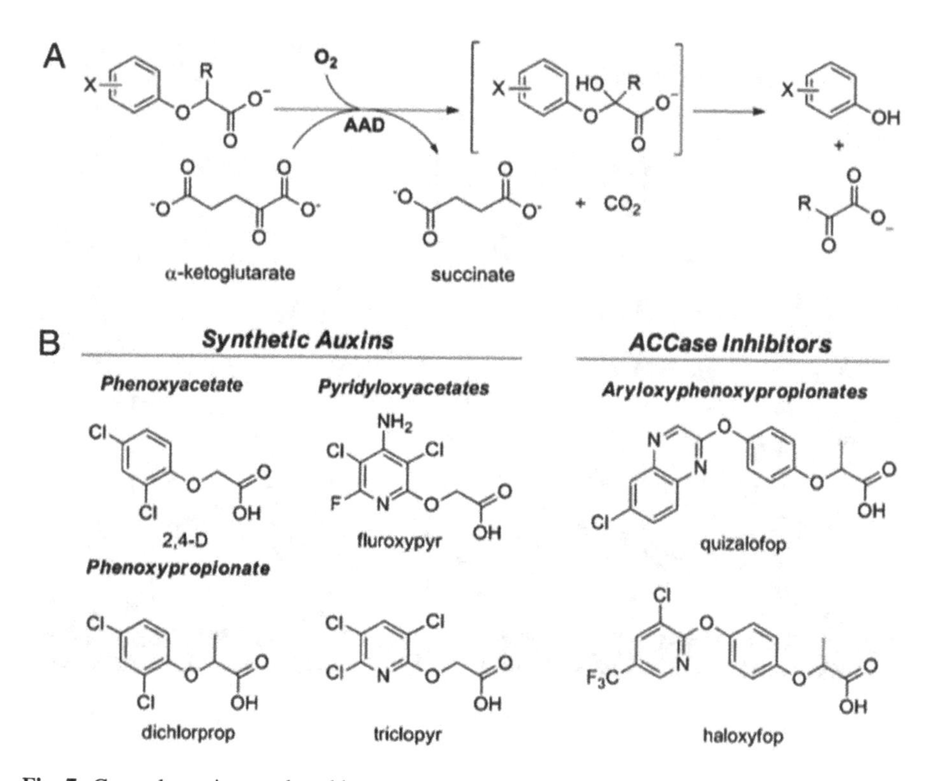

Fig. 7 General reaction catalyzed by enzyme AAD-1 (**a**) and herbicide compounds that are substrates for AAD-1 (**b**). (**Source:** Adapted from Wright et al. 2010)

producers in relation to products that can be applied in corn with this technology (Jhala et al. 2021).

With the resistance of *D. insularis*, ryegrass (*Lolium multiflorum*), and goosegrass (*Eleusine indica*) to glyphosate, one of the few groups of herbicides efficient in the management of these weeds is the inhibitory herbicides of ACCase. In *D. insularis* plants in a stage from 1 to 2 byes, the application of haloxyfop promoted satisfactory control (Melo et al. 2017). Barroso et al. (2014) observed that isolated applications of haloxyfop or quizalofop provided satisfactory control of *D. insularis* in the four-leaf stage. However, in plants of 3–4 leaves, the application of graminicides alone was not efficient. According to these authors, even though *D. insularis* plants are resistant to glyphosate, when associated with quizalofop with glyphosate (ammonium salt or potassium), the control was greater than 90%.

In the control of goosegrass in the stage of one child, applications of haloxyfop, fluazifop, or quizalofop promoted a control of 100% of the plants (Takano et al. 2018). On the other hand, in the stage of four tillers, the control above 90% was observed only in the application of haloxyfop at the dose of 60 g/ha. In the association of graminicides with glyphosate, these same authors did not observe an increase in the control of goosegrass plants.

Fig. 8 Response of corn plants resistant and not resistant to quizalofop applications. (**Source**: Wright et al. 2010)

Fig. 9 Levels of injuries caused by sethoxydim (**a**) applied to 210 g a.i./ha and clethodim (**b**) applied to 119 g a.i/ha in corn Enlist™ resistant to the herbicides of the chemical group of aryloxyphenoxypropionates (FOPs). (**Source:** Striegel et al. 2020)

The efficiency of graminicides in glyphosate-resistant weeds is dependent on the application stage. Thus, a good desiccation associated with the use of residual herbicides is an extremely important tool for effective management, being decisive to achieve control in POST. The herbicides atrazine, S-metolachlor, and isoxaflutole were effective in the control of goosegrass, in the evaluation performed at 80 days (Melo et al. 2017). At the same time, Takano et al. (2018) also observed efficient control of goosegrass in evaluations performed at 60 days after application, using S-metolachlor and trifluralin.

The possibility of applying 2,4-D in POST corn will also assist in the management of dicot weeds, especially those tolerant and resistant to glyphosate, as well as plants outside the stage that make it impossible to use ammonium-glufosinate. Resistant plants, such as *Conyza* spp., are well controlled with the use of 2,4-D, especially when associated with glyphosate (Takano et al. 2013). The authors obtained satisfactory control of *Conyza* spp. up to 15 cm tall by associating 2,4-D and glyphosate. This association was also efficient in the control of *C. benghalensis*, *R. brasiliensis*, *E. heterophylla*, *S. latifolia*, and *I. grandifolia*, weeds with a history of tolerance to isolated applications of glyphosate.

The management of glyphosate-resistant grasses in POST corn has always been a major challenge. This resistance technology, especially FOPs, can help solve these problems; however, care should be taken in the use of this technology, as there are already reports of resistance of weed biotypes to these herbicides in Brazil.

4.7 Dicamba Resistance

To confer resistance to dicamba, the *dmo* gene (demethylase) from the soil bacterium *Pseudomonas maltophilia* (DI-6 strain) was introduced into the genome of soybean, cotton, and corn plants, which expresses the dicamba monoxygenase protein (DMO), responsible for the dicamba resistance characteristic.

The DMO protein catalyzes the addition of a molecular oxygen molecule (O_2) to the methyl dicamba group, which leads to the conversion of this compound into a non-herbicide product, called 3,6-dichlorosalicylic acid (DCSA), formaldehyde (CH_2O), and water (H_2O). MoD acts on the destruction of dicamba herbicide activity before the product can reach toxic levels in treated GM plants (Fig. 10).

In the cultivation of soybean and cotton without this technology, dicamba applications in desiccation and soon after sowing of the crop is harmful, due to the residual left in the soil by dicamba, which may affect the germination and initial growth of soybean and cotton. In this case, application of dicamba in desiccation may make it impossible to sowing sensitive soybean for a period of 30–60 days and may reach 120 days, depending on soil characteristics and environmental conditions. Thus, soybean and cotton resistant to dicamba, in addition to increasing the control spectrum in POST, allows dicamba applications in desiccation, without harming plant growth and development. This creates the possibility of a flexibility in the management of weeds of PRE and POST crops.

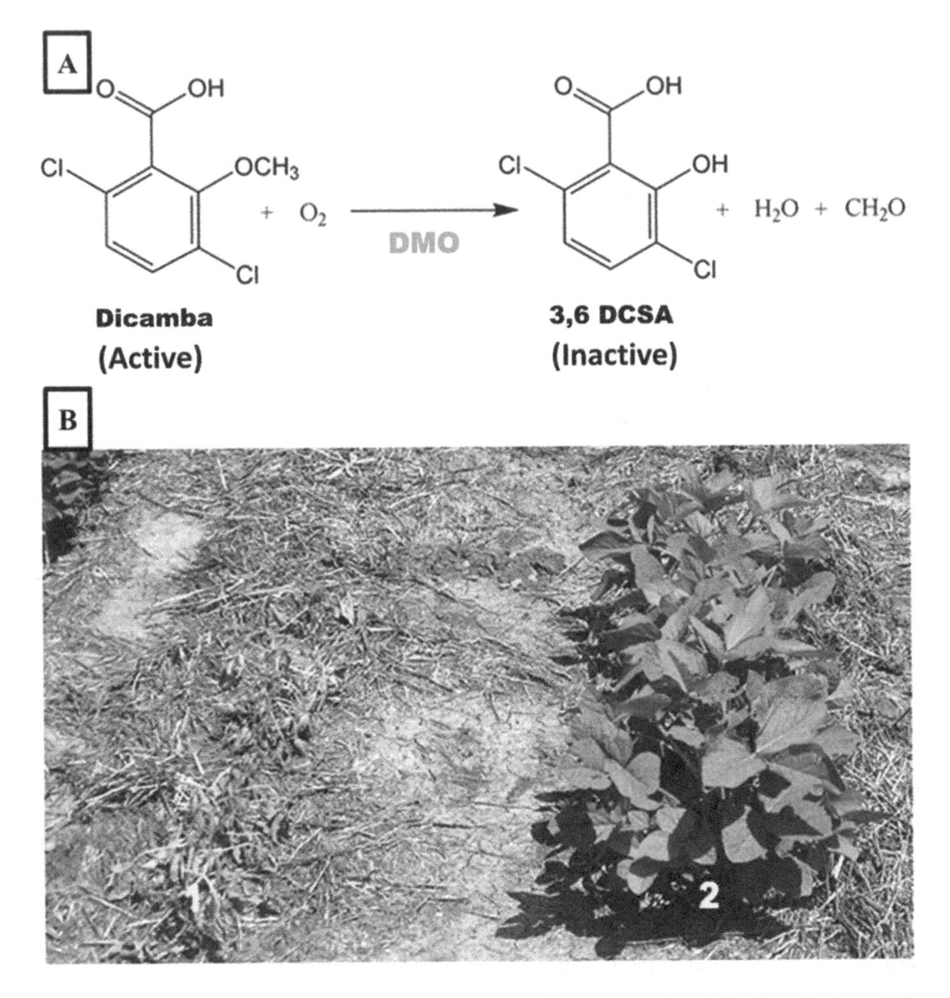

Fig. 10 Conversion of dicamba into DCSA by the DMO enzyme, promoting inactivation of the herbicide molecule (**a**) and effects of dicamba treatments on non-transgenic (1) and transgenic (2) soybean plants containing the genetically modified DMO gene under field conditions (**b**). Dose of 2.8 kg/ha of dicamba at 8 days after application. (**Source:** Behrens et al. 2007)

Even though it is an herbicide recommended only for the control of dicot weeds, dicamba has the ability to cause damage in non-transgenic corn plants, applied both before and after emergence (Cao et al. 2011). The damage occurs with greater intensity in the root system, promoting the appearance of atrophied roots, reduction of root volume, and reduction in the formation of adventitious roots, making it difficult to sustain the plant by taking them susceptible to the occurrence of tipping (Fig. 11).

The possibility of applying dicamba alone, or associated with glyphosate in POST soybean, cotton and corn, is an important tool in the management of dicot weeds. Dicamba has high efficiency in the control of fleabane (*Conyza* spp.) plants, besides presenting good effect on the management of *A. palmeri* and morning glory

Fig. 11 Commutations of corn plants expressing the MOD protein containing the transit peptide of the Arabidopsis Rubisco gene (**a**) or the transit peptide of the Rubisco gene of corn (**b**) with non-transgenic lines B73 (**c**) and Hi II (**d**), 2 weeks after dicamba treatments at 27 kg/ha. (**Fonte:** Cao et al. 2011)

species (*Ipomoea* spp.) (Merchant et al. 2013; Flessner et al. 2015). The use of dicamba, especially when associated with glyphosate, also performs well in the control of weeds such as *R. brasiliensis*, *C. benghalensis*, *C. bonariensis*, and morningglory (*Ipomoea nil*) (Osipe et al. 2017), plants that have a history of tolerance to glyphosate.

The application of dicamba will be an important tool in weed control. However, dicamba applications deserve some care, to avoid the injury of sensitive cultures. As the use of technologies and care with weather conditions at the time of application, avoiding the possibility of volatilization and drifting of the product, as well as performing the correct cleaning of the sprayers after application, with the aim of eliminating possible residues in the tank.

5 Herbicide-Resistant Voluntary Weeds

Plants from grain losses and harvest remains may germinate in crops in succession. In this case, even though they originate from cultivated species, these plants will behave like weeds, being called raccoon or voluntary plants, which become weeds in the crop in succession (Adegas et al. 2014).

Voluntary plants are able to compete with crops of interest in water, light, nutrients, and space, interfering in operations of cultural tracts and harvesting, in addition to serving as a host of pests and diseases. In the presence of voluntary corn plants in soybean cultivation, losses can reach close to 70%, if uncontrolled (López-Ovejero et al. 2016). Corn as a voluntary plant can be highly competitive, the presence of one plant per m^2, can generate yield losses close to 20%, when present in soybean crops (Adegas et al. 2014) (Fig. 12a, b).

Fig. 12 Voluntary plants in crops. Volunteer corn in soybean (**a**, **b**), voluntary soybean in corn (**c**, **d**), volunteer— cotton in soybean (**e**, **f**). (**Source:** Jhala et al. 2021. **Photos:** Lucas Heringer Barcellos Júnior and Guilherme Piccolotto Feltes dos Santos)

In the corn harvest, when losses of pieces of ears or even whole ears occur, these plants end up emerging clusters. Under these conditions, corn can be more competitive and cause greater crop yield loss in succession compared to individual plants (Chahal and Jhala 2016). Andersen et al. (1982) reported a reduction of 31% and 83% in soybean yield in the presence of corn clumps, with densities of one and four clumps spaced every 2.4 m of soybean line. In cotton, volunteer corn also has highly competitive capacity, and Clewis et al. (2008) observed loss of cotton fiber yield in the range of 4–8% for each 0.5 kg in voluntary corn biomass per meter of cultivation line.

Soybean plants also interfere in the corn crop as present. Four soybean plants per m^2 can generate losses of 4%, and can reach losses of 40%, with density of 32 plants per m^2 (Adegas et al. 2014) (Fig. 12c, d). In the case of the cotton, the problem is beyond interference as a competing plant (Fig. 12e, f), since the elimination of voluntary plants, as well as the destruction of the clumps is essential for compliance with the sanitary void, ensuring the absence of plants available for the feeding of the *Anthonomus grandis*, the main pest of the crop in Brazil (Grigolli et al. 2015).

With the adoption of herbicide-resistant GM crops, herbicide options for the control of voluntary plants have become more restricted. Glyphosate-resistant voluntary plants are one of the main problems, as glyphosate is one of the most widely used products in pre-sowing desiccation and POST applications. Volunteer corn presents a more marked complication, compared to soybean, since in the harvest there are losses of free grains, grains attached to the stem (cob) and grains attached to the stem surrounded by straw. This loss feature provides different emergency flows during the crop cycle in succession/rotation, making it difficult to control.

The choice of crops with resistance to more than one herbicide molecule in the same cultivar can generate greater problems in succession crops, such as the choice of a corn cultivar resistant to 2,4-D herbicides; FOPs; glyphosate and ammonium-glufosinate, reduces control options in soybean or cotton grown in succession, by restricting a high number of herbicides with the possibility of control.

The main management strategy of voluntary plants is in the regulation of harvesters and operator care to avoid tipping plants and reproductive structures during harvest. In the case of corn, the choice of hybrids with good stem quality can also mainly avoid plant tipping (Jhala et al. 2021).

In areas of voluntary corn interfering in soybean and cotton crops, the alternative use of ACCase-inhibiting herbicides in desiccation and POST management is one of the main alternatives. The applications of clethodim (108 g/ha), in POST of soybean, showed good control of voluntary corn (López-Ovejero et al. 2016). The herbicides of the chemical group of FOPs also present satisfactory control, with haloxyfop and fluazifop-p-butyl (Petter et al. 2015). However, in the case of corn resistant to the FOPs group, the availability of graminicides decreases, leaving only the use of herbicides belonging to the chemical group of DIMs for the control of voluntary corn (Striegel et al. 2020).

For the control of voluntary corn, technology rotation is important. In the case of voluntary glyphosate-resistant corn in POST soybean, the choice of a soybean cultivar with ammonium-glufosinate resistance technology is an interesting alternative. Ammonium-glufosinate applied in a single or sequential application promoted 85–97% of corn volunteers in ammonium-glufosinate-resistant soybean (Chahal and Jhala 2015); on the other hand, if voluntary corn is resistant to glyphosate and ammonium-glufosinate, an ACCase-inhibiting herbicide needs to be applied (Chahal and Jhala 2015; Jhala et al. 2021) (Fig. 13).

In the case of voluntary soybean interfering in corn crop, the use of atrazine may be an important alternative (Dan et al. 2011), but the rotation of technologies may also present very interesting options. In the case of voluntary cotton plants and plants from the regrowth clumps, the main problem is the use of cotton cultivars

Fig. 13 Symptoms of ammonium-glufosinate injuries in glyphosate-resistant volunteer corn (**a**) in ammonium-glufosinate-resistant soybeans, and sethoxydim in glyphosate-resistant volunteer corn and ammonium-glufosinate (**b**) in dicamba-resistant soybean and glyphosate. (**Source:** Jhala et al. 2021)

resistant to 2,4-D. 2,4-D is one of the most efficient herbicides in the control of cotton clumps, with this, the choice of soybean cultivars resistant to 2,4-D will be an important alternative for the control of clumps and voluntary plants in POST soybean; however, if cultivated cotton also presents this technology, there will be problems in the control. Some herbicides applied in PRE are also efficient in the control of cotton from cotton from sowing, such as sulfentrazone, metribuzin, and diclosulam (York et al. 2004; Minozzi et al. 2017).

The control of herbicide-resistant voluntary plants depends mainly on the planning and strategic use of different technologies in succession crops. The choice of crops with the same technology grown in succession restricts the use of some products, but when chosen intelligently can increase the possibility of applied products. An example of this would be in a soybean/cotton succession system, in which the adoption of a glyphosate-resistant soybean cultivar would help control certain weeds, while in succession the best option would be the choice of a cotton cultivar resistant to another herbicide molecule, such as ammonium-glufosinate or 2,4-D, which helps in the control of voluntary soybean in POST of cotton (Braz et al. 2011a, b).

6 Relationship Between Genetically Modified Crops and Weed Resistance in Brazil

While some researchers point out that the adoption of herbicide-resistant GM crops will reduce herbicide use, others suggest that herbicide use will actually increase. Nevertheless, the number of herbicides applied tends to decrease, because farmers opt for more specific herbicides, according to the acquired technology, in addition to the use of broad-spectrum herbicides, such as glyphosate and ammonium-glufosinate. These conditions may imply ecological changes, mainly related to biodiversity in cultivated areas. These changes in biodiversity can generate population changes in the weed community and favor the evolution of herbicide-resistant biotypes.

Another controversial question among weed science researchers is: Does the use of GM crops influence the emergence of herbicide-resistant weed biotypes? There is no scientific evidence to confirm this information. In order to relate the reports of resistant weeds in Brazil and the release of GM crops, Fig. 14 shows the increase in cases of resistance and the year of insertion of GM technologies in Brazil.

Currently in Brazil, 53 cases of weed resistance to herbicides are reported. The herbicides that most contribute to resistance are ALS inhibitors, involving 30 of the total cases, that is, 53.6% of resistant cases are related to this mechanism of action. When relating this to ALS-resistant crops, only soybean resistant to the chemical group of imidazolinones (imazapyr and imazapic) was approved in 2009, when 13

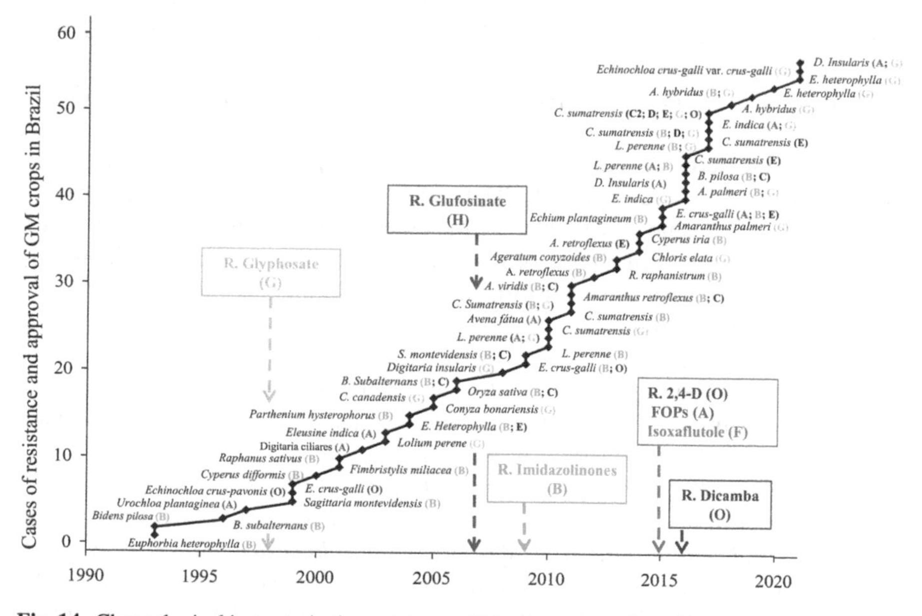

Fig. 14 Chronological increase in the resistance of biotypes of weed species in Brazil and CTNbio approval of GM crops resistant (R.) to herbicides (classified by letters according to the Herbicide Resistance Action Committee, HRAC). (**Source:** HEAP 2021 and CTNBio 2015)

cases of resistance to ALS had already been reported. Moreover, this soybean cultivar did not have a high acceptance in the Brazilian market, being little used.

Glyphosate, because it was the first molecule with resistant cultivars approved in Brazil, may report the influence of the use of resistant GM cultivars on the evolution of weed biotype resistance. Glyphosate-resistant soybean was approved for commercialization in Brazil in 1998, however, only from 2003, that soybean planting began to be allowed in commercial areas in Brazil. In the same year, a glyphosate-resistant ryegrass (*Lolium perene*) biotype was reported, reaching 21 cases of glyphosate resistance in Brazil today. In the case of glyphosate-resistant technology, there was a high adoption by producers, this associated with the high rate of application and the broad spectrum of glyphosate action may have favored the emergence of resistant weed biotypes.

A curious case is with ammonium-glufosinate, because it is a broad-spectrum herbicide, used in weed control from pre-sowing desiccation to pre-harvest desiccation, there are no reports of weed biotypes resistant to this herbicide in Brazil. In 2007, when the first corn cultivars followed by cotton and soybean, resistant to ammonium-glufosinate, obtained an increase in the use of this herbicide. The non-appearance of resistance may be related to the pattern of use of ammonium-glufosinate, because it is a contact herbicide and with low efficiency in more developed stage weeds, ammonium-glufosinate then lost a lot of space, especially for glyphosate. This low usage pattern may have reduced herbicide selection pressure and minimized selection pressure for resistance evolution. In addition, more specific and highly efficient herbicides such as ALS and ACCase inhibitors have several reports of resistant weeds; on the other hand, herbicides that inhibit the enzyme protoporphyrinogen oxidase (PPO or PROTOX) and GS inhibitors, despite the intense use in agriculture, have few reports of resistance, as they have a lower pressure of plant selection, which tends to select individuals in a slower way (Inoue and Oliveira 2011).

In relation to herbicide-resistant GM crops belonging to the mechanism of action of synthetic auxins, ACCase and HPPD, the commercial release is recent in Brazil; in this sense, there is no way to relate to the emergence of resistant weed biotypes. However, there have been reports of weed biotypes resistant to ACCase-inhibiting herbicides and synthetic auxins in Brazil, which can hinder the adoption and efficiency of these technologies.

The use of resistant GM cultivars is an alternative for the control of resistant weed biotypes; however, some herbicides have been losing effect on certain species. An example of this is the perch of glyphosate efficiency. Today in Brazil, there are 12 species of glyphosate-resistant weeds (Heap 2021), which are: *L. multiflorum*, fleabane (*Conyza bonariensis*), horseweed (*Conyza canadenses*), *D. insularis*, tall fleabane (*Conyza sumatrensis*), *C. elata*, *A. palmeri*, *E. indica*, and recently the *A. hybridus*, *E. heterophylla*, barnyardgrass (*Echinochloa crus-galli* var. *crus-galli*), and *D. insularis* (Fig. 14). Even with the presence of resistant weed biotypes, the use of glyphosate in POST soybean is indispensable, given the wide spectrum of action in the control of sensitive weeds.

In the case of management of resistant biotypes, one of the alternatives is the association of glyphosate with ACCase inhibitors for the control of narrow leaf weeds; and in the case of broad leaves, the association would be with auxinic herbicides. In POST of soybean and cotton, the association with auxinic herbicides with 2,4-D and dicamba can only be performed in cultivars resistant to these herbicides, since in the case of corn, the use of ACCase-inhibiting herbicides can only be made in cultivars with resistance technology to the FOPs group.

The adoption of herbicide-resistant GM cultivars helps producers to comply with the most important principle of the management of resistant weed biotypes, which is the prevention of survival and dissemination of resistant populations. However, weed management should not depend only on herbicides and especially not on a single herbicide, as happened in GM crops resistant to glyphosate. In this sense, producers will need to use a series of methods associated with chemical control in integrated weed management systems. Some management strategies are represented in Table 4, describing their benefits and risks in the management of resistant weed biotypes (Green and Owen 2011).

7 Future Perspectives of Genetically Modified Crops Resistant to Herbicides

In Brazil, one of the great prospects of herbicide-resistant GM crops is in the development and approval of technologies for other crops, in addition to soybeans, corn, and cotton. Crops such as canola, potato, rice, alfalfa, wheat, beetroot, sugarcane, tobacco, linseed, and eucalyptus present research and even commercial events with resistance to herbicides; however, they do not have commercial interest or are not approved and available in Brazil (Rao 2014; Krishnan and Preston 2018).

Many doubts still focus on the use of crops resistant to hormonal herbicides in Brazil, such as 2,4-D and dicamba, in addition to the resistance of corn to the group of FOPs. Many questions are asked about the issue of weed control efficiency, injury problems in neighboring crops due to problems of volatilization and drift, mixture of herbicides in tank, and problems in the control of voluntary plants, such as cotton snooze.

Another important issue related to hormonal herbicides and FOPs is in the perspectives in relation to weed selection pressure, because of the greater possibility of herbicide application such as 2,4-D and dicamba, and the chemical group of FOPs, at different times of the crop cycle, since there are reports of weed biotypes resistant to these herbicides in Brazil.

In the case of hormonal herbicides, some species have resistance to only a single widely dominant allele, which means that resistance can be selected more easily (Preston and Malone 2015). On the other hand, there are also examples of species with recessive genetic inheritance or even multigenic inheritances, in which both tend to delay the evolution of resistance (Sabba et al. 2003; Weinberg et al. 2006). In addition, and with the increased applications of these herbicides before and during crop development, it can accelerate the appearance of resistant weed biotypes.

In this sense, the need for research related to herbicide-resistant GM crops in Brazil is evident, especially in a continental country such as Brazil, which presents

Table 4 Evaluation of strategies commonly used for the management of herbicide-resistant weed biotypes

Strategy	Benefits	Risks	Potential for control
Herbicide rotation	Reduction of selection pressure, control of resistant biotypes	Lack of herbicides with new mechanism of action, potential to cause plant injuries, high cost, limited spectrum of alternative weeds	Excellent
Herbicide mixtures	Reduced selection pressure, improved control, broad control spectrum	Poor activity in resistant weed species, increased cost, potential to cause injuries in plants	Excellent
Variable application (dose and time)	Better control of resistant biotypes, use of more efficient herbicides	Lack of residual effect of the herbicide, the timing may be too late to protect the yield potential, more applications	Good to excellent
Soil revolving	Reduction of selection pressure, consistent efficacy, depletion of the weed seed bank	Increased time required, increased soil erosion, increased costs, additional strategy required	Good to excellent
Planting configuration	Improved competitive capacity for culture, reduced selection pressure	Unavailability of mechanical strategies, emphasis on herbicides, equipment limitations	Good
Seed bank management	Reduction of pressure of resistant weed biotypes, reduced selection pressure	Lack of understanding about the dynamics of the seed bank, requires aggressive soil preparation, emphasis on late herbicide applications, high level of management skill required	Reasonable to good
Crop rotation	Alteration of the agroecosystem, allows applications of different herbicides, reduces the selection pressure	Economic risk of alternative rotation crop, lack of culture of adapted rotation, similar rotation crops, and, therefore, minimal impact on the weed community, herbicides required, lack of research base, inconsistent impact on populations of resistant weed biotypes	Reasonable to good
Adjusted herbicide doses	Better control of target species	Increased target site selection pressure with higher rates, increase in non-target location with lower rates (polygenic resistance)	Poor to reasonable
Mechanical control	Reduced selection pressure, consistent effectiveness, relatively inexpensive	Increased time required, high level of management skill required, additional strategy required, potential to cause plant injury	Poor to reasonable
Fit in the sowing season	Improved effectiveness in target weed control, reduction of selection pressure	It requires alternative strategies (crop or herbicide), potential for loss of yield, need to increase rotation diversity	Poor to reasonable

(continued)

Table 4 (continued)

Strategy	Benefits	Risks	Potential for control
Precision in herbicide application	Reduction of herbicide use, reduction of selection pressure	Increased application cost, unavailability of weed population maps, poor understanding of weed seed bank dynamics, increased control variability	Poor
Adjustment in seeding rate	Reduced selection pressure, improved competitive capacity for culture	Increased seed cost, potential increase in pest problems, increased intraspecific competition, reduction of potential yield	Poor
Cover crops, dead cover, Crops association	Improved competitive ability, reduced selection pressure, improved system diversity, allelopathy	Inconsistent effect on resistant weed biotypes, lack of understanding about the systems, limited research base, potential crop yield loss, herbicide need to manage the cover crop, lack of good cover crops	Poor
Adjustment in nutrient use	Improved competitive capacity for culture, efficient use of nutrients	Lack of research base, inconsistent results, potential loss of crop yield	Poor

Source: Adapted from Green and Owen (2011)

different climatic conditions within the territory. In addition, Brazil, because it is a country on the rise, has many flaws in terms of access to technologies and also information. Thus, one of the country's biggest future challenges with the intensification of these technologies is to find ways to ensure that producers use them correctly for efficient weed control.

8 Concluding Remarks

The evolution of resistant weed biotypes, the difficulty in discovering new herbicides or new mechanisms of action and the non-selectivity of herbicides for certain crops, caused the industry to intensify its research with herbicide-resistant GM crops already available on the market, with the objective of solving weed control problems.

The recently approved 2,4-D and dicamba resistant soybean, cotton, and corn cultivars will help control broadleaf weeds, such as *Conyza* and *Amaranthus*, in POST, in addition to facilitating the rotation of herbicide use. As an alternative for the control of narrow leaf weeds, such as *Digitaria insularis* and *Eleusine indica*, the possibility of applying ACCase inhibitor herbicides (FOPs) in POST corn will be a very important tool. In addition, cultivars tolerant to more than one herbicide will enable the association of herbicides of different mechanisms of action, increasing the weed control spectrum.

One of the major challenges will be to maintain the longevity of these herbicides in the efficient control of weeds, which is being threatened by the selection of resistant weed biotypes. In this sense, conducting research will be important to increase the longevity of technologies, maintaining product efficiency for as long as possible.

Herbicide-resistant GM cultivars will also require greater knowledge of technicians and farmers, related to the different transgenic events and herbicides that can be used in each of them. In addition, these new technologies will require farmers to planning more rigorously for production within the property, not only to improve control efficiency, but primarily to avoid problems of controlling voluntary plants in succession/rotation crops.

No weed control technique used in isolation is sustainable because weeds adapt to any single strategy (Green 2012). With this, in addition to strategies involving the sustainable use of herbicide use, the adoption of integrated weed management will be indispensable to maintain efficient weed control and longevity of technologies available in the Brazilian market.

References

Adegas FS, Gazziero DL, Voll E (2014) Interferência da infestação de plantas voluntárias no sistema de produção com a sucessão soja e milho safrinha. Paper presente at the XXVIX congresso brasileiro da ciência das plantas daninhas, SBCPD, Gramado, 1–4 September 2014

Adoryan ML, Novo MCSS, Favoretto P et al (2002) Eficácia de isoxaflutole no controle de plantas daninhas na cultura da batata. Rev Bras Herb 3(2–3):133–138

Albrecht LP, Albrecht AJP, Biazoto FS et al (2018a) Soja transgênica tolerante a imidazolinones: passado, presente e futuro. J Agro Sci 7(Suppl):24–32

Albrecht LP, Albrecht AJP, Mundt TT et al (2018b) Soja transgênica Liberty Link® e o seu manejo. J Agro Sci 7(Suppl):33–42

Andersen RN, Ford JH, Lueschen WE (1982) Controlling volunteer maize (*Zea mays*) in soybeans (*Glycine max*) with diclofop and glyphosate. Weed Sci 30(2):132–136

Barroso AAM, Albrecht AJP, Reis FC (2014) Interação entre herbicidas inibidores da ACCase e diferentes formulações de glyphosate no controle de capim-amargoso. Planta Daninha 32(3):619–627

Behrens MR, Mutlu N, Chakraborty S (2007) Dicamba resistance: enlarging and preserving biotechnology-based weed management strategies. Science 316(5828):1185–1188

Braz GBP, Constantin J, Oliveira Júnior RS et al (2011a) Controle de solanáceas por herbicidas utilizados em algodoeiro. Rev Bras Herb 10(3):190–199

Braz GBP, Oliveira Júnior RS, Constantin J et al (2011b) Herbicidas alternativos no controle de *Bidens pilosa* e *Euphorbia heterophylla* resistentes a inibidores de ALS na cultura do algodão. Rev Bras Herb 10(2):74–85

Braz GBP, Constantin J, Oliveira Júnior RS et al (2012) Performance of cotton herbicide treatments for *Amaranthus lividus* and *Amaranthus hybridus*. Rev Bras Herb 11(1):01–10

Cahoon CW, York AC, Jordan DL et al (2015) Palmer amaranth (*Amaranthus palmeri*) management in dicamba-resistant cotton. Weed Technol 29(4):758–770

Cao M, Sato SJ, Behrens M et al (2011) Genetic engineering of maize (*Zea mays*) for high-level tolerance to treatment with the herbicide dicamba. J Agric Food Chem 59(11):5830–5834

CERA Center for Environmental Risk Assessment (2011) A review of the environmental safety of the CP4 EPSPS protein. Environ Biosaf Res 10(1):5–25

Chahal PS, Jhala AJ (2015) Herbicide programs for control of glyphosate-resistant volunteer maize in glufosinate-resistant soybean. Weed Technol 29(3):431–443

Chahal PS, Jhala AJ (2016) Factors affecting germination and emergence of glyphosate-resistant hybrid maize (*Zea mays* L.) and its progeny. Can J Plant Sci 96(4):613–620

Clewis SB, Thomas WE, Everman WJ et al (2008) Glufosinate-resistant maize interference in glufosinate-resistant cotton. Weed Technol 22(2):211–216

Constantin J, Raimondi MA, Franchini LHM et al (2011) Asociação de amônio glufosinato e pyrithiobac-sodium para o controle de picão-preto e trapoeraba em algodão Liberty Link®. Paper presented at the 8th cotton expo congresso brasileiro de algodão, Embrapa Algodão, Campina Grande, 19-22 September 2011

Correia NM, Resende I (2018) Response of three *Chloris elata* populations to herbicides sprayed in pre-and post-emergence. Planta Daninha 36:e018176117

CTNBio Comissão Técnica Nacional de Biossegurança (2015) Plantas geneticamente modificadas aprovadas para Comercialização. http://ctnbio.mctic.gov.br/liberacao-comercial. Accessed 30 Feb 2021

Dan HA, Procópio SO, Alberto LDL, Dan LGDM, Neto AMO, Guerra N (2011) Controle de plantas voluntárias de soja com herbicidas utilizados em milho. Revista Brasileira de Ciências Agrárias 6(2): 253–257

Everman WJ, Burke IC, Allen JR, Collins J, Wilcut, JW et al (2007) Weed control and yield with glufosinate-resistant cotton weed management systems. Weed Technol 21 (3):695–701

Ferreira ACDB, Bogiani JC, Sofiatti V et al (2018) Chemical control of stalk regrowth in glyphosate-resistant transgenic cotton. Rev Bras Eng Agríc Amb 22(8):530–534

Flessner ML, Mcelroy JS, Mccurdy JD et al (2015) Glyphosate-resistant horseweed (*Conyza canadensis*) control with dicamba in Alabama. Weed Technol 29(4):633–640

Freitas RS, Ferreira LR, Berger PG et al (2006) Manejo de plantas daninhas na cultura do algodoeiro com S-metolachlor e trifloxysulfuron-sodium em sistema de plantio convencional. Planta Daninha 24(2):311–318

Gazziero DLP, Vargas L, Roman ES et al (2004) Manejo e controle de plantas daninhas em soja. In: Vargas L, Roman ES (eds) Manual de Manejo e Controle de Plantas Daninhas. EMBRAPA Uva e Vinho, Bento Gonçalves, pp 595–635

Gonçalves Netto A, Nicolai M, Carvalho S et al (2019) Control of ALS-and EPSPS-resistant *Amaranthus palmeri* by alternative herbicides applied in PRE-and POST-emergence. Planta Daninha 37:e019212505

Green JM (2012) The benefits of herbicide-resistant crops. Pest Manag Sci 68(10):1323–1331

Green JM, Owen MDK (2011) Herbicide-resistant crops: utilities and limitations for herbicide-resistant weed management. J Agric Food Chem 59(11):5819–5829

Grigolli JFJ, Crosariol Netto J, Izeppi TS et al (2015) Infestação de Anthonomus grandis (Coleoptera: Curculionidae) em rebrota de algodoeiro. Pesqui Agropecu Trop 45(2):200–208

Heap I (2021) The international herbicide-resistant weed database. http://www.weedscience.org/Home.aspx. Accessed 30 Nov 2021

Inoue HM, Oliveira RS Jr (2011) Bancos de sementes e mecanismos de dormência em sementes de plantas daninhas. In: Oliveira RS Jr, Constantin J, Inoue HM (eds) Biologia e Manejo de Plantas Daninhas. Omnipax, Curitiba, pp 37–66

I.S.A.A.A. Global Status of Commercialized Biotech/GM Crops (2019). Biotech Crops Drive Socio-Economic Development and Sustainable Environment in the New Frontier. https://www.isaaa.org/resources/publications/briefs/55/executivesummary/pdf/B55-ExecSum-English.pdf. Accessed 20 may 2022.

Jhala AJ, Beckie HJ, Peters TJ et al (2021) Interference and management of herbicide-resistant crop volunteers. Weed Sci 69(3):257–273

Krenchinski FH, Cesco VJS, Castro EB et al (2019) Ammonium glufosinato associated with post-emergence herbicides in maize with the *cp4-epsps* and *Pat Genes*. Planta Daninha 37:e019184453

Krishnan M, Preston C (2018) Genetically engineered herbicide tolerant crops and sustainable weed management. In: Korres NE, Burgos NR, Duke SO (eds) Weed control sustainability, hazards and risks in cropping systems worldwide. CRC Press, Boca Raton, FL, pp 191–212

López-Ovejero RF, Ferreira AC, Crivellari A et al (2014) Culturas Geneticamente Modificadas Tolerantes a Herbicidas. In: Monquero PA (ed) Aspectos da Biologia e Manejo das Plantas Daninhas. RIMA, São Carlos, pp 285–306

López-Ovejero RF, Soares DJ, Oliveira NC et al (2016) Interferência e controle de milho voluntário tolerante ao glifosato na cultura da soja. Pesqui Agropecu Bras 51(4):340–347

Melo MSC, Rocha LJFN, Brunharo CADCG et al (2017) Alternativas de controle químico do capim-amargoso resistente ao glifosato, com herbicidas registrados para as culturas de milho e algodão. Rev Bras Herb 16(3):206–215

Merchant RM, Sosnoskie LM, Culpepper AS et al (2013) Weed response to 2,4-D, 2,4-DB, and dicamba applied alone or with glufosinato. J Cotton Sci 17:212–218

Minozzi GB, Christoffoleti PJ, Monquero PA et al (2017) Control in soybean pre plant of volunteer glyphosate and ammonium glufosinate tolerant cotton and *Eleusine indica*. Rev Bras Herb 16(3):183–191

Oliveira Neto AM, Guerra N, Almeida Dan H et al (2010) Manejo de *Conyza bonariensis* com glifosato+ 2,4-D e amônio-glufosinato em função do estádio de desenvolvimento. Rev Bras Herb 9(3):73–80

Osipe JB, Oliveira RS Jr, Constantin J et al (2017) Spectrum of weed control with 2,4-D and dicamba herbicides associated to glifosato or not. Planta Daninha 35:e017160815

Petter FA, Sima VM, Fraporti MB et al (2015) Volunteer RR® maize management in Roundup Ready® soybean-maize succession system. Planta Daninha 33(1):119–128

Preston C, Malone JM (2015) Inheritance of resistance to 2,4-D and chlorsulfuron in a multiple-resistant population of *Sisymbrium orientale*. Pest Manag Sci 71(11):1523–1528

Quadros AS, Bandeira L, Kasper N et al (2020) Associações de herbicidas na dessecação pré-semeadura de soja. Rev Bras Herb 19(2):1–9

Raimondi MA, Oliveira Junior RS, Constantin J et al (2012) Controle e reinfestação de plantas daninhas com associação de amonio-glufosinato e pyrithiobac-sodium em algodão Liberty Link®. Rev Bras Herb 11(2):159–173

Rao VS (2014) Transgenic herbicide resistance in plants. CRC Press, Boca Raton, FL. 480 p

Sabba RP, Ray IM, Lownds N et al (2003) Inheritance of resistance to clopyralid and picloram in yellow starthistle (*Centaurea solstitialis* L.) is controlled by a single nuclear recessive gene. J Hered 94(6):523–527

SINDIVEG (2020) Sindicato Nacional da Indústria de Produtos para a Defesa Vegetal. Defensivos Agrícolas. https://sindiveg.org.br/wp-content/uploads/2020/08/SINDIVEG_Paper_REV_FINAL_2020_bxresolucao.pdf. Accessed 10 Feb 2021

Skelton JJ, Simpson DM, Peterson MA et al (2017) Biokinetic analysis and metabolic fate of 2,4-D in 2, 4-D-resistant soybean (*Glycine max*). J Agric Food Chem 65(29):5847–5859

Spaunhorst DJ, Bradley KW (2013) Influence of dicamba and dicamba plus glyphosate combinations on the control of glifosato-resistant waterhemp (*Amaranthus rudis*). Weed Technol 27(4):675–681

Striegel A, Lawrence NC, Knezevic SZ et al (2020) Control of glyphosate/glufosinate-resistant volunteer maize in maize resistant to aryloxyphenoxypropionates. Weed Technol 34(3):309–317

Takano HK, Dayan FE (2020) Glufosinato-ammonium: a review of the current state of knowledge. Pest Manag Sci 76(12):3911–3925

Takano HK, Oliveira RS Jr, Constantin J et al (2013) Efeito da adição do 2,4-D ao glifosato para o controle de espécies de plantas daninhas de difícil controle. Rev Bras Herb 12(1):1–13

Takano HK, Oliveira RS Jr, Constantin J et al (2018) Chemical control of glifosato-resistant goosegrass. Planta Daninha 36:e018176124

Takano HK, Beffa R, Preston C et al (2020) A novel insight into the mode of action of glufosinate: how reactive oxygen species are formed. Photosynth Res 144(3):361–372

Weinberg T, Stephenson GR, Mclean MD et al (2006) MCPA (4-chloro-2-ethylphenoxyacetate) resistance in hemp-nettle (*Galeopsis tetrahit* L.). J Agric Food Chem 54(24):9126–9134

Wright TR, Shan G, Walsh TA et al (2010) Robust crop resistance to broadleaf and grass herbicides provided by aryloxyalkanoate dioxygenase transgenes. PNAS 107(47):20240–20245

York AC, Stewart AM, Vidrine PR et al (2004) Control of volunteer glyphosate-resistant cotton in glyphosate-resistant soybean. Weed Technol 18(3):532–539

Index

© The Editor(s) (if applicable) and The Author(s), under exclusive license to Springer Nature Switzerland AG 2022
K. F. Mendes, A. Alberto da Silva (eds.), *Applied Weed and Herbicide Science*,
https://doi.org/10.1007/978-3-031-01938-8

D

E

F

P

R

X

Z